LOCAL AGRARIAN SOCIETIES IN COLONIAL INDIA

Japanese Perspectives

Centre of South Asian Studies,
School of Oriental and African Studies,
University of London

COLLECTED PAPERS ON SOUTH ASIA

10. Institutions and Ideologies
D. Arnold & P. Robb

11. Local Agrarian Societies in Colonial India
P. Robb, K. Sugihara & H. Yanagisawa

COLLECTED PAPERS ON SOUTH ASIA NO. 11

LOCAL AGRARIAN SOCIETIES IN COLONIAL INDIA
Japanese Perspectives

Edited by

Peter Robb, Kaoru Sugihara

and

Haruka Yanagisawa

CURZON

First published in 1996
by Curzon Press
St John's Studios, Church Road, Richmond
Surrey, TW9 2QA

Printed in Great Britain by
TJ Press (Padstow) Ltd, Padstow, Cornwall

British Library Cataloguing in Publication Data
A catalogue record for this book is available from the British Library

Library of Congress in Publication Data
A catalog record for this book has been requested

ISBN 0–7007–0471–X (Hbk)

CONTENTS

v

CONTRIBUTORS

Toshie Awaya	Faculty of Letters, University of Tokyo
Yukihiko Kiyokawa	Institute of Economic Research, Hitotsubashi University
Tsukasa Mizushima	Institute for the Study of Languages and Cultures of Asia and Africa, Tokyo University of Foreign Studies
Nariaki Nakazato	Institute of Oriental Culture, University of Tokyo
Akihiko Ohno	Faculty of Economics, Osaka City University
Peter Robb	Department of History, SOAS, University of London
Osamu Saito	Institute of Economic Research, Hitotsubashi University
Kaoru Sugihara	Department of History, SOAS, University of London
Shinkichi Taniguchi	Faculty of Economics, Hitotsubashi University
Kohei Wakimura	Faculty of Economics, Osaka City University
Haruka Yanagisawa	Institute of Oriental Culture, University of Tokyo

ACKNOWLEDGMENTS

This volume originated in a conference held in the School of Oriental and African Studies (SOAS), in July 1992. We are grateful to SOAS and to the Japanese Ministry of Education, Science and Culture for financial support, for the conference and this publication. We are also grateful for assistance to Miss Janet Marks, until June 1995 Executive Officer of the Centre of South Asian Studies, and to Mrs. Catherine Lawrence (for maps and figures). Dr Michael Hutt has assisted as the chairman of the Centre's Publications Committee, responsible for the series in which this volume appears. Camera-ready copy for the volume has been prepared in SOAS.

Peter Robb, Kaoru Sugihara, and Haruka Yanagisawa

Chapter 1.1

DISTINCTIVE ASPECTS OF RURAL PRODUCTION IN INDIA: THE COLONIAL PERIOD[1]

Peter Robb

This volume collects papers on the rural economy of India, all written by Japanese scholars. They were originally presented in London at the School of Oriental and African Studies, as part of an extensive investigation of agriculture and economic organisation, undertaken by historians and economists from India, North America and Europe. The proposition was that there is a need for a new vocabulary to describe Indian conditions. We kept the Japanese papers separate then, and do so again in this book, because our inquiry was partly into concepts and different ways of understanding data. And these papers do reveal some distinctive features. First, their implicit model derives from what happened in Japan, not in the West. This aspect is discussed in most of the papers, in the second introduction, and in special comparisons. Secondly, the papers show an eagerness to collect, adjust and then calculate from quantitative data. This contrasts with the scepticism often shown, at least by Western-trained historians of India, about the value of such records.[2] Thirdly, the papers tend to focus on the actuality of production

[1] I am grateful for comments and suggestions by other contributors to the volume, especially during a day-long seminar on a draft of this introduction at the University of Tokyo in November 1993.

[2] This aspect speaks for itself in the papers which follow, but I will also make a few comments about particular cases. In my view, elaborate computations are problematic because the data are often suspect, and debates about figures can quickly become sterile or self-generating; there is no substitute for fully contextual, empirical, historical analysis—for political economy. It would be ironic to urge economic historians to take the figures more seriously, at a time when, among economists, econometrics and ahistorical theory seem somewhat in retreat (for example in new institutional economics, for which a broader understanding of *political* economy, such as will be proposed here, is required). And yet it is obvious that the concrete edge provided by statistics (if only for internal comparisons) may lead to questions which otherwise cannot be answered, and thus provide some valuably different hypotheses. I believe these papers do this.

decisions and processes—often, that is, on micro- rather than macro-
economic aspects.[3] Finally, the papers disaggregate decisions and identify roles at different levels of the society and in production.[4] The last
two of these characteristics together comprise one common aspect of
the papers on which my introduction will concentrate. They provide, I
suggest, important perspectives especially on the connection between
social and institutional structures and economic development. This
introduction offers a personal view of the findings: it will not cover all
aspects or try to treat all of the papers in equal depth. It is one reading
of the texts, and not intended to preclude others.

One way of considering these papers is in the context of debates
about the nature of economic change. Central themes of the debates, to
present them in caricature, include the independent role of the market in
the 'modern' era, as against the predominance of social and political
forces in pre-capitalist societies. Though production, specialisation and
exchange always occur, according to some theories it is only in some
places and times that they occupy an autonomous sphere, governed by a
market rationale. Thus trade for profit is seen as an incubus which
breaks open societies, a cuckoo outgrowing the pre-capitalist nest. The
argument generalises many explanations of development *from outside*—

---

[3] This is necessary and welcome, though it may imply methodological
conclusions that these scholars (and social scientists generally) do not find
congenial: namely that, in economic history, the way through the 'smoke-
cloud' of ideology, concepts and generalisation, is to attempt specific studies of physical contexts, material culture, social organisation, and production, doing so in contact with broader issues but without the expectation of
reaching any one precise and detailed answer to large questions, like a single law of physics. See Peter Robb, 'British rule and Indian "improvement"', *Economic History Review* 34, 4 (November 1981), pp.507-23.
Thus we might consider, say, national economies only as analytical con-
structs, which is what they are—without hope of a definitive 'why', but with
a good chance of showing 'what' and 'how' within the limits of the model
applied. The limitation is relevant to the comparisons between India and
Japan which arise here as elsewhere.
[4] See Peter Robb, 'Peasants' choices? Indian agriculture and the limits of
commercialization in nineteenth-century Bihar', *Economic History Review*
45, 1 (February 1992), pp.97-119, notably pp.98-101 and 108-9. Taniguchi's paper in this volume in particular might be read as a companion-
piece to this article, but similarities will also be demonstrated with other
essays.

feudal Europe from medieval cities, pre-capitalist Asia from European expansion. There are contrary views which attribute change also to non-economic or to indigenous factors, but the end-product usually remains a 'capitalist' society.

The debates about these processes are particularly pertinent at the moment with regard to South Asia, because of recent studies which are transforming our view of the eighteenth century and hence of the nature of nineteenth- and twentieth-century developments.[5] The papers in this volume do not reflect directly upon these questions: as said, they are written against a different set of theories and preoccupations. But for that reason they may offer new insights. My own position is that, although economic developments obviously occur, the trajectory is not from a world without markets to one in which the market determines the whole of economic life.[6] In particular there are, in all societies, independent roles at different levels of political and economic processes. There are, in 'pre-modern' societies, specialist intermediaries operating between producers and consumers, including merchants but also various kinds of managers and agents; moreover, there are also individual profit-motives rather than only priorities of state or society (that is, change-seeking as well as equilibrium-maintenance). On the other hand, in 'modern' societies, there are non-market forces, including some working through intermediaries; there are political and social strategies which qualify the maximising of profit. Thus though there may be exchange without profit (or trade without the market), these features are not confined to particular kinds of societies.

Among what I take to be current arguments, I endorse the following.[7] (1) The market is a social as well as an economic arena. (2) It is necessary to disaggregate within as well as between economic roles. (3) Social transformation does not occur from influences of a single type—whether merchants and demography (that is, economic forces, à la Pirenne) or autonomous developments in politics and class structure

[5] See Sanjay Subrahmanyam (ed.), *Merchants, Markets and the State in Early Modern India* (Delhi 1990), especially the introduction.

[6] See Robb, 'Peasant choices'.

[7] They are adapted from Subrahmanyam, *Merchants, Markets*, pp.1-17.

(following Brenner). (4) The distinction is not absolute between 'amateur' and 'specialist', for example in the case of merchants; however there are specialist brokers in all complex societies. (5) Extensive market relations pre-date the 'capitalist' age, and extensive 'pre-capitalist' forms persist during it. The last proposition, it may be noted, is a re-statement of the first.

Let us assume that the advent of private property, the political independence of commercial interests, and hence the accumulation of capital needed for economic development and industrialisation, all depended upon the achievement of stability and power (as in city states) and upon the growth of public institutions of law and civil society, including those of money. In fact there are examples of large merchants existing without all of these conditions,[8] but let us accept their importance. We will then expect to base our periodisation on a divide between times when such conditions did or did not exist. But then we find, from the new politico-economic history, that European conquest did not mark such a divide between periods, thus defined, in India, as used to be assumed, and as we would expect from the increase of foreign trade which accompanied colonial rule. The anomalies of the pre-colonial age include extensive trade independent of the state and its revenue needs, the socio-political resources of some commercial interests, the possibility of a continuity of merchant houses over time, and occasions of state-merchant alliances bringing economic dynamism to local areas. My concern has been to add parallel (that is, of course, opposite) caveats upon the role of the market and the state in the colonial era.[9] Some reflect incumbrances imposed by colonial rule, and some inheritances of Indian society. Thus, I conclude, socio-economic change consists in the shifting weight and role of various components, rather than in a movement from one complete and coherent system to another—the growth of the market, say, driving out pre-capitalist relations of production. This introduction will reflect on the papers in the volume from this perspective.

The earliest example here of production and exchange is one based

[8] See ibid, pp.242-65.
[9] Robb, 'Peasants' choices?'.

on shares of the output, the so-called *mirasi* system, as described by Mizushima, in south India and particularly Changalpatti (Chingleput) district, in the late eighteenth century. This is a kind of modified 'village community' model. Importantly there were shares in land; Muzishima concentrates on the different types of tax-free holding. But the system's focal point was the harvest, at which transactions were designed mostly to meet needs collectively. On the other hand the community was neither unchanging nor confined to the village. It was not egalitarian or self-sufficient. The 'village' was after all not the community. Rather, as Mizushima presents it, the community was a continuum of local and external forces, expressed through but not limited to harvest shares. This was a society with specialised roles, penetrated by political and administrative elites and the state. (I will come back to what is meant by the state.) The system was adaptable but also capable of unravelling in the face of commercial development, of new forms of dominance, or of the advent of individual property rights.[10]

Mizushima distinguishes four types of harvest share: the first two calculated gross, whether (1) from the crop before threshing or (2) as grain, and the second two calculated net, from the remainder of the grain heap which was divided between cultivators and the state; they were

[10] The account should be read in conjunction with that of David Ludden, *Peasant History in South India* (Princeton, 1985), which shows that, in Tirunelveli district, the *ur*, encompassed one or more villages, later complicated by a superior organisation, supported by Vellalas, in the form of Brahman settlements (*brahmadeya*), which in their extended form represented a kind of sub-regional structure and arena of production (pp.34-41). After about 1400 military adventurers and migrating peasant communities began further to develop political structures and regional patterns of agriculture: notably village magnates and officials controlling dry-zone villages (see pp.81-4), and closest to Mizushima's account, share-holding elites in the wet-zone (pp.84-94). Even in the latter case village office eventually became an important asset (after 1800), while landed Brahmans and Vellalas, mainly through control of irrigation, controlled 'sub-regional networks of state and market activity' (p.90). Ludden also, in the remainder of the book, discusses at length the impact of colonial rule and commodity production, including the rise of village headmen and the development of new concepts of rights, including 'individual entitlement to property' (p.199).

deducted either (3) from both cultivators and the state, or (4) from the state's share alone. Mizushima's evidence is mainly statistical, and his reasoning strongly and directly functionalist: these harvest shares (and their distinctive forms) are each supposed to have a rational purpose. For example, because payments to cultivators' servants (dependent employees) appeared *inter alia* among category (4), it is concluded that they were used to secure public works or services to the state. One might equally suggest that, if the state's share was paying for services to the local community or the cultivator, then this meant the locals were effectively reducing what they were paying to the state, thereby having the same effect by another means as concealment or avoidance. From the state's point of view one might call this a compromise conceded in order to gain leverage over the cultivators or village elites.

Differences of perspective, with their implication of negotiation, are central to Mizushima's account. His emphasis on shares will be familiar to students of south India,[11] but the mechanics and minutiae of the system do throw up some interesting points. As described here, the system must be an ideal which was not always achieved in practice; also its characterisation and terminology show some influence from European assumptions, most particularly about property rights. But plainly the system had a distinctive character: it represented one kind of bargain between the cultivator and all those with whom he had relationships. The purpose was to secure labour and services, meeting obligations and demonstrating status: the *mirasi* system was a method embodied in ways of calculating payments. It had behind it the force of custom, and physical, social and political power, and it did not reflect any one or

---

[11] See Burton Stein, 'Idiom and ideology in early nineteenth-century south India', in P. Robb (ed.), *Rural India. Land, Power and Society under British Rule* (London 1983 or Delhi 1992). Note that the shares were usually held not by individuals but by households, joint families or institutions— as has often been argued, this constituted another inhibition of change, one only gradually attacked by colonial law and administration; see D.A. Washbrook, 'Law, state and agrarian society in colonial India', *Modern Asian Studies* 15, 3 (1981), pp.649-721. In this introduction, taking that point for granted, such collectivities will be termed 'landholders', 'intermediaries', 'cultivators' and so on, as the smallest units identified for our purposes in these roles.

sectional interest and rationalisation, or meet the requirements of any party individually. It was pre-eminently decentralised, though focused on the threshing floor. It was at once general and specific, the result of local conditions, negotiation or dispositions of power. The extent to which norms prevailed may be measured from the remarkable consistency (at least in the record) of the deductions made from the net shares of the cultivators and the state jointly. On the other hand, we may note the variations firstly between villages and in the total share for any recipient or class of recipient. Though the data are not fully presented here, it seems too that adjustments necessitated by high payments for, say, labour in some villages, or to landholders in others (no doubt reflecting local circumstances), were not made through all kinds of harvest shares equally. Rather they were sometimes made through different rates of deduction from the state's proportion alone, and sometimes in the allocation of the gross product after threshing. In the case of landholders, as for temples and other services, they were also made by the grant of tax-free land. Differences occurred secondly in the proportions retained by the cultivators and in the state. Mizushima thinks that a lower share for the state reflected higher costs to the cultivators or a lower yield (though presumably the latter would alter the *proportions* of harvest-division only when necessary to secure a minimum return for the cultivator). Probably discounts to secure allies, attract cultivators or encourage reclamation were also involved.

The underlying question should be why cultivators, or at least localities, agreed to pay anything at all to outsiders. Mizushima's analysis shows that part of the means of enforcement, indeed of state-building, was an interdependence expressed within the revenue and employment systems. He reminds us that bargains imply agreement. These *mirasi* bargains relied first on the continuum between the local and the external, marked by the extra-local beneficiaries, and secondly on the penetration into the locality of external forces (norms, temples, the 'state'), a linkage made possible because the locals who benefited from harvest shares became the potential allies of the outsiders. It would have been difficult, in the *mirasi* system, for the cultivator to have had the idea of not paying out the shares, or, that is, of an exclusive ownership in the product of his land and labour. He needed workers and services, and secured them through harvest shares, collectively, rather than through separate master-servant relations as in slavery, wage-labour or *jajmani* arrangements. (Mizushima sees the last, interestingly, as an imperfect transition from collective to individual relations, or a kind of concealed continuation of the *mirasi* system at a time when it was crumbling in face of various challenges, including market forces.) The cultivator had to take the collective system as a whole, and therefore, once the state had entered into it, and, provided the state remained effective, the cultivator was bound to pay the state as well as well as local clients. The latter ensured the state's share so as to secure their own.

What was this state? Mizushima is reticent. But it was surely not a unitary bureaucracy or a centralised oligarchy, swollen by its take of one quarter of the agricultural produce of the region, and responsible for a commensurately large range of public functions. Like the village community in Mizushima's account, this state was a continuum of linked forces. Mizushima puzzles over the mysterious non-appearance of landholders' rents in his major source; but what is identified here as the state's share must have helped support local elites whom we may regard either as the tip of the state's tentacles or as local powers allied to intermediary and superior ones. The lords, poligars (*palaiyagar*; military chiefs), revenue-collectors, accountants, temples and Brahmans that benefited from shares of the harvest also took directly from their own lands and from the 'state'; indeed, alongside urban, courtly, and particularly military functionaries, they constituted the state, which was a *role*, at once local and central.

II

Village elites and intermediaries form an important part of this story; they were rediscovered by British officials in the later nineteenth century.[12] They do not seem to have been fully integrated into descriptions of Indian agriculture and rural industries. In this volume we confront

[12] Compare C.J. Dewey, 'The rehabilitation of the peasant proprietor in nineteenth-century economic thought', *History of Political Economy* 6, 1 (1974).

them most directly when Taniguchi defines and describes them in an analysis of late eighteenth-century Bengal. He is concerned with village elites, principal raiyats (cultivators), identified as those possessing larger than average holdings, and as distinct from smaller self-sufficient farmers and a larger body of dependent landholders or workers. These principal raiyats did not only have land; they had privileges and influence. They managed rent and revenue collection; they provided agricultural loans, facilitated reclamation, and operated in the grain market; and they led local protest movements. Their privileges were a price paid to ensure their support, by superiors whose agents they were. Conflict occurred when that reward was insufficient or support was withheld. Hidden here, as in the *mirasi* system, is a model of rural society which sees it as defined by trials of strength, or according to negotiation between allies—whether for revenue and rent collection, or in encouraging commercial production. This interdependence between internal and external forces in the Bengal countryside was expressed in the prevalence of indirect management of estates, and in general in a repeated subdivision of tasks and authority which presumed the continual interposition of mediators, a feature which was undoubtedly connected to political structures and to the caste system. It is not clear whether this supports or contradicts Mizushima's 'local society', but it certainly suggests a congruence of procedures and methods from one end of India to another.

Similarly, in Malabar, Awaya shows that it was only in the colonial period (and to a limited extent even then) that the great estates became well-enough managed as to be more than a scattered congeries of diverse landholdings, rights and dependants under a vague, or rather a socio-religious, overlordship: real power, she argues, lay with local agents, and even in colonial times there was little central knowledge of how land control or tax liabilities operated in the localities. Thus Awaya records the key roles played by middlemen, including *taravads* (matrilineal joint families) and Brahmans, in the land management of Malabar. In specific examples, she shows how certain tenants, by sub-letting to cultivators, were able to corner a large share of the agricultural output on holdings of between 5 and 135 acres, a share larger than that of the superior landlords, the *janmis*. These tenants remained intermediary leaseholders, their *kanam* leases being medium-term (12-year) renewable agreements associated with a usufructuary mortgage,[13] comparable to those whereby land was held by some of the *thikadars* (sub-proprietors) of north Bihar and elsewhere.

For Bengal, recognition of the importance of village elites and village-level intermediaries was pioneered by Ratnalekha Ray, using the term 'jotedar'; Taniguchi acknowledges her contribution and largely reinforces her findings. Rajat Datta's studies also support the case for the ubiquity of village elites.[14] But the origin and longevity of these social formations is a matter of some controversy, indeed an aspect of one of the great divides in the interpretation of India's modern agrarian history, between what we might call the school of 'remarkable structural stability' and the believers in progressive differentiation and socio-economic revolution.[15] Where Ratnalekha Ray described continuities, Taniguchi sees at least two different kinds of jotedar as having arisen in special circumstances—one in the 'disorder' of Mir Kasim's government (1760-4), and another in response to the market opportunities of the late nineteenth century. With regard to the first of these, we can see that in Bengal the period from the 1740s until at least the 1790s was, as a whole, one of increasing revenue demands and expanding cultivation. At the village level this probably sharpened economic differentiation by reinforcing the instruments of power in the hands of village heads and others. The extensions of cultivation and commercial

[13] They still paid rent but at reduced rates—the reduction paid off the mortgage for the superior landlord.

[14] For references to the works by Ray and Datta, and further discussion, see Taniguchi's essay below.

[15] In the relative stability camp, on the whole, are Dharma Kumar, Ludden, Baker, Morris, Rajat and Ratna Ray, Abu Abdullah, and so on; the 'revolutionaries' include R.C. Datta, the Thorners, Bhatia, and Binay Chaudhuri. Some of the differences are merely about timing; others are more fundamental. Compromises are suggested in terms of regional and sub-regional differences (Stokes, Stone, Sugata Bose), or by disaggregations (such as Chaudhuri tracing the dispossession of occupancy raiyats in the absence of basic or structural change in the peasant economy). See the discussion and references in Sugata Bose, *Agrarian Bengal. Economy, Social Structure and Politics, 1919-1947* (Cambridge, 1986), pp.146-8.

cropping were often encouraged by the granting of privileges and incentives, not to all, but to those with local power and prestige who could organise the changes. Similarly, the farming-out of state revenue (and of local rent-collection) was resorted to as a way of maximising short-term returns while minimising direct management. This too meant opportunities for some village leaders as well as for interlopers. As a result the permanent settlement of the land revenues (1793) and its grant of exclusive property rights, were partly designed to restore the position of the superior rent-collectors and landholders, the zamindars, who had supposedly lost out to such large tax-farmers and their petty local allies.

But by that time, Taniguchi finds, a selection of successful raiyats had achieved spectacular upward mobility. Though there may have been elements of some kind of Chayanovian cycle, there was also, in Taniguchi's view, a general and rapid polarisation made possible because new circumstances altered the value of certain resources. Taniguchi thinks there was a general labour shortage (a thesis presumably supported by the losses of population during the great famine of 1769-70). This shortage encouraged and somehow enabled richer raiyats to depress their smaller, barely self-sufficient neighbours. Other commentators have noted the relatively favourable rent-rates apparently allowed to so-called non-resident raiyats in the aftermath of the famine;[16] this too may be attributed to population decline, but it may also have meant that enterprising people were able to gain larger holdings and participate in the grain, oilseeds and fibre markets which were then growing as a result of European-led exports and urbanisation. Such a line of reasoning used to be ruled out by macro-economic notions about, first, late-Mughal economic decline and, second, the East India Company's 'drain of wealth' from Bengal. But recent scholarship makes it seem more plausible.[17] It may even suggest a way of reconciling Datta's idea of indigenous development with Binay Chaudhuri's insistence on

[16] See W.W. Hunter, *Bengal Mss. Records, 1782-1807*, vol.1 (London, 1894), which contains a valuable but neglected essay on agrarian structure, rent, revenue and property rights in eighteenth-century Bengal. Non-resident (*pahikasht*) raiyats are discussed further below.

[17] See Bayly and Datta; the references given by Taniguchi below.

externally-stimulated commercialisation.[18]

The nature of these intermediaries and of the changes during colonial rule provides one of the major themes of this collection. We shall see that there are two main scenarios. One analyses the changing strength of those who *managed* production without participating in it directly. The other implies a polarisation *among the producers*, with advantages for an elite and impoverishment for the majority. The choice is between the broker and the independent peasant. Taniguchi is primarily concerned with the first of these issues. His argument—in common with that of several of the other papers—is that there were radical changes in the *structure* of society. Concentrating on landholding rather than labour or other factors of production, he observes a new role for credit intermediaries, attributable (and Nakazato would agree) to the establishment of property rights in land and the commercialisation which increased land values. Similarly, Mizushima describes a society newly based on individual rights, and household units of production: the raiyat's plot.

Yet it is not clear that the stratification Taniguchi demonstrates should be associated with any particular circumstance or period. There are after all, very many customary categorisations of status within agrarian society, in the revenue terminology and otherwise, and all of them claim ancient origins. A derivation from the late eighteenth century or earlier is convincing for the terms found in the nineteenth-century record, where they persisted as local usages against a strong tendency towards standardisation on the part of the rulers and their revenue administration, a point to which we shall return. It may be instructive, in this connection, to consider the history of one of the most widespread of distinctions, that between *pahikasht* (non-resident, unprivileged) and *khudkasht* (original resident, owner-cultivating) raiyats.[19] This is more complex and controversial a matter than might be thought. Mizushima's paper in this volume also introduces the distinction, in discussing residents and non-residents as part of the continuum

[18] Ibid., for references to Chaudhuri's work.

[19] See B.H. Baden-Powell, *The Land-Systems of British India*, vol.1 (Oxford, 1892), pp.599-600. The *khudkasht* raiyat was in many ways equivalent to the *kaniyatchikkarar* discussed by Mizushima below; *kaniyatchi* rights could be bought and sold by the start of colonial rule.

he calls 'local society'. I raise it in order to suggest that, whatever the complexity or changes of form, there are some important continuities of function in Indian society. My view is that in the late eighteenth century there were privileged and unprivileged tenures (and their holders) in most villages or estates, but no such rigid distinction between operative holdings: many households, especially the more powerful, held land in a number of guises, in villages where they had *khudkasht* status and in others as *pahikasht* raiyats; there are illustrations of this in Taniguchi's evidence. Households might even be 'resident' in respect of some but not all land held in the one village. The terms should not be taken literally: they reflect a kind of right or obligation, rather than, necessarily, places of residence or even ancestry.[20] The distinctions no doubt had been blurred over many centuries, whatever the original meaning of the terms; by the start of Company rule they were used with regard to revenue matters, but not, it seems, in popular parlance, except that it was common for *pahikasht* raiyats (like Mizushima's non-residents) to pay lower rents (that is, to retain a higher share of their output), presumably because of a need to attract labour and extend cultivation, or because of their exclusion from privileges and income available to those with *khudkasht* status. In the nineteenth century, however, it came to be more generally held that *khudkasht* implied a proprietary right held by the descendants of original settlers, whereas the *pahikasht* raiyat was, by the same argument, either a non-resident outsider or the resident descendant of one who had not been among those who had originally cleared the land. This interpretation may have owed more to then-current European theories about the origin of property rights than to any observation of Indian understanding. The latter, for example, probably gave more weight to social status, for a necessary though not sufficient condition for *khudkasht* status seems usually to have been membership of a recognised cultivating caste. In any case, during the nineteenth century the distinction, and its link to caste, were

[20] See also Meena Bhargave, 'Perception and classification of the rights of the social classes: Gorakhpur and the East India Company in the late eighteenth and early nineteenth centuries', *Indian Economic and Social History Review* 30, 2 (1993), pp.215-37.

eroded by legal changes: the idea of the *khudkasht* raiyat was replaced by
that of the 'settled' or 'occupancy' tenant, a different concept, enshrined
in Western legislation, though one which nonetheless had drawn on
what was taken to be the spirit of the original. A brief resumé of this
aspect is provided here by Nakazato. I think the moral of this story may
be that there was perpetual stratification, but that its basis could change
radically over time. The same conclusion might be drawn from the
ubiquity of the concept (under many different names) of village headman
or principal raiyat.

III

Before returning to discuss this issue more fully, we should raise the
second kind of question: the impact of general forces and trends. Several
are illustrated in this volume. I suggest that we approach them with
caution. Again there are rather a lot of arguments in favour of continuity. In particular, in considering the stratification of the nineteenth century, one should not imagine (as might be read into Taniguchi's
account) a steady shift from non-market to market relations, or an
unqualified acceptance of the familiar arguments about the impact of
changes in the legal incidents of landed property, and hence in regard to
credit. First, there was an influence from the market in the eighteenth
century—the jotedars wanted to capture surplus for sale, as well as to
profit from their local lending and agrarian management. Secondly,
redistributions of land were possible in the eighteenth as well as the
nineteenth century. The burden of Binay Chaudhuri's interpretation of
the late nineteenth and early twentieth centuries is that there was
'depeasantisation', but what Taniguchi argues is that 'peasants'—self-
sufficient farming households—were already in quite a small minority
in the eighteenth century. He records instances, which do not seem
dependent on Company laws or records, whereby poor tenants were
ousted and then brought back to labour on their former holdings.
Thirdly, at the end of the nineteenth century local dominance still rested
upon social as well as market forces.

Operating in the land market was far from the only means whereby
elites could gain power; the point is made (though not illustrated) by

Nakazato. On the other hand, high status in relation to land and labour involved social expenditure, which therefore was higher among higher castes and the wealthy. This and other costs connected with status (for example, the high castes employing ploughmen or secluding their womenfolk) suggested a degree of redistribution within agrarian society. Of course it is often argued that this interdependence diminished during the period of colonial rule, resulting in a polarisation of the rich landed and the impoverished landless. But social costs, as nineteenth-century observers often complained, continued to be incurred—and habitually, at most levels of society, in amounts above what could be readily afforded. Consistently, a certain strain on resources was expected at times of public celebration or family ceremony, as a sign of pride, hospitality, generosity, or social ambition. Wealthier households had higher costs and often became indebted to meet them, and so too did families of middling rank. How then could a greater polarisation occur? It did precisely because, as Nakazato confirms, most land purchases were made by people of roughly the same legal and caste status as the sellers (though some petty officials and traders accumulated property). The law facilitated differentiation, without transforming roles: more land was purchased or acquired by dominant groups, and the numbers of landless grew disproportionately, but this did not seem to result in fewer and larger units of production. Village elites gained a firmer grasp over dependants —but treated them as, say, share-croppers, not fellow-tenants.

Another feature to be considered is the difficulty of generalising. Nakazato most explicitly addresses the question of regional variation and the different trends perceptible in different areas. He groups the districts of Bengal into three regions; expressed in a kind of shorthand, they were the west (with strong landlords), the north and centre (with village elites) and the east (with small peasants). For the end of his period Nakazato finds the registered transfer of holdings least common in the central districts, but higher in the east and west. He suspects that the east's prominence was because of economic development, and that the west's figures were deflated by the virtually-landless majority of the population. The *numbers* of sales must relate to the total numbers of holdings, and, other things being equal, would be higher in a smallholding culture than where land had been consolidated. But the increases were not steady and general. In the late nineteenth century, registered transfers affected the largest proportions of land (according to Nakazato's maps) in Hooghly, close to Calcutta, in Midnapur, with its many government estates and forest zones, and in Tippera and Noakhali, on the eastern frontier. By the early twentieth century, though most returns were up by at least 100 per cent, the highest figures were shown first for these same districts and some of their neighbours with similar conditions—most spectacularly Chittagong in the east—but secondly also for newcomers (the west Bengal districts, and Dacca and Rangpur). It follows that, in all cases, the pattern of registered transfer has to be related to other, often earlier bases for stratification, as well as to changing conditions in particular places. The figures show (Nakazato's Table 4) that throughout Bengal, by 1945, only a minority of households owned enough land to support themselves; and there is some evidence that the proportion declined sharply in the 1940s; certainly it fell between 1920 and 1950. Those owning more than five acres in the west made up 10 per cent of households, with 65 per cent of the land; they formed 20 per cent with 70 per cent of the land in central districts, and 11 per cent with 53 per cent in the east. In the west 53 per cent owned no agricultural land at all, which was true for 31 per cent in central districts, and 32 per cent in the east. We must remember that these figures do not reveal anything much about production units. By concentrating on ownership (*khas* land) we miss share-cropping and other short-term or legally-invisible sub-letting arrangements. Nonetheless we can see that in Bengal the centre had a somewhat larger 'elite' (owner-cultivators or landowners), while the west was more sharply polarised, having a smaller 'elite' and very many landless labourers; the east, with an equally small 'elite', had more land owned in very small holdings by many households.[21] However this situation arose—in reducing the

---

[21] Those owning some but less than 5 acres of agricultural land represented 60 per cent of the holders in the east, with over 45 per cent of the land; such holdings accounted for around 30 per cent of the land in both the other regions, and for 50 per cent of the owners in central districts and 32 per cent in the east. For the delineation of these regions see Nakazato's essay below.

average size of holding, partition must have been as or more significant than transfer—it is clearly the proper starting-point for investigations of change. Indeed one would hope to go further by disaggregating to the district level, or, preferably, to ecological regions.

Certainly the continuing variety of conditions at any one time must always be noted. For example, Table 9 of Taniguchi's essay shows how unstable the situation was in the late eighteenth century. Mizushima and Yanagisawa are both aware of the mobilities of labour. and resources. For Bengal the position of zamindars both before and after 1793 is over-standardised in most descriptions, and the same may be said for most other sectors of the agrarian structure. There were different kinds and varying sizes of estate, indeed various levels of quasi-zamindari control. Some differences pre-dated Company rule, and others developed under it. Most notably the number recognised by the state as having zamindari rights (that is, landownership) greatly increased after 1793, and some of these smaller zamindars resembled large jotedars; but the increases were not evenly distributed. Some estates were consolidated and preserved; others partitioned. We have here, in short, a series of recurring elements plus contingency and rapid fluctuation. Similarly, in late nineteenth-century Bihar, I found that the value of resources depended on circumstances, and that, within a studied group of 'surplus' raiyats, there could be remarkable annual fluctuations of landholding and output.[22] Dominant villagers would often be difficult to dislodge, as would zamindars, but among the 20 per cent or so whom Taniguchi regards as (at least potentially) small local leaders, there was surely opportunity for both successes and failures.

For all this, there do seem to be, as presented here, some recurrent features or related models of nineteenth-century change. One key argument is provided by Yanagisawa, from his analysis of the landholding statistics of Tiruchirapalli district (Trichinopoly) in the Madras presidency about the turn of the nineteenth century. He reveals a reduction in the area of land owned by Brahmans and in the number of Brahmans

----

[22] Peter Robb, 'Hierarchy and resources. Peasant stratification in late nineteenth-century Bihar', *Modern Asian Studies* 13, 1 (1979), pp.97-126, especially pp.119-26.

among the large landholders, and the acquisition of small amounts of land even by some low-caste people. The change marked the declining rural and growing urban interests of the Brahmans, and suggests *inter alia* a reduction in the number of relatively small holdings which employed non-family labour—another example (in addition to the one cited by Taniguchi and several others to follow) of the shift of advantage away from rent-receivers and towards owner-cultivators and direct managers. It is more difficult to judge the significance of the minute plots showing up in the record as being in the hands of the lowest castes. Low-caste people, including bonded labourers, in practice had long had tiny plots on which they could grow a few crops: it was a way of keeping down the payments which had to be made from their employers' output. But if a few of these people had effective and not just notional possession of these lands (that is, if the change was not merely administrative or formal), then such people must be assumed to have acquired an additional asset, provided always that one presumes the lower castes previously to have been invariably poorer and weaker than those of higher social status.[23] Most important for the questions I am now taking up is, Yanagisawa argues, that the relative advantage of the larger farm decreased and greater rewards accrued to intensive, careful cultivation. Given the shifts in occupational patterns—Brahmans using their education to gain jobs in towns, and others also migrating further afield, supporting their families by remittances—it was not surprising that there should have been more leasing-out of land. One may imagine that elsewhere comparable imperatives might have been met by the evident extension of both the small family farm and share-cropping. We should note the primacy given here, as by Mizushima or Kiyokawa and Ohno, to the role of economic efficiency or maximisation.

The question of labour is a difficult one. Yanagisawa considers the demand to have changed in character and also regards the shift from permanent to daily working in this period to be advantageous to labour,

[23] Washbrook has suggested that 'pariahs' were in a stronger position in the eighteenth than the nineteenth century in south India; 'Land and labour in late eighteenth-century South India: the golden age of the pariah', in P. Robb (ed.), *Dalit Movements and the Meanings of Labour in India* (Delhi, 1993), pp.68-86.

in that the tenant-cum-labourers who were engaged in daily work may have gained in status as small farmers with a sense of independence. He stresses the new opportunities for work. Emigration, for example, provided some lower-class people with the funds to buy land, and also reduced the dependence of some of the high castes on rural incomes. Other scholars, for example Jan Breman, have thought that the casualisation of labour weakened its position, while Ramachandran found a persistent variety of labour relations.[24] The advantages of different systems must depend in part on total income over a whole year or longer: daily rates tended to be higher, but daily agricultural labourers were usually unemployed and unpaid for significant periods. Rather paradoxically, Yanagisawa suggests that there was a growth of independence among labourers stimulated by emigration, while at the same time hired labour became less productive, leading to an increase in share-cropping. In Bengal a similar increase had been attributed to population pressure and the loss of viable landholdings by cultivators.

Mizushima offers another version of change though not one confined to the colonial period. His argument about south India starts from the perception of pre-colonial developments (especially, he thinks, in broadening the involvement of intermediaries with the villages); but he adds three related developments which accelerated or began during British rule. First, village leaders emerged with power as individuals (rather than through collective mechanisms). This power, like the notion of landed property, was at variance with the *mirasi* system. The village heads, often bought off by the Company with large revenue-free holdings, began to claim new dues and to support activities formerly paid for through harvest shares. Secondly, and subsequently, the individual plot became the unit of revenue administration, breaking up a system which had united revenue-free land, harvest shares and the provision of taxes, labour and services in a single socio-economic process. The outcome was ultimately to marginalise the headmen, and (more controver-

---

[24] See Jan Breman, 'Labour migration and rural transformation in colonial Asia', *Comparative Asian Studies* 5 (Amsterdam, 1990), and V.K. Ramachandran, *Wage Labour and Unfreedom in Agriculture. An Indian case study* (Oxford, 1990).

sially) to remove all effective intermediaries between the state and the plot-holder or raiyat. Thirdly, (once again) growing commercial activity placed new resources in the hands of classes which had not participated in the *mirasi* system, or provided them outside the system for those who had been involved: headmen, local intermediaries, traders and government all found that the alternative sources of income and influence were being enhanced.

Particular attention is being paid in all these accounts to landholding. This is also the subject of Nakazato, who reassesses the record of registered transfers of tenancies (raiyati holdings) in Bengal between the 1880s and the 1940s. We have here as well, incidentally, a case study for the evaluation of quantitative methods. Interpretation of the remarkable but uneven increase in transfers is made difficult, as Nakazato explains, by the known under-reporting of the earlier period—he quotes but considers too high an estimate that some 30 per cent of transfers were recorded in the 1890s; my view is that 10 per cent (the very proportion in which registrations for the 1880s stand to those of the 1940s) would not be implausible. Subsequently, various pressures and legal and administrative changes greatly increased registration by making it easier and more advantageous. In these circumstances, it requires a certain act of faith, which, however, many have been prepared to make, to believe that the record actually shows something about changes in the total numbers of holdings being sold. Interpretation based on the numbers of transfers is further complicated by re-transfers—the problem of double-counting never seems to be resolved, for example when calculating the percentage of all land or holdings transferred. Interpretation is further complicated, though Nakazato does not accept this point, by the undoubted subdivision of holdings under conditions of demographic and economic strain—or, for that matter, in search of social or economic advantage. At its simplest, holdings became smaller on average, though this did not preclude regional variations or consolidations of production units in some cases. This actual fragmentation, which Nakazato puts at about 25 per cent over several decades but which it is difficult to measure precisely, needs to be added, therefore, to the increased registration of smaller transactions discussed in the paper. Some indications can be obtained of the influence which this could have had on the registrations. From the statistics in Nakazato's Table 4, and others he has provided, we can gauge very roughly that at the start of his period the average occupancy holding whose sale was registered was large by the standards of Bengal, and that the variations between districts corresponded to some extent with those for price.[25] In the conditions of the times, we would have expected registrations to have been made more commonly for larger and more valuable holdings—because of lack of knowledge, uncertainty of the advantages, inconvenience and cost (including a fee to the landlord), and disputes about whether or not tenancies could legally be sold by their occupants. One can safely assume that the average holding was very much smaller in the enormously larger number of registered transfers in the 1940s. Apart from demographic change and alternative employment (which permitted families to hold on to or acquire very small plots), there had been, as Nakazato reminds us, a huge increase in the number of registration offices, a rise in the value of land, and a tendency to sell part rather than entire holdings.

These are the kinds of difficulty which bedevil the use of statistics And yet what can Nakazato's calculations show us? First, the increase of registration measures that intrusion, already mentioned, of public institutions into agrarian relations, and the standardisation of practice and rights; particularly it charts state influence over the securing and enforcing of mortgage contracts. As landownership was increasingly defined and recorded, credit was increasingly secured through a mortgage on land rather than on moveable property. Such formal arrangements clearly spread between 1880 and 1940 so as to affect very large proportions of all holdings. Secondly, just as clearly, they provided mechanisms for the use of those who had either high status or particular advantages. As the man:land ratio worsened, such people were enabled to resist a trend towards fragmentation; as commercial profits grew, they were often able to gain in independence. One feature of Nakazato's interpretation then takes on importance: it is the evidence that those gaining in land were predominantly 'raiyats', except in some districts around Calcutta where the urban middle classes were very significant

[25] See the table appended to this chapter.

purchasers. Of course the spread and increase of the urban classes' involvement in land purchases are important, though the buyers may often have been the children of landholders or those who served agriculture, and though such groups, including professional moneylenders, had always had some interests in land. But the persistence of raiyat purchasers shows the relative slowness of any change; and these raiyats were not predominantly cultivators or self-sufficient producers—they were employers of labour and renters to sub-tenants, including a growing number of share-croppers. In some places they were 'jotedars'. Nakazato wishes to show the impoverishment of many in Bengal; the converse of this, however, is that he also reveals something of the character of those who were gaining in wealth.

IV

As said, the changes are seen in these essays more in terms of landholding than of land use or of labour. But they rest on a picture of considerable upheaval or transformation. The nineteenth and twentieth centuries, these contributors would agree, saw a change in the effective units of production and in labour relations, the development of land markets and a new basis and sources of supply for credit, the commercialisation of production, and the reorganisation of trade and of structures of power.[26] Ecology, disease and demography played an important part too, as did administration, law and politics. The forces of 'modernity' attacked and undermined the pre-existing order. To assess these judgments, in this section I will discuss some aspects of Indian adaptability in the face of such challenges, the purpose being to measure the extent of the changes.

Much in these papers is about stratification, and except for Yanagi-sawa, and other evidence of improvements filtering to the poor (for

---

[26] The individualisation of units and means of production, as described by Mizushima, was gradual and incomplete (perhaps more so than he admits) but the eventual direction was plain enough—towards household farms held individually within fixed boundaries, towards wage and casual labour or rental agreements, towards separate irrigation (wells or government canals rather than village channels and tanks), towards individual credit and marketing arrangements.

example, in alternative sources of income or rising wages),[27] the general impression here is the familiar one of mass impoverishment under the expansion of the colonial state and market relations. The first picture is one of burgeoning and unanticipated consequences. The emphasis is on the unintended victims of economic 'progress', just as it was on the inadvertent consequences of legal changes. Three challenges may be considered. The second and third are colonial rule and commercialisation. The first is the worsening context in which production took place: the instance given here is Wakimura's discussion of famine and disease. He is anxious to show the relationship between malaria and 'excess' mortality in north India, as others have done for other regions.[28] The essay contributes to debates in demography, but it also casts light more obliquely on questions of social and economic change, which is Wakimura's ultimate concern. This paper adds to the catalogue of external forces—all examples of interference or competition—which affected the Indian countryside. It is also valuable in reinforcing our theme of Indian conditions and the way they reacted with the external inputs.

Wakimura's problem is to explain high mortality rates and demo-

---

[27] Like Yanagisawa, Christopher John Baker, *An Indian Rural Economy. The Tamilnad Countryside* (Oxford, 1984), notes (pp.152-3 and 174-5) some signs on the plains, and even in the valleys, of wages edging upwards and of subordinate groups taking on small amounts of land (enough so as to raise government complaints at the spectre of 'pauper raiyats').

[28] The point is made by Dyson; see Wakimura. For an explanation of the role of malaria in 'famine' mortality in Madras (1877-8) and Punjab (1896-7 and 1900-1), see Elizabeth Whitcombe, 'Famine mortality', *Economic and Political Weekly* 26 (5 June 1993), pp.1169-79. The reference to social dislocation is also taken up in Arup Maharatna, 'Malaria ecology, relief provision and regional variation in mortality during the Bengal famine of 1943-44, *South Asia Research* 13 (1), May 1993, pp.1-26; but his main point—that excess 'famine' deaths due to malaria were greatest in areas of least malaria endemicity raises some difficulties for connecting 'famine' malaria with canal-building (which also arguably increases endemicity). Hence the large increases in fever mortality near Calcutta may suggest another factor rather neglected in these accounts, namely famine migration— whether to the city of Calcutta, or to irrigated districts as in Wakimura's U.P. However, I suspect there will be a closer correlation between increased death rates and the availability of work, than between death rates and *prices*, and hence the direct availability of food. Maharatna shows the latter not to be significant, measured either in prices or gratuitous distributions; see also Robb, *Rural India* (Delhi, 1992), pp.124-31 and 149-50.

graphic mobility in nineteenth-century India. He stresses the importance of epidemic disease rather than famines (in so far as the two can be distinguished), the role of malaria in 'famine' mortality, and in particular the important role of irrigation canals in furthering these epidemics. There are important lessons here with regard to the means of understanding the changing context of the colonial period. Wakimura concentrates on the thirteen districts which comprise the Ganges-Jumna *doab*.[29] A refurbished West Jumna Canal reopened in 1820, and the East Jumna Canal ten years later. The huge new Ganges Canal was fully operational by 1857. The Agra Canal opened in 1874, serving Agra and Muttra, and the Lower Ganges Canal was completed in 1878, taking over the lower part of the former Ganges canal; that in turn was later greatly extended in the north by new branches through the centres of Saharanpur, Muzaffarnagar, Bulandshahr and Aligarh.[30] Wakimura finds that in the 1870s malaria mortality, for example, was highest of all in Aligarh and Bulandshahr in the central *doab*, next highest in the adjacent districts (Meerut, Etah, Muttra, Mainpuri and Budaun, only the last of which was outside the *doab*), and high, though somewhat lower again, in another ring of districts around these ones. The links between the disease and the canals, as discussed by Wakimura, are complex, even if one accepts the mortality statistics (which may exaggerate the incidence of malaria). The connection with the canals is undoubted but the geographical and chronological fit is incomplete. Though canal-building between 1868 and 1878 may well have increased the incidence of malaria, the highest mortality in 1879 was in the *doab* towns, not the countryside, and in districts which had had canals for more than a generation. (The further large extension of the canals in these most affected districts dated from after 1879.) Nonetheless, there clearly is a significant coincidence in this case, as elsewhere in India *but not invariably*, between famine and malaria; and this needs to be explained, possibly in ecology, or socio-economic conditions, or state policy.

[29] From north to south these are Saharanpur, Muzaffarnagar, Meerut, Bulandshahr, Aligarh, Muttra, Etah, Agra, Mainpuri, Farrukhabad, Etawah, Kanpur and Fatehpur.

[30] See Ian Stone, *Canal Irrigation in British India* (Cambridge, 1984), pp.144-57.

There is a related point mentioned but not discussed by Wakimura which is important for the volume as a whole—that to illustrate the link between famine mortality and malaria, he has to appeal obliquely to a notion of social breakdown at times of crisis (especially the movement of population). This implies a more resistant system functioning in 'normal' conditions. It is not precisely described in the paper, but certain assumptions about it are apparent. This introduction and many of the papers in this volume suggest ways these might be tested. But where Wakimura's account has the Indians rather *subject* to outside forces, other papers stress the *adaptability* of Indian responses to changing conditions. Nakazato and, less obviously, Yanagisawa alert us to the importance of the legal framework and the ways Indians could make use of it.

This brings us to the second point: any structural changes consequent upon colonial rule. The example given here, and the major component in most of these papers, though not in all recent scholarship, is the traditional one which also worried the colonial authorities in the nineteenth century. The argument is that the socio-economic structures changed with the advent of enforceable, transferable, individual property rights, for example as provided by law in Bengal after 1793; they worsened the consequences of debt by permitting mortgages to be secured against real property, and making it easier for the insolvent to be deprived of land. After 1819 this liability was extended to sub-proprietors or *patnidars*, as it was to occupancy tenants after 1859 or, less, ambiguously, 1885. By the early twentieth century, as Nakazato argues, a land market was sufficiently developed as to permit a certain redistribution of land. In particular this was the case as population rose, holdings were partitioned, and commercial interests intervened.

As suggested, however, the use of the new laws and institutions in making these changes might be seen as another instance of adaptation by Indians. And, arguably, the suggestion of social breakdown is then qualified by evidence of re-built structures and allegiances. Awaya concentrates on this question. She argues that European misconceptions of the nature of landholding shaped not only subsequent law and the historiography, but also popular perceptions and tenancy agitations in Malabar. The error consisted in assuming that there were clear-cut divisions among the agricultural population on the basis of rights over land, and in turning these distinctions into legal and economic power. British law more often sought to protect Indian institutions than to transform them, as in Awaya's example of the restrictions on the partition of the property of the *taravad* (the matrilineal joint family). But, in fact, the introduction of European principles inevitably altered or disrupted pre-existing practices framed according to different understandings. With regard to *janmi* or landlords and *kudiyan* or tenants (who were further divided into categories according to the nature of their 'leases'), Awaya illustrates the hardening and standardising, under colonial rule, of legal rights which previously had not been so definite or mutually exclusive: nor had their holders constituted homogeneous classes, though there were classes which might have been discerned—for example, great overlords, middlemen, and rich independent 'tenants'. Yet, Awaya argues, it was the common 'point of reference' provided by the colonial terminology which permitted 'tenancy' agitations to be organised against 'landlords'. In general, whatever the elements which produced the change—in this case resentment at the social and economic dominance of Nambudiri Brahmans, *taravad* quarrels, educated and higher-caste tenants, Mapilla revolts, modern political ideologies, and so on—it is certainly true that protest movements were differently articulated and rationalised in the twentieth century, and particularly in terms of rights and classes. In this form—the interests of 'tenants'—agrarian questions came to the attention of legislatures and governments, in Malabar as elsewhere, during the 1920s and 1930s. The precursors were the nineteenth-century debates, especially over the 1885 Tenancy Act in Bengal.

Our third subject, the growth of export crops, and of ancillary industries, is clearly another of the other major agents of economic change. Here too there are instances of adaptability. It was the growth of the jute industry which helped impart a particular character to the eastern districts of Bengal, as discerned in the pattern of registered land-sales discussed by Nakazato. Others have noted the dynamism of dry areas of south India, and the consequent social adjustments, with the expansion of cotton and groundnut production.[31] All these cases suppose a capacity to *react* on the part of the indigenous system—whether or not one follows Mizushima in thinking that it had thereby to destroy itself. An interesting perspective on this aspect is offered in this volume by the example, one of apparent stagnation in the Indian sugar industry, that is investigated by Kiyokawa and Ohno. Here the adaptability worked to permit the survival of what seemed to be an uncompetitive industry. My view is that the story has a pre-history. Though this paper doubts that it was relevant to the period studied, there can be little question but that Indians resisted factory-refined sugar on grounds of taste and religious fears, at least initially, in the nineteenth century. These objections help explain the barriers in the Indian market, prior to protective tariffs, to the international competition brought about by cost-cutting in sugar production and transport. Gradually an increasing difference in cost led, for example, to the mixing of refined sugar with *gur* (well-boiled undried cane-juice) for the making of sweets. This 'substitutability' increased in the present century, though total sugar consumption also clearly grew over the same period. At this stage Kiyokawa and Ohno begin their inquiry. Having measured the backwardness of the modern factory sector in terms of production-efficiency (attributed to intrinsic weaknesses), they find that indigenous manufacture survived precisely in so far as it was efficient or competitive. It was enabled to compete through technological adaptation. Though this is a volume primarily about agriculture, there should be no doubt of the general relevance of such findings on economic performance, based as they are on an assessment of the effectiveness of production methods. It enables us to focus on a question which is also crucial for agriculture: what were the conditions and limits of adaptability? In the modern sugar factories, one major weakness, as suggested by Kiyokawa and Ohno, was the absence or inadequacy of direct or vertically-integrated manage-

---

[31] Baker regards the development as a monetisation and commercialisation growing out of earlier, co-existing strategies of 'minimal' and 'intensive' agriculture, the change spearheaded by village and middle-level elites who had formerly managed and creamed off the surplus from an interdependent *mirasi*-style system of production and exchange; Baker, *Rural Economy*, pp.137-68.

ment of production. The indigenous manufacturers and suppliers (*khand-saris*) allowed increased output and even technical innovation and new crops, but perhaps they prevented rather than facilitated radical change.

V

Though there are always difficult issues of categorisation to be resolved, the implication of these papers is that there were various versions of common trends. However, it is plain both that intermediaries continued to be important, and that new circumstances did not merely benefit an unchanging agrarian elite. Different conditions favoured different skills and resources, but of the persistence of a certain kind of role and relationship, from the *mirasi* system to mid-twentieth-century Bengal and beyond, there should however be no doubt. It is striking how often change is seen as taking the form of stratification, and in particular as emphasising the middling roles. In Bengal and Bihar some smallholders from among the 'sturdy agricultural castes' and artisans may have gained in independence, and taken on additional land, benefiting from commercial agriculture: this is the echo in Taniguchi of Mizushima and Yanagisawa. There are also, as we have seen, arguments for a decline of the zamindar and the rise of the jotedar—a level above those prospering agriculturists—in late nineteenth-century or twentieth-century Bengal.[32] Similarly the essay in this volume by Awaya notes the significance of increasing evictions in fomenting tenancy agitation in Malabar, and suspects that they may have been used to bring in what she calls 'more resourceful' tenants, a category which therefore may be supposed to

[32] See, for example, Partha Chatterjee, *Bengal 1920-1947. The land question* (Calcutta, 1984), which suggests a declining class of proprietors, a rising rich peasant and moneylender-trader class, and a growing class of poor peasants and share-croppers. But see also Akinobu Kawai, *'Landlords' and Imperial Rule. Change in Bengal agrarian society c.1885-1940* (2 vols.; Tokyo, 1986 and 1987), for an argument that zamindars and their rental incomes did not decline significantly until the 1930s, and did so then because of the depression and not the rise of the jotedars. For an important argument, which somewhat reconciles these differences, about a degree of resilience and adaptability of property relations but also new kinds of landlordism growing up at the lower levels of the zamindari system (associated with share-cropping), see Nariaki Nakazato, *Agrarian System in Eastern Bengal, c.1870-1910* (Calcutta, 1994).

have been gaining land.[33] These tenants were, in effect, land-controlling groups. Later Awaya emphasises the political importance of such dissatisfied intermediate land-controllers in securing tenancy legislation; these are the counterparts of the jotedars as defined by Ratnalekha Ray. However, there is an ambiguity to be resolved between such accounts and those which stress the advance of a section of smallholding agriculturists. It may seem that independent smallholders are middle or rich peasants, while jotedars are brokers of some kind. The advance of the one may imply a process of 'peasantisation' alongside one of growing landlessness, but higher status for the other suggests a closer hold by village elites mediating between labour the market, or between the taxpayer and the landlord or state. We need to clarify this possible confusion of categories and changes.

There are elements of a kind of middle-peasant thesis, suggestions pointing to the relative success of small independent cultivators, to the growing profitability of production rather than rent-receiving. In Taniguchi's examples such developments apply somewhat to the later eighteenth century but more especially one hundred years later. A similar point has been made about peasants in west and central Bengal, and might at some stages have applied to jute cultivators in eastern Bengal. On the other hand, the argument about jotedars is really about the importance of dominant villagers—people with large lands, employers of labour. (The role might also be performed by small zamindars or intermediary landholders; it did not depend on the legal niceties of tenure.) Baker describes, on the dry plains of Tamil Nadu, 'a shift from the old system in which "intensive" farmers commanded the labour services of the sub-subsistence cultivators and rewarded them with grain, to a new system in which the "intensive" cultivator parcelled out surplus land, often in units which could be worked with one pair of bullocks'.[34] The latter system still involved the landholder very closely

[33] The increases were not steady or very dramatic: from Awaya's figures, decrees, differing percentages of which were executed, seem to have numbered 1,000 to 2,000 a year in mid-century, 2,000 to 3,000 in the 1880s, 4,000 to 5,000 in the 1890s, around 2,500 in the 1910s, and 4,000 to 5,000 in the 1920s.

[34] Baker, *Indian Rural Economy*, p.153.

in details and inputs of production, and still rewarded the actual cultivator poorly. And, though in this example the cultivator made a fixed cash payment for the land, it was analogous in other ways to the share-cropping which existed, and indeed notably spread, elsewhere, including Bengal: the tactics were often designed to intensify cultivation for the market while keeping down the rewards for those who actually laboured in the field. It seems that, as the precise social disposition differed from place to place, so too did the reactions to change. Even if we consider just one factor, the increased value of production for the market, we may still find areas where middle tenants of agricultural caste took advantage, areas where population increase and subdivision of holdings prevented any benefits, and areas where landlords profited, as in parts of Bengal and Bihar where they retained or extended large demesne lands for commercial production. It is quite possible therefore for our central actor to be the middle peasant in one place or level, and the jotedar in another: both might be carrying out roles in cultivation, moneylending and trade, acting on behalf, or commonly at the expense, of their weaker neighbours and dependants.

The privileged raiyat provides a useful model for understanding the nature of those of middling rank, the society's many brokers, and the extent to which the role was more significant and permanent than those who played it. There is room for significant doubt about how far inter-mediaries and specialists—and for that matter caste—became elaborated during the nineteenth century, partly under colonial influence. But there can be no doubt that these same characteristics existed already in the seventeenth and eighteenth centuries, and presumably far earlier as well. In this sense nineteenth-century changes may be regarded as variations upon a theme, rather than a new departure. The colonial state penetrated into the countryside and into 'private' transactions more than many of its predecessors, and estate-owners gained its support through new laws and administrative or political change; there were adjustments in the balance between different rural sections and resources; there were intru-sions by some broader, indeed international, credit and marketing net-works. But office, credit and social prestige remained important to local power, and production decisions were still strongly influenced by village-level conditions and controllers. Even in a merchant- or state-sponsored system of harvest credit, such as was used to encourage commercial production, the local intermediaries were generally key players. Indeed arguably their position rested as much upon such external links —administrative, political, commercial and socio-religious—as upon their landholding or local lending.

What then is meant by the intermediary in this context? The crucial point is that he is not merely a political or an economic functionary: he is the manifestation of a social cleavage. Clearly there is any number of roles in which there are genuine middle ranks, brokers or agents. These exist in all societies, their position and tasks changing over time. In India there were government or village officers, social elites, military chieftains, landlords, 'professionals' (priests, accountants, vakils, bankers), creditors, labour recruiters, and traders. The middlemen may be interdependent, mutually supportive, overlapping or conflictual. To say that there were intermediaries in this sense is to say very little. Hence my concern is less with personnel or structures, and more with the way in which these necessary roles were performed, with the norms, limits and customs in which they were located.[35]

[35] This is a kind of political economy of the intermediary, not unlike—but wider than—that provided for 'portfolio' capitalists described by Subrahmanyam and C.A. Bayly (see Subrahmanyam, *Merchants, Markets*, pp. 242-65), who demonstrate the varied activities, skills, resources and social sets of specialists operating between producers and consumers (see also ibid., p.2). My argument here is about roles—of government credit, production-management and exchange—performed by various actors. Many were 'amateur' in that they did other things as well (as did Subrahmanyam's and Bayly's capitalists), but this does not, in my view, prove the usual argument limiting trading professionals to a 'capitalist' system. Rather, the entire frame of that argument seems unhelpful: the 'portfolio' of sources and activities enabling individuals to perform these roles was not immutable but entirely 'specialist' and 'professional', and the mix might not be so very different from that with which a 'modern' merchant operates. What differed would be the ideological and institutional context. Subrahmanyam and Bayly propose a hypothesis about change in these terms, taking up Breckenridge's concept of 'social storage' (the accumulation of knowledge in institutions) with some qualifications concerning likely pluralities. Thus commercial skills could be developed alongside political and military resources to produce great merchant entrepreneurs in India in the sixteenth to eighteenth centuries. European power destroyed some of these and changed the character of others; in localities, however, I believe small equivalents

The intermediary, in this discussion, is a powerful local figure who is able to operate in two worlds (though to a limited degree), providing services and drawing profits. He is an intermediary because of his liminal position, but he is neither an absolute controller of production, nor merely an agent for outsiders. His strength and limitations depend upon the multiple sources of his prestige, and the variety of his operations. He is revealed in such terms as 'malik', 'patel', 'dominant lineage', 'principal cultivator' or 'village banker', for commonly he is a social leader, a village office-holder, a rich peasant, an independent village moneylender, or a local agent of trade. These roles did not imply functionaries of identical type or size, but rather a range of lords and brokers of society, labour and trade. Where the roles were not concentrated in one household or group of associates, they were allocated across a number of rivals who were, however, paradoxically, mutually-dependent or complicit. Their ultimate alliance is illustrated by the resilience of the systems of control. Among the mass of poorer cultivators and share-croppers, some could at times choose their creditors or masters, and play middlemen off against each other, or even appeal past them to their principals in the state or market—the greater intervention of both law and trade into the countryside could have this liberating effect. But, by thus changing one creditor or patron for another, the poor seldom reduced their underlying subordination. Also the upwardly mobile or the outside official, planter and entrepreneur could join the system but found it difficult to undermine it. Intermediaries survived and gained in strength because of the range of their resources, in inheritance, status, wealth, office, information, and because they were necessary to a society and economy marked by perpetual delegation and fragmentaion in the processes of authority, production and exchange.

The critical question, uniting the discussions of structure and change in this volume, is whether Indian institutions, which we have agreed are subject to pressures and are adaptable, could develop on an uninterrupted trajectory, or whether they reached some limit or impediment after a certain point. For example, Ronald Toby has shown how one family transformed itself in late-Tokugawa Japan, from moneylending headmen

still thrived in the nineteenth century.

into components of a rural interbank network. From a base as substantial peasants in a village already enmeshed in a wider economy, the family became receivers of deposits, entrepreneurs and specialised lenders to more substantial clients.[36] There are certainly examples of similar developments in nineteenth-century India, but it may be that they were more typical of rural Japan. One reason for India's shortfall is probably colonial rule—for example, a credit system focused on Europeans and a few export commodities, or a legal system initially supportive but increasingly restrictive of moneylenders. But other reasons may include persistent norms defining the ways tasks were grouped or divided. For this reason I do not altogether follow Mizushima when he proposes, in effect, the demise of middling or managerial roles in favour of a new individualism. I believe that revolution still to be unfinished in India.

VI

Two issues are particularly important in regard to intermediaries; the first is how far they were a persistent feature of Indian society, and the second is the impact they had on production and hence 'development'. It should be noted that the precise evolution of the intermediaries may not be material to their impact on production decisions. Why should they have a particular character in India? The explanations include caste, for example the aloofness from agriculture of the ritually high, or the tendency of operations within a single process to be fragmented among many hands. A favourite candidate in the past has been climate, and it *is* clear that a much subdivided system of production and exchange posits a certain level of ready surplus. (This is, for example, a tempting argument, added to social conservatism, to explain the longterm decline of Bihar.) Another related possibility is the culture of particular crops and production methods—transplanted rice, multiple harvests, mixed cropping, localised small-scale processing, and so on—which in some circumstances might invite the bifurcation of production and exchange into particular specialised tasks. Or there is law, the fact that the

[36] Ronald P. Toby, 'Changing credit: from village moneylender to rural banker in protoindustrial Japan', in Gareth Austin and Kaoru Sugihara (eds.), *Local Supplies of Credit in the Third World, 1750-1960* (London, 1993), pp.55-90.

colonial power favoured individual ownership while restricting the dispossession of landholders (a recipe for encouraging agents and, through inheritance, fragmentation). Finally, with regard to sugar factories but also applicable to cultivation, there may be yet another possibility: labour absorption, the dispersal of work across the population, made possible by the relative efficiency and profitability of small-scale operations, or conversely the relative lack of competitiveness in large-scale, 'modern' enterprise.

Ultimately, the economic characteristics must be assessed at the micro-level. There the subdivision of function and personnel could have the effect, in brief (as in the sugar industry), of dissociating decision-making and profit-taking from production. The cultivator made subsistence decisions, and to a large extent (subject to custom and other restraints) determined agricultural methods. But commercial decisions and profits tended to be located with non-producers—officials, merchants, zamindars, jotedars. Cultivators had limited direct access to markets and commercial intelligence, while merchants were similarly isolated from most cropping decisions and production processes. Obviously jotedars (for example) did sponsor agricultural improvement, but only to a limited extent; and they too tended to become users and creators of further intermediaries once they had achieved a modicum of success. This is not to say that these modes of operation *explain* differences in economic performance, for it is difficult to single out any one factor. But certainly economic development requires some higher mobilisations of capital and labour, and that need must focus attention on intermediary or managerial levels. How far can unified decisions be taken on production—whether by farmers having independent access to necessary resources and information, or by local controllers intervening in factors of production, or by supralocal sponsors and consumers? The social structure and economic stratification must have affected the answers to these questions (and will allow us to make deductions about production decisions).

We should consider agricultural practice in more detail. First, we can conclude that there was a degree of specialisation. If we take larger holdings in Bengal, as Taniguchi does on the basis of Buchanan's observations early in the nineteenth century, we find that winter rice was very generally grown, but that varying amounts were produced of other valuable crops, such as oilseeds, sugarcane and *khesari* (a pulse often used for fodder). The recorded holdings each produced more of one at least of these crops than could have been intended for household consumption.[37] The choice of cash crop would obviously have reflected ecology (soil, access to water and so on), know-how and seed, labour supply (for troublesome crops such as sugarcane or transplanted rice), and demand (the expected market price or the preference of a creditor). Secondly, however, from the same evidence we can deduce that considerable effort was expended, even on larger holdings, on a range of crops (presumably to spread risk and to match a variety of agricultural conditions), and in quantities which suggest that they were intended to provide only for consumption. Taniguchi attempts some detailed analyses of presumed household budgets from 1814; these are calculations of the kind usually looked at with scepticism, but they do give broad indications of the terms of decision-making. Taniguchi's Table 10 is necessarily schematic (for example, in ignoring mixed cropping, which was widespread), but from it we can suggest that, in most holdings, a certain proportion of production was motivated by subsistence needs, in the sense that small quantities of low-value crops may be presumed to have been intended for consumption. For example, on 21 bighas of cultivated land (including double cropping on 5 bighas) we find 5 bighas devoted to summer rice, apparently *bhadai dhan*, or rice sown in April or May and harvested in July, August or September, usually called *aus*. The growing of this crop implies a distinct purpose, which was perfectly apparent a hundred years and more after this example—indeed was implicit in the existence of kinds of rice and their seasons.

Though there were hundreds of varieties, and differences from one district to another, and several schemes of classification, broadly speaking there were three main categories of rice according to harvest in Bengal, their importance differing from one area to another: winter rice

---

[37] We do not need to concern ourselves with the detailed calculations of production costs and profits, which would differ according to crop, soil, location, expertise, tenure and labour regime.

(*aman*); the summer *aus*, of lesser importance; and the richer, wet spring rice (*boro*) sown from December to February, harvested from April to May.[38] The first, found in the delta and on irrigated or inundated lands, was wet-cultivated, transplanted, high-quality and widely-marketed (though possibly smaller proportions of the total output were sold, as jute-growing expanded). It was a standard item of trade, performing the role which was also attributed to other high-value crops such as sugarcane, oilseeds, cotton and so on. The second kind of rice, found on higher, drier land, especially in west and north Bengal, was poorer in quality and regarded as a subsistence grain. The last, often high-yielding and a food for the poor, was also grown on wet deltaic lands, commonly as a double crop.[39] Put briefly, it follows from this account that rice-cultivators did not always intend to maximise the output of the most valuable grain, despite the existence of a vigorous rice market.[40] We can trace different motives in the use to which agricultural land was put; in particular the subsistence motive is apparent in the smallholding culture, whereby modest or, often, tiny units of production produced small amounts of a wide range of crops. The allocation of effort between them, as between these three kinds of rice, was determined by external demand, market opportunities, soil and water, labour-requirements and subsistence need. Cultivators' priorities and non-commercial considerations still held considerable sway under

[38] See the entry under *Oriyza sativa* in G.E. Watt, *Dictionary of the Economic Products of India* (1890).

[39] Ibid. For a brief description, indications of the regional differences, and some twentieth-century changes, see also Bose, *Agrarian Bengal*, pp.38-58, 71-2 and 94-5.

[40] It certainly could have been intended to maximise and spread the *total* output, but not necessarily that of the highest-value crops. Thus, though the best classes of winter rice might be grown by themselves (between May and October or November), it was often the case that even good-quality *aman* crops were transplanted into lightly-flooded fields in which *aus* rice was already established, having been sown broadcast in April or May, and weeded well into June. Similarly, the summer *boro*, a coarser grain eaten by poorer people, might include fast- and slow-growing varieties sown together but harvested separately. Local variations on cropping patterns were remarkable, but so was the possibility for different strategies between seasons, given that so many types of rice were available and five or six different harvest seasons feasible.

the semi-detached control of the intermediaries who linked them to but insulated them from the market.

The difference in production may be seen, then, in types of crop. But the story is more complicated than this. Taniguchi assumes that much of the winter rice output would have been sold to meet cash requirements (and we might be misguided enough to think of it therefore as commercial in character). He also suggests that the average holding had a small output of valuable crops in household gardens. His analysis of household budgets for 'surplus' raiyats assumes a kind of stability within these arrangements, excluding the effect even on these middling families of poor seasons, family cycles or extraordinary expenditure. Sometimes one has to add, on the positive side, the fluctuating possibilities for non-agricultural income. But the long-term effect of household crises was surely to reduce the independence of the majority, and to make (at least in the richer and more secure areas, paradoxically) for a sharp gap between the few rich households and a further-stratified majority. One reason for this is that there is more to the consideration of production decisions than the distinction just proposed on the basis of crops and land-use. We need to add the credit and trading systems into our considerations. Though their features are well-known, their general economic significance is not fully appreciated.

I want now to go back to the example of rice, with credit and marketing in mind. A late nineteenth-century estimate was that a little over 3 per cent of the rice produced in Bengal was available for net exports, and that some 79 per cent went for 'local consumption'. In a global sense then much of the rice was grown for subsistence, and, though large quantities (albeit the equivalent to less than 4 per cent of this estimated total) were imported from beyond its immediate environs to feed Calcutta, and other substantial amounts of grain were marketed more locally, yet one can postulate any number of concentric circles of local exchange, from the region to the individual village or even household; in the words of A.P. MacDonnell's celebrated *Food-Grains in Behar and Bengal*, 'a quantity...equal to the absolute wants of the people during the [short] intervals between harvests, is always in the district'.[41] The

[41] Quoted by Watt, see note 31. The preceding figures were calculated from

aus crop was thought to represent less than 20 per cent of the output, and it was exchanged mainly within the narrowest of these circles. But the bulk of the *aman* crop too was also exchanged on a similar basis, despite the 'large business' of the traders from outside each district. To call this subsistence production is not to imply a body of self-sufficient peasant households: I have in mind an extended meaning of the term, to distinguish it from production oriented to the market. In particular, as (though differently) in the *mirasi* system, there were localised (or personalised) relations of interdependence whereby the grain which was produced was exchanged immediately at harvest to meet immediate demands (including advances received for seed and production costs, or for food between harvests). MacDonnell saw this perfectly well; he wrote of subsistence production (in the usual sense) and the equally imperative disposal of a proportion of the crop to pay for rent and other outgoings to others. What shows the character of this exchange, and hence of the production which made it possible, is that (though minor purchases might be made by drawing on stores of grain exchanged at more favourable rates) a large proportion of the payments of rent and for credit was made forcibly on the threshing floor, or at least at the moment of harvest, and hence at the time when prices were lowest. (In addition, the prices allowed to the cultivator were commonly at a further discount upon even the cheap rates then prevailing in the wholesale markets.) Here too is the true significance of the intermediaries to production decisions. It does not seem to have declined in most areas despite the changes of the colonial period.

It should be stressed that this argument is not merely another re-invention of the 'self-sufficient village' model defined by colonial administrators and popularised by Marx. Rather, it is a way of reconciling the contemporary observations of subsistence strategies with the growing evidence in the recent historiography of dynamic trade, well beyond the needs of the state and the military, in the pre-colonial period. The argument is for the co-existence of subsistence and commercial strategies, even in the 'capitalist' era. It is, moreover, for a continuing influence in both cases (subsistence and commercial) from

the same source.

and upon the institutions and ideologies of the so-called village. It is finally in favour of an abandonment of schematic explanations of change.[42] It addresses some of the same questions from a different perspective, as those who have contrasted forced or distressed selling and buying by peasants, as a survival strategy, with 'normal' commercialisation for profit on a kind of vent-for-surplus model.[43] The argument here stresses differentiation (the mixed roles of intermediaries) and also the co-existence of subsistence strategies on the part of the producers with profit motives on the part of elites, controllers, lenders and traders. The difference is important for development. A division of function and motive in decision-making may possibly encourage the extension and even a distress-led intensification of production, but it is less likely to promote an absolute increase in productivity per acre or per worker.

It is a matter of judgment and for further investigation how far these features also pertained in the eighteenth-century *mirasi* system already discussed. We may note however that if there were (as Mizushima suggests) some cultivation where there was insufficient margin to pay the state at the full rate, it would follow, conversely, that state revenues and village costs were more dependent on higher-value cultivation or irrigated land. The distinction is apparent in the detail of Mizushima's account of irrigated villages, and would be even more apparent in the

[42] This argument, on the *co-existence* of different modes and motives of production, contrasts with the apparently similar idea of subsistence-oriented trade, familiar in the African literature though less often discussed in regard to India. That concept is useful in describing the exchange of goods within a social system which gives priority to subsistence and kinship, which lacks universal monetary value, and in which the factors of production are not commoditised. In such literature, however, it should be noted that, despite the identification of some 'transitional' situations, the underlying idea is still of a process *from* subsistence *to* market-orientation, because of the incompatibility of various forms of 'tradition' with market-relations. See, for example, Richard Gray and David Birmingham, eds. *Pre-colonial Trade. Essays on trade in central and eastern Africa before 1900* (London, 1970), especially pp.1-6.

[43] Amit Bhaduri, 'The evolution of land relations in eastern India under British rule', *Indian Economic and Social History Review* XIII (1976) and 'Class relations and commercialisation in Indian agriculture: a study in the post-independence reforms of Uttar Pradesh', in K.N. Raj *et al.* (eds,), *Essays in the Commercialization of Indian Agriculture* (New Delhi, 1985).

more stark distinction made, particularly by scholars of the south, but replicated everywhere, between wet and dry lands. It implies that there was some specialisation of production drawing output on to the market, but directed towards meeting external charges over and above the costs of production and without any implication of a profit motive on the part of the cultivator. By MacDonnell's account, if the raiyat were only partially indebted, then he could retain some grain for his family's food in addition to the proportion needed for these outgoings. If the raiyat were more completely in thrall, then the whole would go to the *maha-jan* (the moneylender, who, however, had to ensure that the family survived, lest he lose future income). Theoretically there could be a tripartite division in which the final portion was a residue or profit for the raiyat—and then market production would be possible. But, as Mac-Donnell noted, the third part was generally in the hands of moneylenders, and it was this which permitted the export of grain from the countryside which produced it. It is important to note that in this description the role of 'moneylender' might be performed by village elites or zamindars as well as by traders and bankers—that is, the picture was complicated but not fundamentally altered by village-level bosses who, though themselves indebted and owing rent, could lend to their poorer neighbours and dependants.

If we take the minimum developmental inputs to be labour (its skill and care), irrigation and fertiliser, then we have already noted that small agriculturists or managing zamindars might be relatively efficient. Up to a point, household labour would be more productive than hired workers or share-croppers, particularly where land rights and markets permitted only modest economies of scale. And, though extensive canal or tank irrigation, requiring systems of co-operation, control and distribution, would probably tend to assist regional and village elites, yet in some circumstances wells might require little capital and be within the reach of smallholders. Thus, in Yanagisawa's account we find indications that irrigated land was mostly beyond the means of poorer cultivators however industrious, and that large areas of dry land were equally out of their reach for want of capital. But we also find that smallholders could gain some surplus by growing garden-produce on small plots irrigated by wells. It may be this fact which explains the slightly odd pattern of increased landholding revealed by Yanagisawa—that is, that the beneficiaries included those apparently without much capital, either social or economic. But a strategy of more intensive small-scale cultivation, commonly reported among agricultural castes, was restricted in its impact: it was limited to plots of a size which could be cultivated by family labour; it depended on ample and appropriately-timed rains, or a relatively high water-table and soil conditions that permitted the digging of simple wells; it required that cultivators could avoid providing forced or hired labour for others, and that their tenurial and credit conditions allowed them to retain any extra income gained by extra effort. The perpetuation of these limits upon increased profitability in agriculture depended, it seems, on the socio-economic control placed between village and region, producer and trader, subject and ruler.

Let us now return, for a final example of the changing circumstance against which these continuing limits were played out, to the background against which Wakimura discussed the role of malaria in 'famine' mortality. Many analysts have implied that increased exports, for example of wheat in the 1870s, drew upon a steady level of production which otherwise would have been consumed within India (specifically in this case that wheat production contributed to a kind of national foodgrain stock, even for rice- or millet-eating areas and classes affected by drought), and moreover that this marketed output was deducted from produce potentially consumed by the individual cultivators or communities which produced it. From what we know of production practices we can see that this conclusion requires a good deal of qualification. In the economy as a whole, we would be concerned with foodgrain prices, which indeed were taken as the measure of scarcity, but would want to calculate them not just in absolute terms but with reference to incomes, disposable surplus, reliance on the market and so on. Similarly we might try to assess total foodgrain availability, taking into account imports as well as exports, and regional differences. But there seem too many unknowable variables to be confident that such a computation would be convincing. The consumption patterns, output per acre, and 'exchange entitlement' all seem unquantifiable. At the micro-level, we

should remember the picture of many smallholders habitually contri-
buting fairly low proportions of their output to the market for their
own commercial reasons, and consuming produce from others of their
lands, or from elsewhere, while also being caught up in exchange to a
larger degree through rent and debt. At the other extreme we should also
note the independent farmer who was able to make cash profits from
increasing output and rising prices. (Some households would have fluc-
tuated between one condition and the other, and fortunes also changed
between areas.)

The variety of possible responses indicates that any increasing polarisation in society depended on the prior but strategic strengths of particular sections, especially those we have called intermediaries. This situation helped define the consequences for different sections of the population. If exports did reduce the food-stock available to the poor or during
famines, then one would expect to find either an increased involvement
in the market in terms of land-use, or a decreasing access to resources
on the part of poorer sections of the population. These increased vulner-
abilities do not depend on demonstrable changes at the macro-level.
Given a free market, any village economy would obviously be more
vulnerable to drought-induced famine the more it was subsistence and
localised in character: broader market involvement implies reserves in
money or valuables rather than in grain, and the operation of the market
(as McAlpin suggests)[44] enables grain to be purchased as a distance.
Famine could occur in areas which did not import food while exporting
produce (or tribute—rent and revenue), a lopsidedness of market
involvement which may explain some of the harsh famines in the last
decades of the nineteenth century. Indeed, production and employment
patterns may affect the nature and incidence of famine. A trigger for suffering among all those who had to sell their labour to others would be
the drying-up of credit, or climatic or other conditions, both of which
conditions would reduce the demand for labour, in agriculture and processing. In addition, the availability of casual labour reduced the need
for holders of grain or money to distribute it charitably in order to

[44] M.D. McAlpin, *Subject to Famine. Food Crises and Economic Change
in Western India, 1860-1920* (Princeton, 1983).

secure future production (the 'moral economy' argument). Against that suggestion we should note first the increased demand for labour in towns, factories, plantations, other commercial production and irrigated areas, and for transport and through emigration (though whether there was a net increase ahead of population growth is controversial, given the decline in military and local service occupations, and the collapse of some local industries under foreign competition—the allegation of 'deindustrialisation'). However, we should take note of the effective labour shortage in many areas and for many would-be employers because of the socio-political controls which continued to be exercised over their workers and dependants by the local elites. The older, more collective order survived in part, it seems, and was able to shape the consequences of new law, trade and production.

In this volume, first, there is a working assumption that economic advance was disappointing. Secondly, we agree that nonetheless there *was* economic change, including intensification of production, some technological innovation, the introduction of new commercial crops, and new export markets. Moreover, consequential changes are generally discerned within Indian agrarian society. Yanagisawa remarks on the decreasing advantage of the larger farm—perhaps what has long been termed the 'inverse relationship' between farm size and productivity,[45] and Mizushima notes the growing importance of individuals managing their own production. Nakazato, in quantifying the land sales to richer cultivators, presumably shows another aspect of this process, while the upheavals suffered by some are mirrored more obliquely in the patterns of malaria incidence discussed by Wakimura. All these are aspects of the school of thought which sees colonial rule and commercialisation as disruptive of an older (and usually more collective and supportive) social order. On the other hand, some attention is paid—especially by Taniguchi, and indirectly by Awaya and by Kiyokawa and Ohno—to the structures or relations of production, and some of these arguments

---

[45] See Terence J. Byres, 'Charan Singh (1902-87): an assessment' in D. Arnold and P. Robb (eds.), *Institutions and Ideologies. A SOAS South Asia Reader* (London, 1993), pp.296-8; and see also the discussion of fragmentation, pp.301-3. The article is reprinted from *Journal of Peasant Studies* 15, 2 (1988), pp.139-89.

invite attention to pre-colonial trends, or to indigenous responses to changing conditions. There is a possible discrepancy between this emphasis and the certainty of radical change. Two issues might be singled out: peasant independence and the role of brokers. This introduction has suggested that, in some senses, they are closely related.

A pattern of partial and uneven agricultural development in India is related to the subdivision of production and of management associated with incomplete specialisations and commercialisation, and the role of intermediaries. Some people prospered; and some major changes occurred, especially in the amounts of produce exported, in the cultivated and irrigated acreage, and probably in the extent of double-cropping. These were manageable without structural changes. But changes would be needed to achieve greater, more rapid and widespread improvement, through consolidations of land, investment of capital and mobilisation of labour. These arguments, derived from an interpretation of agricultural conditions and practices, may also apply more generally—to secondary industries and to broader economic and policy issues, the context or impact of which may be defined or characterised according to these features of the agrarian system. The market was not fully sovereign: the analogy is with states, whose jurisdiction, though never unqualified, may be even, unified, and definite within fixed borders, or else, even on matters of crucial importance, may be uneven, fragmented, vaguely-bounded, incomplete. We reach this conclusion by concerning ourselves less with forms or structures, and more with the way roles were performed in practice.

Table: *Estimate of average size of occupancy holdings transferred, with average value (Rs.) per bigha, in Bengal districts, 1891-5*

	Bighas	Rupees		Bighas	Rupees
Burdwan	3.98	22.3	Dinajpur	4.89	14.2
Bankura	4.98	12.7	Rajshahi	9.34	9.4
Birbhum	5.7	15.1	Rangpur	7.58	12.7
Midnapur	3.17	21.9	Bogra	4.95	13.0
Hooghly	3.14	33.8	Pabna	2.93	27.2
24-Parganas	6.58	11.3	Dacca	3.24	22.1
Nadia	9.37	7.7	Mymensingh	4.97	16.9
Jessore	8.08	5.9	Tipperah	3.72	18.9
Khulna	11.94	6.5	Chittagong	3.31	20.9
Murshidabad	2.69	11.3	Noakhali	6.74	9.0

Note: Prices for Dinajpur are for 1891-4. My thanks to Nariaki Nakazato for providing the figures from which this table has been calculated. The figures represent the total numbers of recorded transfers divided by the total area transferred. Unfortunately comparable data are not available for the period after 1904. It is interesting that the variation between districts in presumed total prices paid for each holding is far smaller than between either the average area or the average price; but too much should not be read into the figures. Among other variations to be considered is the different proportion of land in occupancy holdings in the different districts.

Bibliography

Amin, Shahid, *Sugarcane and Sugar in Gorakhpur. An inquiry into peasant production for capitalist enterprise in colonial India* (Delhi, 1984).
Baker, Christopher John, *An Indian Rural Economy 1880-1955. The Tamilnad countryside* (Oxford, 1984).
Bandopadhyay, Arun, *The Agrarian Economy of Tamilnadu, 1820-1855* (Calcutta, 1992).
Banerji, Himadri, *Agrarian Society of the Punjab, 1849-1901* (Delhi, 1980).
Bayly, C.A., *Rulers, Townsmen and Bazaars: North Indian society in the age of British expansion* (Cambridge, 1983).
Bhaduri, Amit, 'The evolution of land relations in eastern India under British rule', *Indian Economic and Social History Review* XIII (1976).
Bose, Sugata, *Agrarian Bengal. Economy, social structure and politics,*

1919-1947 (Cambridge, 1986).

Breman, Jan, *Labour Migration and Rural Transformation in Colonial Asia* (Comparative Asian Studies 5; Amsterdam, 1990).

Charlesworth, Neil, *Peasants and Imperial Rule. Agriculture and agrarian society in the Bombay Presidency, 1850-1935* (Cambridge, 1985).

Chatterjee, Partha, *Bengal 1920-1947. The land question* (Calcutta, 1984).

Chowdhury, Benoy [B.B. Chaudhuri], *The Growth of Commercial Agriculture in Bengal, 1957-1900* (Calcutta, 1964).

Chaudhuri, K.N. and Clive Dewey (eds.), *Economy and Society. Essays in Indian economic history* (Delhi, 1979).

Desai, Meghnad, *et al.* (eds.), *Agrarian Power and Agricultural Productivity in South Asia* (Berkeley, 1984).

Dewey, Clive (ed.), *Arrested Development in India. The historical dimension* (New Delhi, 1988).

— and A.G. Hopkins (eds.), *The Imperial Impact. Studies in the economic history of Africa and India* (London, 1978).

Frykenberg, R.E. (ed.), *Land Tenure and Peasant in South Asia* (New Delhi, 1984; first ed. 1977).

Goswami, Omkar, *Industry, Trade and Peasant Society. The jute economy of eastern India, 1900-1947* (Delhi, 1991).

Guha, Sumit, *The Agrarian Economy of the Bombay Deccan 1818-1941* (Delhi, 1985).

Hill, Polly, *Dry Grain Farming Families* (Cambridge, 1982).

Islam, Sirajul, *The Permanent Settlement in Bengal. A study of its operation 1790-1819* (Dacca, 1979).

—, 'Bengal Land Tenure. The Origin and Growth of Intermediate Interests in the 19th Century', *CASP* 13 (Rotterdam, 1985).

Kawai, Akinobu, *'Landlords' and Imperial Rule. Change in Bengal agrarian society c.1885-1940*, (2 vols.: Tokyo, 1986 and 1987).

Kumar, Dharma (ed.), *The Cambridge Economic History of India*, vol.II (Cambridge, 1983).

Ludden, David, *Peasant History in South India* (Princeton, 1985).

McAlpin, M.D., *Subject to Famine. Food crises and economic change in western India, 1860-1920* (Princeton, 1983).

Mitra, Manoshi, *Agrarian Social Structure. Continuity and change in Bihar 1786-1820* (New Delhi, 1985).

Nakazato, Nariaki, *Agrarian System in Eastern Bengal c.1870-1910* (Calcutta, 1994).

Pandian, M.S.S., *The Political Economy of Agrarian Change. Nanchilnadu 1880-1939* (New Delhi, 1990).

Pouchepadass, Jacques, *Paysans de la plaine du Gange: le district de Champaran, 1860-1950* (Paris, 1989).

Prakash, Gyan, *Bonded Histories: Genealogies of labour servitude in colonial India* (Cambridge, 1990).

— (ed.), *The World of the Rural Labourer in Colonial India* (Delhi, 1992).

Radhakrishnan, P., *Peasant Struggles, Land Reforms and Social Change. Malabar 1836-1982* (New Delhi, 1989).

Ramachandran, V.K., *Wage Labour and Unfreedom in Agriculture. An Indian case study* (Oxford, 1990).

Raj, K.N. *et al.* (eds.), *Essays on the Commercialization of Indian Agriculture* (Delhi, 1985).

Ray, Ratnalekha, *Change in Bengal Agrarian Society, 1760-1850* (New Delhi, 1983).

Robb, P. (ed.), *Rural India. Land, power and society under British rule* (London, 1983 or Delhi, 1992).

—, 'Peasants' choices? Indian agriculture and the limits of commercialization in nineteenth-century Bihar', *Economic History Review* 45, 1 (February 1992), pp.97-119.

Rothermund, Dietmar, *Government, Landlord and Peasant in India. Agrarian relations under British rule 1865-1935* (Wiesbaden, 1978).

— *et al.* (eds.), *Urban Growth and Rural Stagnation. Studies in the economy of an Indian coalfield and its hinterland* (New Delhi, 1980).

Satyanarayana, A., *Andhra Peasants under British Rule. Agrarian relations and the rural economy 1990-1940* (New Delhi, 1990).

Sen, Asok, Partha Chatterjee and Saugata Mukherjee, *Perspectives in Social Sciences 2. Three studies on the agrarian structure in Bengal 1850-1947* (Calcutta, 1982).

Stein, Burton (ed.), *The Making of Agrarian Policy in British India 1770-1900* (Delhi, 1993).

Stone, Ian, *Canal Irrigation in British India* (Cambridge, 1978).

Stokes, Eric, *The Peasant and the Raj* (Cambridge, 1978).

Washbrook, D.A., 'Law, state and agrarian society in colonial India', *Modern Asian Studies* 15, 3 (1981), pp.649-721.

Whitcombe, Elizabeth, *Agrarian Conditions in Northern India. The United Provinces under British rule 1850-1900* (Berkeley, 1972).

____, 'Famine mortality', *Economic and Political Weekly* 28 (5 June 1993), pp.1169-79.

Yang, Anand, *The Limited Raj. Agrarian relations in colonial India. Saran district, 1793-1920* (Berkeley, 1989).

Chapter 1.2

INTERNAL FORCES OF CHANGE IN AGRICULTURE: INDIA AND JAPAN COMPARED[1]

Kaoru Sugihara with Haruka Yanagisawa

Introduction

This chapter discusses some aspects of the histories of Indian and Japanese agriculture in an Asian comparative perspective. Its focus is the identification of internal forces of change in Asian agriculture, and its explicit yardstick is the Japanese experience. In particular, three aspects stand out as crucial for the development of Japanese agriculture or, more generally, Asian agriculture. First, small farmers acted as internal forces of change. Even when they did not own land, they were often actively involved in technological and managerial improvements. There is no clear-cut answer to the question of whether Asian agriculture should have opted for large-scale farming, or had good economic reasons to remain small-scale, but it is clear by now that small-scale production does not always imply backwardness, nor does large-scale farming necessarily mean 'capitalistic'. A rough guide to the case of the cultivation of rice, the single most important crop in Asia, is that it is more or less scale-neutral.[2] Second, the strength of small-scale production arose from both the development of labour-intensive technology and its ability to absorb labour, particularly family labour. Manageri-

[1] This chapter was written by Sugihara on the basis of a series of intensive discussions between the two authors. Yanagisawa is primarily responsible for the interpretations on India, Sugihara for those on Japan and the presentation of comparative perspective. The views expressed in this chapter are our own, and are not necessarily shared by other contributors of this volume. But we are grateful for their comments and support. In the following footnotes, attempts were made to cite English-language references, in preference to Japanese, wherever possible. However, we felt it appropriate to refer to several important Japanese-language works on India which had not been made accessible to the English-language reader.

[2] Francesca Bray, *Rice Economies: technology and development in Asian societies* (Oxford, 1986).

ally-independent farmers had better incentives and opportunities than large-scale farm managers to fully absorb labour and adapt new tools, seeds and methods. This stronger incentive and the more detailed co-ordination among workers more than compensated for disadvantages. And it acted as a major force of change in production methods. Whenever land became scarce, small-scale production had a chance to push this 'technology choice' further, and was able to retain its competitive edge.[3] Third, Asian agriculture responded to Western impact, not by shifting to Western-style farming, but by incorporating Western technology and organisations into small-scale agriculture. The integration of Asian peasant producers into the world economy meant, in many cases, the destruction of traditional social structure and the reorganisation or emergence of economic and social stratification. Nevertheless small-scale production appears to have persisted, along with the development of some large-scale farming.[4]

A brief look at the history of Japanese agriculture suggests that these characteristics were all present. The managerial independence of small farmers during the second half of the sixteenth and the seventeenth centuries has been identified as a crucial watershed in Japanese history, signalling a decisive break from its locally or regionally determined social texture in which larger units dominated decision-making processes. The Tokugawa regime (1603-1868) created a strict occupational division between warrior-rulers, farmers, craftsmen and merchants, and required the peasant to be tied to the land. Contacts with foreign countries were highly controlled and restricted, although the traditional understanding of the seclusion policy has recently been questioned, especially with regard to the earlier period. Thus when land

[3] For a theoretical discussion, see Shigeru Ishikawa, *Essays on Technology, Employment and Institutions in Economic Development: comparative Asian experience* (Tokyo, 1981), ch.1.

[4] The Chayanovian model has had some influence on the Japanese discussion which had otherwise been dominated by the Eurocentric assumption of the superiority of large scale farming. See, for example, Osamu Saito, 'Peasant families' work patterns in a proto-industrial setting: the case of early Meiji Japan', in Richard Wall and Osamu Saito (eds.), *The Economic and Social Aspects of the Family Life-cycle: Europe and Japan, traditional and modern* (Cambridge, forthcoming).

became particularly scarce in the eighteenth century, a full absorption of labour was attempted by a tightening of the family system and the village community. The influences of people outside the village were of course important. Grass-roots agricultural technologists helped diffuse new seed varieties and agricultural tools, and advised on their adaptation to local conditions, while merchants helped bring in commercial fertilisers and market cash crops. Also, the state effectively approved of the idea of independence of small farmers and village political autonomy, in exchange for securing land tax. By and large, however, the main actors that carried out these changes were managers of production, that is, the household and the village community. When such a community was in charge of implementing technological and organisational changes rather than being dominated by external forces, it can be concluded that internal forces of change were at work.

Furthermore, internal forces continued to lead Japanese agriculture after Japan opened its ports to foreign trade in 1858. A new government with a will to build a modern state came into power in 1868. The modernisation of agriculture, the basis of the economy, was thus attempted. The outcome of this developmental policy presents a classic example of institutional and ideological modernisation without Westernisation. The early Meiji craze for things Western meant the introduction of Western-style larges-cale agriculture, but without success. With the continued development of labour-intensive technology, the labour productivitiy of traditional agriculture continued to rise. There was a development of landlordism from the Tokugawa period, and landlords became politically powerful after 1868. The tenancy rate rose to nearly 50 per cent by the end of the nineteenth century. However the smallness of the management unit of production was maintained, and the ratio of land tax to agricultural output tended to decline over time. Agriculture eventually became internationally uncompetitive during the interwar years when the indigenous technology began to fail to compensate for severe land scarcity. However small-scale production persisted, partly because of agricultural protection, but also because it incorporated the benefits of the interactions with modern industry, through the use of electric pumps, artificial fertilisers and part-time employment in industry into traditional economic and social settings. Thus traditional agriculture provided the basis for both Japan's proto-industrialisation from the second half of the eighteenth century and her industrialisation from the late nineteenth century, by channeling the bulk of capital and labour into industry and providing a market for industrial goods. Land productivity had surpassed that of all other Asian countries by the early nineteenth century at the latest, and has continued to rise to this day.[5]

The technological development based upon small-scale production originated from China, and Japan was certainly not alone in exploiting its advantages further (Chapter 7.2 discusses the successful case of the growth of the traditional Taiwanese sugar industry). Were there comparable Indian experiences? If we take the period from the late nineteenth century, an emphasis on 'internal' forces of change has not been a strong feature of Indian historiography. On the one hand, the development of transportation networks, the growth of exports of agricultural produce, commercialisation, and population changes heavily affected by famines and epidemics, all implied significant changes in the patterns of production and demand. At the same time, agricultural output and land productivity are thought to have grown only slowly, if at all, during this period, and many features of local agricultural societies remained essentially unaltered.[6] Thus it often looks as if forces of change had come from outside. Where internal forces of change were identified, they were usually represented by the activities of business elites, landlords or rich peasants, and it was often alleged that the benefits of these initiatives had not been shared with the rest of the population. Also, the nature of economic change was measured, though not always consciously, by the degree to which it was comprehensible to observers of agricultural development in Britain and other Western

[5] For a fuller discussion, see Yujiro Hayami and Saburo Yamada, *The Agricultural Development of Japan: a century's perspective* (Tokyo, 1991), and Kaoru Sugihara, 'Agriculture and industrialisation: the Japanese experience', in John Davis and Peter Mathias (eds.), *Agriculture and Economic Growth* (Oxford, forthcoming).

[6] Sumit Guha (ed.), *Growth, Stagnation or Decline?: Agricultural productivity in British India* (Delhi, 1992); Peter Robb, 'Peasant Choices? Indian agriculture and the limits of commercialization in nineteenth-century Bihar', *Economic History Review* 45, 1 (1992), pp.97-119.

countries.

However there are some findings which echo Japanese historiography. It has been noted that those merchants, moneylenders, landlords and officials who had dominated local and regional economic activities during the eighteenth century did not disappear when British rule was established, but reorganised themselves into key intermediaries during the first half of the nineteenth century.[7] In any case the direct impact of the British rule on producers may have been limited. At the same time, the 'rationality' of Indian peasants has been identified in various contexts. In western India, for example, peasants took initiatives to respond to market opportunities offered to them as a result of the development of transport networks.[8] In South India the increase in the demand for labour in the late nineteenth century weakened labourers' traditional dependence on the landlords.[9] In Bengal agriculturalists responded to export opportunities from the mid-nineteenth century on, including producers' initiatives, with the result that major social changes occurred.[10]

However were the internal forces of change in India of a similar nature to those seen in Japan? If not, what were the key differences? In this chapter we present some of the basic findings of the papers included in this volume (and some related works by Japanese scholars) relevant to these questions, and relate them to the Japanese experience. The intention behind this exercise is, firstly, to draw out some of the methodological background and thematic foci of the papers in this volume, in order to help the non-Japanese reader understand the broader

[7] C.A. Bayly, *Rulers, Townsmen and Bazaars: north Indian society in the age of British expansion* (Cambridge, 1983).

[8] Neil Charlesworth, *Peasants and Imperial Rule: agriculture and agrarian society in the Bombay Presidency, 1850-1935* (Cambridge, 1985).

[9] Dharma Kumar, 'Agrarian relations: south India', in Kumar (ed.), *Cambridge Economic History of India,* vol. 2 (Cambridge, 1983).

[10] Sugata Bose, *Peasant Labour and Colonial Capital: rural Bengal since 1770* (Cambridge, 1993). B.B. Chaudhuri acknowledges the important role the richer section of the peasantry played in the reclamation of the waste lands. B. Chandhuri, 'Rural power structure and agricultural productivity in eastern India, 1757-1947', in Meghnad Desai, Susanne Hoeber Rudolph and Ashok Rudra (eds.), *Agrarian Power and Agricultural Productivity in South Asia* (Berkeley, 1984).

implications of their arguments.[11] The second aim is to locate the three short chapters (4.2, 6.2, 7.2), which concentrate on the more specific comparative aspects, in a general context. Finally, this chapter should supplement the introductory essay by Peter Robb by introducing a Japan-centred comparative perspective to Indian agrarian history.

Small farmers as forces of change

Breaking up the locally determined social structure. An obvious starting point is the question of how independent small farmers appeared. Mizushima's paper discusses this, though indirectly, with respect to South India. According to his account of the Barnard Report, local society in South India in the late eighteenth century had at least three levels of 'formations', or socio-politico-economic units; individual, communal and state. The state took a large proportion of surplus from peasants, an amount comparable to the early Tokugawa practice, although its role in the management of production does not appear to have been important, as was the case in Tokugawa Japan.

There was then a communal formation which was a larger unit than the village, comprising around nine villages.[12] This unit provided the village with various key occupational activities without which village life would not have functioned. In other words, it was the minimum self-sufficient economic unit, although state intervention, commercialisation and credit networks all played a part in modifying the nature of this community. The activities of professional people in the community were defined, recognised and secured through the concept of 'shares'

---

[11] Although most authors have worked within the Indian historiographical context and differ substantially in their views, there has also been a Japanese academic community which provided them with some shared methodological and factual understanding. Most relevant to this volume are the publications of three edited volumes which resulted from group research activities: Toru Matsui and Toshio Yamasaki (eds.), *Indoshi ni okeru Tochi Seido to Kenryoku Kozo* (Agrarian Relations and Power Structure in Indian Society) (Tokyo, 1969); Toru Matsui (ed.), *Indo Tochi Seidoshi Kenkyu* (Studies in the History of Indian Agrarian Relations) (Tokyo, 1971); Noboru Karashima (ed.), *Indoshi ni okeru Sonraku Kyodotai no Kenkyu* (Studies in the Village Community in Indian History) (Tokyo, 1976).

[12] See Tsukasa Mizushima, *Nattar and the Socio-economic Change in South India in the 18th-19th Centuries* (Tokyo, 1983), pp.409-56.

which referred to the right to produce. These shares were freely sold and mortgaged. Because these rights were called *mirasi*, this was a society dominated by the 'mirasi system'.[13]

Mizushima's understanding of the *mirasi* system echoes Kotani's work on medieval Deccan.[14] Kotani argued that a system similar to that of South India existed there in the form of *watan* (various economic, religious and political rights or shares), and that, although the state later intervened, its involvement reinforced rather than destroyed the *watan*-based society. To some extent his view is an interpretation of the works by Fukazawa who was highly cautious of generalisation.[15] It is clear that Kotani's idea of *watan* society closely relates to the literature on medieval Japan where 'shiki' was recognised as a concept which transcended village boundaries and widely prevailed.[16]

Finally, there was the individual formation (village) which functioned as the smallest local unit. Mizushima's thesis is that by the end of the eighteenth century, the communal formation (central to the *mirasi* system in the case of South India) had disintegrated, and was replaced by the emergence of a society with a much stronger element of individualism. He interprets the emergence of village leaders differently from Kotani who held that village activities were submerged in, if not incorporated into, the *watan* society. The difference between the two writers affects our understanding of the state of economy and society immediately before and after the British arrival. Mizushima argues that the introduction of the raiyatwari system simply capitalised upon, rather

[13] The functions of *nattars*, the representatives of the local society, were discussed in Mizushima, *Nattar and the Socio-economic Change*. See also Noboru Karashima, *Towards a New Formation: South Indian society under Vijayanagar rule* (New Delhi, 1992).

[14] Hiroyuki Kotani, *Indo no Chusei Shakai: Mura, Kasuto, Ryoshu* (India's Medieval Society: village, caste and domainal lords) (Tokyo, 1989). A glimpse of this can be seen in his 'The Vatan-system in the 16th-18th-century Deccan: towards a new concept of Indian feudalism', *Acta Asiatica* 48 (1985), pp.27-50.

[15] See Kotani's introduction to Hiroshi Fukazawa, *The Medieval Deccan: peasants, social systems and states, sixteenth to eighteenth centuries* (New Delhi, 1991).

[16] Yoshihiko Amino, *Nihon Chusei Tochiseido-shi no Kenkyu* (Studies in the History of Medieval Japanese Agrarian System) (Tokyo, 1991) pp.547-81. *Shiki*, however, was strictly related to vertical, rather than communal, rights or shares.

than created, this change.

His emphasis on internal change echoes the great transformation in Japanese agriculture from large-scale to small-scale management during the second half of the sixteenth and the seventeenth centuries. During the period of internal warfare in the fifteenth and sixteenth centuries, land in Japan was divided amongst various manorial lords as well as court nobles, religious authorities and regional warlords. Large-scale farming with the use of servile labourers under the multi-layer ownership of land was common. During this period feudal control gradually loosened, and some servile labourers became independent peasants in the more advanced regions. Independent cities functioned as places for exchange, and farmers in surrounding villages became involved in cash-crop production. Small-scale production by motivated independent peasants became increasingly competitive. Meanwhile, the strengthening of warlord power over manorial lords led to the gradual dissolution of the local power structure. Warlords attempted to deny the share of surplus taken by various intermediaries, including manorial lords. Warlords wanted to become the only owners of land who could collect tax, and the peasant, who was identified as a tax-payer for a plot of land, was in turn to be given a strong right to cultivate. The unification of Japan at the end of the sixteenth century was closely related to the successful completion of a series of cadastral surveys which identified the tax payer for each plot of land, as well as the official discouragement of a society based upon multi-layer ownership.

By the time the unification was completed and the Tokugawa regime was fully established in the late seventeenth century, intermediaries were eliminated, and a clear two-layer land ownership and holding pattern emerged, in which land belonged to the ruler, while the peasant household was responsible for cultivating it.[17] The key rural social institutions were *ie* (household: for a more precise definition see below) and *mura* (village). The former was primarily responsible for

[17] This is another way of stating the independence of servile labourers. T.C. Smith's classic account dates this transformation at a later period, reflecting the literature of the 1950s. See Thomas C. Smith, *Agrarian Origins of Modern Japan* (Stanford, 1959).

the management of production, and the latter for the payment of tax; it acted as a supporting management unit of production. The level of tax, particularly in times of bad harvest, was a frequent cause of peasant discontent and uprising. However the rulers did not live in the village, and village affairs were usually left to villagers themselves. Thus there was no 'communal formation' in Tokugawa Japan. It appears, from our perspective, that this was the most important difference between Japan and India during the seventeenth and the eighteenth centuries.

The managerial independence of small farmers. The presence of a smaller economic unit such as the household and the village is very much assumed by Yanagisawa who also discusses South India with regard to the period from the late nineteenth to the early twentieth centuries. He finds evidence of the development of intensive cultivation, and identifies the rise of low-caste farmers, that is, an increase in the opportunities for low-caste agricultural labourers to become tenants or very small-scale farmers. Dharma Kumar's overall picture of Madras was that the pattern of landholding either remained unaltered or moved only slightly toward the large-scale in the period from 1853 to 1946.[18] Yanagisawa, however, pointed out that Kumar's finding might be a statistical reflection of two changes taking place simultaneously in opposite directions: a gradual disappearance of Brahman landholdings, with an extensive use of low-caste agricultural labourers, was countered by a gradual stratification of non-Brahman peasants.[19] Building on this hypothesis, his chapter in this volume concentrates on the latter process, in an attempt to identify internal forces of change, and, in particular, the upward mobility of agricultural labourers as part of these changes.

Crucial to this upward mobility was the opportunity for earning extra income through temporary emigration to work in plantations in

[18] Dharma Kumar, 'Landownership and inequality in Madras Presidency, 1853-54 to 1946-47', *Indian Economic and Social History Review* 12, 3 (1975), pp.229-61.

[19] Haruka Yanagisawa, 'Mixed trends in landholding in Lalgudi Taluk: 1895-1925', *Indian Economic and Social History Review* 26, 4 (1989), pp.405-35. For a full presentation see Haruka Yanagisawa, *Minami Indo Shakai Keizaishi Kenkyu: Kasomin no Jiritsuka to Noson Shakai no Henyo* (Studies in the Socio-economic History of South India: the emancipation of low-caste folks and the change in the agricultural society) (Tokyo, 1991).

Ceylon or Malaya or paddy fields and rice mills in Burma. Though it was hard work and involved the risk of contracting diseases, higher wages were offered. Peasants-turned-workers could utilise their extra time in off-peak seasons. Alternatively, women and children might have covered agricultural work during their absence, which was another form of labour absorption. Social ties with the village could be secured through the *kangani* system in which work gangs were recruited and organised under the *kangani* (headman) within the locality. In his own comparative chapter (4.2), Yanagisawa points out the possibility that the difference in the character of employment-opportunities between South India and Japan might have been important. In Japan the major source of labour absorption outside rice and other cash-crop production came mainly from female by-employment in proto-industry (such as weaving) or male involvement in food processing industries. The emphasis in Madras was more on extra work in plantations and mines, involving long-distance travel and direct contacts with the outside world, not on the opportunity for accessing proto-industrial work more closely interwoven with economic changes in rural life.

There were also other, more fundamental differences. It must be pointed out that Japanese agriculture since Tokugawa times has never had a large proportion of agricultural (landless) labourers. By the end of the seventeenth century village boundaries were fairly strictly observed, and the village community was held jointly responsible for dealing with those households who were unable to pay land tax. The richer household may have lent money or rice, and, by the return of debt becoming more like rent, may eventually have acquired *de facto* ownership of land from the 'defaulted' household. By and large, however, village solidarity was maintained, and the total collapse of the village community was usually avoided. Because of the strong authority of the household and its effective control of family size (see below), the number of households in the village was reasonably well controlled. The village and the household thus acted as a crucial barrier to proletarianisation.

The 'typical' Japanese village was rather small. According to a record prepared for the revenue collection in 1734, it consisted of 120 people, 24 households, 5 horses and arable land, half wet and half dry, capable of producing 500 *koku* of rice (1 *koku* was approximately 4.36 bushels). Each household cultivated an average of two acres. About 10 per cent of the village population might have been craftsmen and merchants. Villagers used communal land and controlled water together, and the households were organised in groups of five for administrative purposes. Commonly there were three main village officials, including one representing the ordinary peasants. Based on the predominantly subsistence rice-growing with some cash-crop production, peasants had access to regular local fairs and employment opportunities in nearby towns.[20] After the second half of the eighteenth century the opportunity for proto-industrial work increased significantly.

Another difference was the penetration of the caste system into agrarian relations. There are important similarities between India and Japan in terms of strict caste-like occupational divisions, but the Japanese system was essentially concerned with the distinction between the four main classes as mentioned above in which all farmers were classified into one class. There was also a class of outcastes (called *eta* or *hinin*) below this four-tier system who were engaged in special trades, and the discrimination against this class of people bears strong resemblance to the Indian case. However their number is thought to have been relatively small. The main system of social hierarchy within the village was based on the differentiation of the status of *ie*, and there was little caste-like division among villagers. By contrast, the proportion of outcastes in South India amounted to some 20 per cent of the total agricultural workforce.

Technology and labour absorption. The managerial independence of small farmers is a necessary condition, but not a sufficient one, for the development of Asian agriculture. In order to generate internal forces of change, it is necessary to create the dynamics between the development of labour-intensive technology and a greater degree of labour absorption. Indeed these dynamics were the core of much of Japanese agricultural development. Shigeru Ishikawa argues that at the initial stage of agricultural development in Asia, it is often the case that labour input

[20] Chie Nakane, *Kinship and Economic Organization in Rural Japan* (London, 1967), p.42. The data were taken from the work of Shinzaburo Oishi.

increases, rather than decreases, as the output per land (land productivity) begins to grow. That is to say, labour productivitiy temporarily retreats, until the turning point is arrived at. After that, capital input generally dominates the increase of output per land, and the relative significance of labour input decreases. This is the 'Ishikawa curve', a term recognised in development economics.[21]

Kiyokawa and Ohno argue, with reference to the traditional sugar industry, that these dynamics were certainly present in India. They identify the technological improvement within this sector of industry, and argue for its competitiveness. According to their estimation, this sector alone absorbed a workforce of about two million, a considerably greater number than that which the development of a modern sugar industry would have required. They suggest that this traditional sector's very survival implies its competitiveness. Supposing that this was the case, then, this may sound quite a satisfactory state of affairs. Kiyokawa and Ohno point out, however, that the successful survival of the traditional sector was largely a reflection of the lack of enterpreneurship (and managerial failure) on the part of the modern sugar industry on the one hand, and, on the other, in the period after 1932 when the protective tariffs were set up, of protection from international competition. Elsewhere Kiyokawa has pointed out, with reference to the modern cotton industry, that there was an important difference in the *speed* of technological diffusion between India and Japan.[22] In such circumstances the traditional sugar industry could have survived relatively easily.

There is another dimension to this issue, which is the question of substitutability between *gur*, the traditional Indian product, and modern sugar. Although the authors are reluctant to place importance on this point, consumer preference may well have played a part in distorting the direct competition which is assumed in this paper.[23] But this point

[21] Ann Booth and R.M. Sundrum, *Labour Absorption in Agriculture: theoretical analysis and empirical investigations* (Oxford, 1984), ch. 1.

[22] Yukihiko Kiyokawa, 'Technical adaptations and managerial resources in India: a study of the experience of the cotton textile industry from a comparative viewpoint', *Developing Economies* 21, 2 (1983), pp.93-133.

[23] Shahid Amin, *Sugarcane and Sugar in Gorakhpur* (New Delhi, 1984).

could only add to an impression of lack of open international competition. Local consumer taste could provide a vital platform through which innovation and competition could be generated, for domestic industries could respond to the rise of local purchasing power in the export sector and the trade opportunities offered by improved transport. One of the most distinctive features of Japan's industrialisation was that it exploited this advantage to the full.[24] In addition, Japan took advantage of the relative similarities of Asian consumer taste as distinguished from Western, and exploited other Asian markets including the Indian. Japanese sources in the late nineteenth and the early twentieth centuries confirm in some detail that local consumer taste in India acted as an effective non-tariff barrier to Western, but not Japanese, penetration.[25]

Why was the modernising force within the traditional sector relatively weak in India? One crucial difference was that an acute population pressure was forced upon Japanese agriculture as early as the early eighteenth century. As was suggested earlier, there was little alternative for the Japanese farmer other than the labour absorption scenario. He was tied to land, and occupational division was strict. He was under the strong influence of the village community which required co-operation in terms of water control and labour-sharing during harvest times. Above all, he was expected to pass the plot of land he had inherited from his father to his heir at all costs. The ideology of *ie*, a unique quasi-kinship institution developed in Japan, had it that *ie* existed as a sort of corporate entity which was to go beyond the life-span of the present members of the family, and that the family assets or customary

[24] See Heita Kawakatsu, 'International competition in cotton goods in the late nineteenth century: Britain versus India and East Asia', in W. Fischer, R.M. McIness and J. Schneider (eds.), *The Emergence of a World Economy, 1500-1914*, vol.2 (Wiesbaden, 1986). The presence of keen competition was identified as a major driving force behind the modernisation of the traditional Japanese cotton weaving industry. Takeshi Abe, 'The development of the producing-centre cotton textile industry in Japan between the two world wars', *Japanese Yearbook on Business History* 9 (1992), pp.3-27.

[25] Kaoru Sugihara, 'Japan as an engine of the Asian international economy, c.1880-1936' *Japan Forum* 2, 1 (1989), pp.127-45. Haruka Yanagisawa, 'The handloom industry and its market structure: the case of the Madras Presidency in the first half of the twentieth century', *Indian Economic and Social History Review* 30, 1 (1993), pp.1-27.

rights, including the plot of land, belonged to such an entity, not to its present members. The maintenance of *ie* overrode kinship concerns, so that a very high rate of adoption was common. In the absence of an able male heir, a suitable male would be adopted to assume headship, either as a husband of a daughter of that household or simply as an adopted son. This was an important departure from the kinship-based organisation which is bound to conflict with meritocratic concerns and economically-rational decisions from time to time. *Ie*, by contrast, was capable of accommodating such rational ideologies as learning from new knowledge, co-operating with other *ie* and situating itself within the village hierarchy, and generating positive responses to outside (market) influences.[26]

Japanese scholars have debated why and how the Japanese peasant reacted in this way. Some paint a dynamic picture of an innovative and market-minded peasant with rising expectations of income, education and standard of living. Others hold on to a more gloomy picture of the peasant who, without having an effective political voice, suffered from heavy taxation and a high level of rent, and was forced to work harder and harder for his survival and pride. However, even the latter view, assumes that labour-intensive technology developed and labour absorption occurred. Economic development appears to have been compatible with this latter scenario as well, so long as peasant motivations for the improvement of production methods were maintained. It is here that a more detailed comparison between India and Japan might fruitfully be made in future.

Malthusian checks and population control. The advance of historical demography has made significant contributions to our understanding of the history of Japanese and Asian agriculture in recent years. A picture emerging from Japanese scholarship on this front suggests that the Japanese demographic regime during the eighteenth and the nineteenth centuries could be characterised by low fertility, relatively low mortality and very mild population growth. It is important that a regime

---

[26] Similar concepts of entail existed in India, but the law generally favoured joint inheritance, and the pressure of the nineteenth century was toward individual ownership and the identification of shares through actual partition.

of this kind was found in the pre-industrial society. That is to say, the demographic regime characterised by high fertility and high mortality should not be assumed even in early eighteenth-century Asia.[27]

The issues here are two-fold. First, relatively low mortality was made possible largely by the absence of major epidemics. Unlike India, Tokugawa Japan was geographically isolated from major international trading routes, and escaped several major epidemics such as typhoid, cholera and bubonic plague which travelled the world.[28] There were no wars and relatively few other major calamities. Although we need to know more about the cases of China and Korea before making any generalisations, the Japanese case looks very different indeed from the experiences of other Asian countries. Second, low fertility may have been a result of the combination of infanticide and abortion, and low fecundity. The former factor has so far been emphasised in the literature. Indeed the idea of the presence of long-term family planning is fully compatible with our impression of the stability of Tokugawa society.[29] It is also the case that no ideological distinction was made between abortion and infanticide in the Buddhist teaching of the time. However the evidence that infanticide was numerically important nationwide is definitely missing. An alternative, somewhat gloomier picture might be that a strong pressure on pregnant women to participate in agricultural work may have resulted in lower fertility when labour-intensive technology gained importance in peasant daily life in the eighteenth century. This better explains the disappearance of the rapid population growth which had occurred during the seventeenth century, and certainly fits with the early twentieth-century evidence of the effects of heavy female participation in agricultural work on fertili-

---

[27] Nigel Crook, 'On the comparative historical perspective: India, Europe and the Far East', in Tim Dyson (ed.), *India's Historical Demography: studies in famine, disease and society* (London, 1989), pp.285-96.

[28] Ann Jannetta, *Epidemics and Mortality in Early Modern Japan, 1600-1868* (Princeton, 1987).

[29] Thomas C. Smith, *Nakahara: family farming and population in a Japanese village, 1717-1830* (Stanford, 1977). Susan B. Hanley and Kozo Yamamura, *Economic and Demographic Change in Pre-industrial Japan, 1600-1868* (Princeton, 1977).

ty.[30] Either way, the fact that major calamities neither occurred nor were expected in Tokugawa society remains undisputed.

Wakimura's paper confirms that the Indian scene was dramatically different. Here we have a classic Malthusian case of low population growth, as a result of high fertility checked by high mortality. Historical demography has added insights into the workings of this familiar demographic regime too, by clarifying the relationships between famines, epidemics and population change. With regard to north India during the period from 1871 to 1921, Wakimura found that high mortality was caused largely by epidemics, particularly malaria, and that death tolls were particularly high when epidemic malaria occurred immediately after famine. After a famine raised agricultural prices and weakened human resistance to epidemics, a return of rain would create excellent conditions for the prevalence of malaria. Using the data compiled by Toru Matsui,[31] he argues that there was no clear relationship between the increase of agricultural prices and mortality, and dismisses the importance of a direct link between famine and high mortality. He also suggests that famines affected fertility in the short term,[32] but that this was compensated for by the 'rebound' of fertility in later years. Although its magnitude was extremely small, there were famines and epidemics in Tokugawa Japan, and, as Saito's comparative chapter

[30] Osamu Saito, 'Infanticide, fertility and "population stagnation": the state of Tokugawa historical demography', *Japan Forum* 4, 2 (1992), pp.369-81, and 'Gender, workload and agricultural progress: Japan's historical experience in perspective', to be published in a book in memory of Franklin F. Mendels (Geneva, forthcoming).

[31] Toru Matsui, *Kitaindo Nosanbutsu Kakaku no Shiteki Kenkyu* (An Historical Study of Agricultural Prices in North India) (Tokyo, 1977). In the context of the introduction of quantitative methods to economic history developed in the United States, Japanese scholars have tended to put greater emphasis on a thorough collection of data and their processing than their interpretation and model-building. Matsui's work on agricultural prices represents such a contribution in Japanese studies of South Asia.

[32] In his analysis of monthly data for famine periods, Tim Dyson demonstrates that a fall in the conception rate took place before the rise in the mortality rate. This shows why both death and birth rates reacted to famines, but may also be taken as testimony of how severe the impact of these famines was. See Tim Dyson, 'On the demography of South Asian famines', parts 1 and 2, *Population Studies* 45, 1 (1991), pp.5-25, and 45, 2 (1991), pp.279-97.

(6.2) shows with regard to the Tempo Famine period, there is evidence to suggest that they were sometimes interrelated.

Under such a violent demographic regime, the perception of what would be needed for stable daily life in north India would have been very different from that in Japan. So would the degree to which one could commit oneself to long-term economic planning. In this respect the development of transport networks and the integration of the region into the wider world does not seem to have brought the same kind of opportunities and risks to the two countries. Rather, a simultaneous introduction (or intensification) of market opportunities and epidemic diseases might well have highlighted the difference in the two countries' social capacity to respond to the Western impact. The relative lack of a sense of stability and security must have affected the chances of generating the dynamics between technological change and labour absorption, which in turn must have lessened the amount of emergency resources to which the peasant could have recourse when calamities occurred.

Socio-economic stratification and its consequences

The growth of rich farmers. Both Indian and Japanese historiographies have been deeply concerned with socio-economic stratification within local agricultural societies. It seems easier, though, to identify when and how stratification took place in Japan, so long as we accept the view that during the seventeenth century the autonomy of local and regional communities was denied by the Tokugawa regime, and the managerial independence of small farmers was established nationwide. After this decisive break, the story goes, a new form of differentiation started around the late seventeenth century. It initially occurred in the form of usury turned into *de facto* land ownership, but some acquired land newly opened up, and, in the later period, some land was actually purchased.[33]

It is less clear if such a break can be established with regard to eighteenth-century Bengal. Taniguchi's paper attempts to show a high

---

[33] Ryuzo Yamasaki, *Kindai Nihon Keizai-shi no Kihon Mondai* (Fundamental Issues in Modern Japanese Economic History) (Kyoto, 1989), ch.3.

degree of stratification, and depicts the economic life of each strata of peasantry. His tabulation suggests that nearly two-thirds of the rural population were under-raiyats and agricultural labourers. They did not have independent means of survival, and were dependent upon wealthy raiyats for the supply of agricultural stocks. His thesis is that this growth of wealthy raiyats took place in the second half of the eighteenth century at the cost of zamindar and the poorer peasants. Peasant disturbances reflect this change in the balance of social structure and its adjustment process.

Taniguchi's picture parallels Japanese historiography on *gono,* an established academic term for the Japanese cultivating landlord-cum-merchant-cum-moneylender who was also influential in village politics.[34] And, in the case of Bengal, it is possible to synchronise the two historiographies; Taniguchi's rich peasants were the contemporaries of *gono.* It has been argued that in the 1780s there was a shift in the nature of Japanese peasant uprisings from the struggle for tax reduction to that for rent reduction, reflecting the growing importance of socio-economic stratification within the village. Around this time *gono* began to become a target of peasant uprisings instead of leading them. *Gono* were a rising force of change, undermining some of the assumptions of the Tokugawa regime (like village boundaries and occupational division) on the one hand, and effectively subjecting the poorer villagers to their mercy through economic forces on the other. Although it is difficult to support the suggestion that *gono* were actually behind the overthrow of the Tokugawa regime, there can be little doubt that they represented a major force which favoured an institutional structure more prone to market forces and entrepreneurship.

It is interesting to note that in both the Bengali and the Japanese cases these forces emerged *before* the introduction of European-style property rights. In the Japanese case *gono* grew within the rigid institutional framework, and gradually established their socio-economic status as a force of change. The formal establishment of European-style property rights after 1868 endorsed change, but it did so in a clumsy

[34] Junnosuke Sasaki, *Bakumatsu Shakai-ron* (The Japanese Society in the Last Years of the Tokugawa Shogunate) (Tokyo, 1969), ch.4.

way and only to a limited extent. Attention to ownership and rent-receiving rights was certainly needed to offer security (hence incentives), but that alone would not have ensured the change in the management of production. Thus legal rights remained less important than customary rights in Meiji Japan. In the meantime, the establishment of the British system in India proved difficult and remained uneven. One of the reasons for this seems to be that European-style property rights did not give due regard to the technical and managerial contributions of small farmers. Small rice-farmers were not inherently market-driven; yet they could be extremely innovative. Given the nature of labour-intensive technology and the degree of labour absorption, Asian rice-farmers may well have deserved a greater recognition of their contributions to land management (hence respect for their customary rights) than their Western counterparts. The issue at stake here was not just colonial rule. Whoever attempted to instigate change, had to deal with this difference between European and Asian agriculture.

It might be thought that the traditional social structure in northern Bengal, particularly the presence of zamindar, hampered the growth of rich farmers. However, according to Taniguchi's work,[35] the zamindars do not appear to have been fully in control of local village affairs. Indeed some of their administrative functions were rather close to those of the samurai bureacracy in Tokugawa Japan. Although zamindar imposed an extraordinary range of taxes relating to various aspects of village life, much of the tax-collection was probably subcontracted. In

---

[35] Shinkichi Taniguchi, '18-seiki Kohan Hokubu Bengaru no Nogyo Shakai Kozo' (The structure of agrarian society in northern Bengal in the second half of the eighteenth century), *Hitotsubashi Daigaku Kenkyu Nenpo: Keizaigaku Kenkyu* 31 (1990), pp.193-248, and 33 (1992), pp.83-170. Taniguchi has published a number of articles relating to this subject in Japanese. For an English-language presentation, see his 'Structure of Agrarian Society in Northern Bengal, 1760-1800', Ph.D dissertation, University of Calcutta, 1977. Takabatake, on the other hand, stresses the harmonious relationship between the zamindar and his raiyats, which, he considers, originated from a distant memory of their once having belonged to a group involved in reclamation in common. See Minoru Takabatake, 'Zamindaru Raiyato Kankei no Genkei' (The prototype of zamindar-raiyat relationship), *Hokkaido Daigaku Bungakubu Kiyo*, 18,1 (1970).

other words, the zamindars appear to have maintained considerable political and religious influence in local agricultural societies, without knowing much about the actual production processes. The organisation of agriculture was small-scale based on family labour; typically each family cultivated about five acres. The typical farm size of *gono* was much smaller, but land productivity was probably higher. The partial involvement in commercial production led to peasant dependence upon merchant credit advances to some extent, but subsistence production remained dominant. Thus there was room for the growth of rich farmers, or, to put it another way, for 'genuine' stratification according to luck and merit. So long as the growth of rich farmers reflected social mobility of this kind, stratification was not necessarily a bad sign in itself. On the other hand, the increase of peasant protest reflected a growing conflict of economic interests among the agricultural producers. In these respects a parallel between northern Bengal and Japan appears to hold, in spite of a number of other general differences mentioned in the previous section.

The degree of stratification. Taniguchi's picture of late eighteenth-century Bengal is close to Ratnalekha Ray's portrayal of a highly-stratified Bengal society, and is radically different from Rajat Datta's less stratified one. But neither Taniguchi nor Nakazato, who writes on late nineteenth-century Bengal, is happy with Ray's emphasis on continuity—that is, the persistence of traditional agrarian relations in the colonial period.[36] Nakazato argues that late nineteenth-century Bengal saw a new development of economic and social stratification, in addition to the existing hierarchy. At this time commercial agriculture, particularly jute production, became lucrative as a result of Bengal's integration into the world economy, and also powerful peasant movements occurred, requesting a full recognition of peasant rights, and leading to the enactment of the Bengal Tenancy Act of 1885. As a result of this

[36] Rajit and Ratna Ray, 'The dynamics of continuity under British imperialism', *Indian Economic and Social History Review* 10, 2 (1973), pp.103-28. Rajat Datta, 'Rural society and the structure of landed Property: zamindars, talluqdars and *la-kharaj* holders in eighteenth century Bengal', in Peter Robb (ed.), *Agrarian Structure and Economic Development*, Occasional Papers in Third World Economic History 4 (London, 1992), pp.34-60.

Act, land transfer became much easier and thus more frequent, and this accelerated peasant stratification. On the whole, the increase of agricultural production and the stratification took place simultaneously, and rapidly, in Bengal between the late nineteenth century and 1945.[37] Unlike some other writers,[38] Nakazato sees this new pattern of stratification as an essential element which created in interwar Bengal a society with an extraordinary degree of polarisation. The question is how much of this general process was a 'genuine' stratification, leading to the rise of rich farmers willing to take opportunities and risks, and what factors were responsible for choking such a development.

In this volume Nakazato offers some statistical information relevant to these questions, by analysing regional patterns of land transfer. His findings echo, but do not exactly coincide with, those in the available English-language literature. Fairly large land transfers were already taking place in the early twentieth century, in which raiyats themselves were the main purchasers of land in the majority of districts. The land market was much more active in the relatively prosperous districts of eastern Bengal than in the region suffering from declining economic activities. Nakasato argues that, although the experiences of economic distress in the 1930s and the Bengal famine of 1943 show the importance of economic strain as a cause of land sales, their phenomenal rise cannot be explained by poverty alone. Taking a long-term perspective, the increase of land sales was as much a result of the intensification of economic activities among the peasantry, which took place in more normal years. Nakasato's picture of jute-growing regions reminds us of Japanese sericultural areas which, helped by export booms, outperformed the rest of the country in terms of agricultural production. How-

[37] In addition to Nakazato's chapter in this volume, see his 'Superior peasants of central Bengal and their land management in the late 19th century', *Journal of Japanese Association for South Asian Studies* 2 (1990), pp.96-128, and his *Agrarian System in Eastern Bengal, c.1870-1910* (Calcutta, 1994). Kawai has examined the impact of the integration into the international economy on agrarian change. Akinobu Kawai, *'Landlords' and Imperial Rule: change in Bengal agrarian society, c.1885-1940* (2 vols: Tokyo, 1986 and 1987).

[38] Sugata Bose, *Agrarian Bengal: economy, social structure and politics 1919-1947* (Cambridge, 1986; Indian ed., 1987).

ever the latter's degree of stratification was much smaller.

With respect to Japan as a whole, we do see the growth of land-lordism under an emerging developmental state after 1868. The land tax was monetised, and the relatively high rate of inflation in the 1870s meant that the *effective* tax burden was considerably reduced. Thus landlords and owner cultivators enjoyed a significant increase of their share in aggregate agricultural income at the expense of the govern-ment.[39] In 1881 Finance Minister Masayoshi Matsukata decided to deflate the economy in the face of twin crises; fiscal and in the balance of payments. During this period of deflation (1881 to 1884) a large number of forced sales for non-payment of tax and foreclosures of mortgages occurred, and the tenancy rate (the proportion of land under tenancy to the total arable land) increased from about 30 per cent in 1873 to 40 per cent in 1887. The tenancy rate then stabilised and stayed between 40 and 50 per cent throughout the period before the Second World War.[40]

Although Meiji landlords were larger than Tokugawa ones, and some of them were very large, owning more than 2,500 acres, they were small by international standards. Typically they owned no more than 50 acres. As cultivating landlords, many of them were involved in such improvements of agricultural methods as seed selection, irrigation and the introduction of commercial fertilisers. And it was they, more than anyone, who represented internal forces of change in Meiji agri-culture.[41] In so far as agricultural surplus was transferred from the unproductive samurai class to the more innovative section of the rural population, the Meiji Restoration of 1868 must have had a favourable effect on Japanese agriculture. Some farmers who lost land during the Matsukata deflation went to cities, but most of those who stayed in the countryside became tenant farmers. Urban population grew, but remained proportionately small until the interwar years, and the indus-

[39] James I. Nakamura, *Agricultural Production and the Economic Develop-ment of Japan, 1873-1922* (Princeton, 1966).

[40] Takafusa Nakamura, *Economic Growth in Prewar Japan* (New Haven, 1983), pp.54-8.

[41] Ronald P. Dore, 'Meiji landlords: good or bad?', *Journal of Asian Studies* 18, 3 (1959), pp.345-55.

trial workforce in modern factories was largely recruited from the countryside as temporary migrants. Thus even this most drastic deflationary measure in modern Japanese history did not produce proletarianisation. Instead, after 1885, there was a counter-trend to stratification with a mild increase in the proportion of middle peasants. Instances of land transfer also declined after 1885. It is true that Meiji landlords emerged as an important political force, and attempted to contain the more democratic political ideas which were supported by many peasants. However the landlords were not yet alienated from agricultural production. They retained, and often enhanced, their sociopolitical status in the local agricultural communities. In the meantime, more than a half of arable land continued to belong to the owner-cultivators who enjoyed an increase in agricultural income as a result of inflation. In this way both landlords and owner-cultivators continued to exploit the advantages of small-scale production. Stratification reflected the growth of market forces; but its mildness reflected the fact that market forces did not change the character of those social institutions which governed the relations of production.[42]

Tenancy Acts and beyond. The question of how to protect tenants from unreasonable eviction while securing the modern sense of ownership has become a major issue in the agrarian histories of many countries from the second half of the nineteenth century on. In India the 1885 Bengal Tenancy Act established a format which was followed by other regions. Awaya's paper concerns the Malabar Tenancy Act of 1930, a region with very different socio-economic characteristics from the rest of the Madras Presidency where the raiyatwari system became the norm. The traditional interpretation was that the exploitation by *janmi* (landlords) led to this legal intervention, but Awaya questions their overwhelming socio-economic strength, and points out that such an image was created, rather successfully, by tenancy agitators. Here again, an underlyng thesis is that landlords were not in control of the

[42] This is the position taken by the *Koza-ha* scholars in the pre-Second World War Marxist debate on Japanese capitalism. However, they viewed this as an indication of Japan's backwardness. For an assessment of this debate, see Kaoru Sugihara, 'The Japanese capitalism debate, 1927-1937', in Peter Robb (ed.), *Agrarian Structure and Economic Development*, pp.24-33.

actual management of land, and that cultivators had a considerable degree of autonomy.

Interwar tenancy disputes were major issues in Japanese society as well. Landlords became heavily involved in investment outside the village, while tenant farmers, with better literacy and a better understanding of the market, began to make managerial decisions. It was natural for them to claim a greater portion of the surplus. In spite of the obvious political power of landlords, rent generally declined, and state intervention probably reinforced this trend.[43]

However Awaya is more concerned with how the Act was actually pushed through, and finds a political answer. Unlike the rest of the Madras Presidency, British officials and courts singled out *janmi* as a group which had an absolute right of ownership of land. Despite the fact that this did not reflect the reality, it gave tenancy agitators an excellent 'reference point' against which they organised their movement. Also, landlords' power was often weakened, rather than strengthened, by other types of legal intervention. For example, the consensual rule on the division of family inheritance resulted in the growth of very large (joint) families, with increasing instances of family feuds. The matrilineal system of inheritance also made it difficult for the male head of the family to protect his immediate family's welfare, and put a pressure on him to take a tough line with his tenants, sometimes leading to evictions. Here one is left with an impression of anarchy: neither the consequences of each legal decision nor the connections between them were clearly understood. Largely for fear of resistance, the British government of India did not have a strong will to enforce the law either.

It is true that the identification of the male head as the sole owner of land at the time of the introduction of the Western legal system caused

[43] Ann Waswo, 'The origins of tenant unrest' in Bernard S. Sinberman and Harry D. Harootunian (eds.), *Japan in Crisis: essays on Taisho democracy* (Princeton, 1971), and Richard Smethurst, *Agricultural Development and Tenancy Disputes in Japan, 1870-1940* (Princeton, 1986). The latter is bitterly criticised by Japanese specialists, however. See, for example, Yoshiaki Nishida, 'Growth of the Meiji landlord system and tenancy disputes after World War I: a critique of Richard Smethurst', *Journal of Japanese Studies* 15, 2 (1989), pp. 389-415.

a similar change in the balance of power within *ie* in Meiji Japan. Generally speaking, family fortunes had been perceived to belong to the *ie* as a corporate entity. They now belonged to the individual, at least on paper, regardless of his ability. Still, no one would wish to characterise Meiji Japan as anarchic, in spite of internal war and chaotic economic conditions in the early period, and diverse cultural and ideological developments later. If anything, Western impact gave Japanese society a sense of purpose, and made it more cohesive. On the whole, the government purpose of nation-building was understood, if not always agreed with, by most of the population.

However was the reality of Malabar society close to anarchy, or is this simply a reflection of the limits of our understanding? Some of the points Awaya makes, such as the difficulties of drawing the lines between landlords and tenants, and explaining the presence of richer tenants and poorer landlords, are perhaps better compared with pre-Tokugawa Japan where functional and hierarchical divisions in society were more obscure and more complex. Modern Japanese history is a poor yardstick here, although our limits may well have been shared by British colonial officials and Western scholarship. Elsewhere Mizushima has argued, primarily with regard to South India but also more generally, that the dissolution of the so-called village community should not be seen as an inevitable path for the emergence of 'individual formation',[44] while Awaya variously exposes the absurdity of dealing with the more fundamental issues of Malabar society by means of the legal mentality of the early twentieth century. In order to conceptually come to terms with these diversities, we would ultimately need an India-centred comparative framework, capable of locating both Western and East Asian experiences in a wider historical perspective.

Some policy implications

This chapter attempted to identify internal forces of change in agriculture, and concentrated on the central role small farmers played. Its main

---

[44] Tsukasa Mizushima, *18-20-seiki Minami Indo Zaichi Shakai no Kenkyu* (Studies in Local Society in South India from the 18th to the 20th Century) (Tokyo, 1990), Conclusion.

message is that these forces did exist in India as well as in Japan, though to a lesser extent and under different circumstances, and that it is important to identify them as an element of change in the history of Indian agriculture. Our intention is not to play down the importance of other factors in accounting for the change in Asian agriculture, but to interpret such factors as the role of the state and the impact of external shocks from a new perspective. Below we briefly outline our thoughts on the role of the state.

The Western impact under colonial rule usually meant that most of the economic regions were put under the regime of 'forced free trade'. On the one hand, the development of transport networks and the introduction of Western technology and organisations opened up the possibility of trade and enterprise. In India the colonial code was essentially that of *laissez-faire*, or the application of what was thought to be proper in Britain. In practice the capacity to enforce the government's will was so severely limited by practical considerations of manpower and the understanding of local languages and cultures that it was often synonymous with the lack of control. At the same time, the British administration did try to protect the economic and political interests of Britain and the British Empire. Thus there were interventions in revenue, tariff and exchange rate policy, and labour conditions, both to curtail Indian industries in direct competition with British industries and to encourage the development of Indian agriculture and industry in so far as they were capable of earning foreign exchange.

On the other hand, the Meiji state attempted to build the modern industries necessary for 'enriching the nation and strengthening the military', most of which were in direct competition with the West, and did not become internationally competitive for quite a long time. The state pursued this policy against the logic of market forces. After the early 1880s, the Japanese government allowed competition from Taiwanese sugar and Chinese and Indian raw cotton to ruin domestic agricultural production, in an attempt to secure cheap raw material, and hence the competitive edge of emerging modern industries. Thus the contrast has traditionally been made between the British government of India which discouraged modern industries and the Meiji government which encouraged them. But a judgment on whether or not the latter pursued an economically-correct policy depends on how one perceives its goal. From the point-of-view of resource allocation alone, it might have been better to leave the attempt to catch up with top technology to a later stage.

It is time to rethink the singular emphasis on modern industry, and to broaden the points of comparison into agriculture, traditional industry, commerce and other service sectors. In these areas *direct* state intervention was much less important. Under the regime of forced free trade, the institutional barriers to local and regional trade and services were largely removed, while the unification of currencies, and the development of transport, ports and cities, provided entrepreneurs with new opportunities for trade. In these respects Japan and India both benefited from institutional change and integration into the world economy. For example, if the export of raw cotton or raw silk rose in one region, there might well in that region be an increase in the demand for local grains and ordinary cloth for local peoples which could only be provided domestically. While many of the traditional trade links were disturbed by these developments, it is nevertheless important to see that considerable linkages and local commercial, agricultural and industrial initiatives emerged in both countries. It was necessary to identify market opportunities, get used to new transport, storage and credit facilities, and guide producers to understand and meet new demand. In Japan initiatives of this kind were well recognised by the Meiji government, and *indirect* state support was systematically pursued. This may well have differentiated the two societies' capacity to respond to new opportunities.

The crucial role the Meiji state played with regard to agriculture was that it built a competitive environment in which small farmers could take their own initiatives: it provided them with information, credit, and social and ideological incentives. Commercial information was provided by Japanese consuls overseas and government researchers, and was fed into local governmental offices to be at the disposal of farmers.[45] Agricultural co-operatives and agricultural and industrial banks

[45] Kaoru Sugihara, 'The development of an informational infrastructure in

provided small farmers with credit on relatively easy terms.[46] Agricultural experimental stations and schools advised on new seed varieties and other technological matters. Above all, various exhibitions and contests were held at local, regional and national levels to put social and political emphasis on the value of the improvement of production methods. This was not protectionism. The Japanese interpretation of nineteenth-century British ideas of free trade was that the state should intervene in creating a *competitive* environment for free enterprises if such a circumstance had not existed already. Thus indirect state intervention was essential for free trade.[47]

Similar efforts were undertaken by the British in India, at first in a piecemeal fashion but more systematically with the establishment of Agricultural Departments in the 1880s. But much of the effort was devoted to export products and to a model of agricultural management and production inconsistent with Indian conditions.[48] Also, arguably, throughout the colonial period, inadequate resources were devoted to such efforts, even after 1918 when the Industrial Commission suggested that the British government of India should emulate Meiji economic policy,[49] or when the Royal Commission on Agriculture reported in 1928. Equally, the question of improving local production methods in agriculture had been largely outside the scope of the main arguments put forward by nationalists in India—they associated market competi-

---

Meiji Japan', in Lisa Bud-Frierman (ed.), *Information Acumen: the understanding and use of knowledge in modern business* (London, 1994), pp. 75-97.

[46] As far as local money circuits and credit networks were concerned, India appears to have been extremely well developed. See Frank Perlin, *The Invisible City: monetary, administrative and popular infrastructures in Asia and Europe, 1500-1900* (Variorum, 1991), and Gareth Austin and Kaoru Sugihara (eds.), *Local Suppliers of Credit in the Third World, 1750-1960* (London, 1993), ch.1.

[47] Sugihara, 'The development of an informational infrastructure'. There is a general recognition at the policy level that at least part of Japanese industrial policy today promotes, rather than hinders, competition, that is, indirect state intervention can be justified even in a fully-developed economy. But the theoretical presentation of the case is yet to be made.

[48] Peter Robb, 'Bihar, the colonial state and agricultural development in India, 1880-1920', *Indian Economic and Social History Review* 25, 2 (1988), pp.205-35.

[49] See Atsushi Mikami, *Indo Zaibatsu Keieishi Kenkyu* (Studies in the Business History of 'zaibatsu' in India) (Tokyo, 1993), ch.2.

tion and export trades with impoverishment and a drain of wealth from India. Both sides linked competition with *laissez-faire*, for or against. Such a frame of reference must have seriously hampered the identification and encouragement of internal forces of change in Indian agriculture.

Chapter 2

THE MIRASI SYSTEM AND LOCAL SOCIETY IN PRE-COLONIAL SOUTH INDIA[1]

Tsukasa Mizushima

Introduction

The raiyatwari settlement introduced in South India by the colonial government in the early nineteenth century constituted a revolution. It did so not because of its very excessive tax assessment or its demand for cash payment, though both of these surely had a critical impact; rather it was revolutionary in assigning an autonomous identity to every land-holding. A plot of land, demarcated, numbered, assessed, and, if a raiyat agreed to be registered as revenue-payer, allotted to him, came to the centre of the stage. The plot could now absolutely and independently express the relations among the different forces concerned. In pre-colonial South Indian society, on the other hand, relations in land could never be isolated from other relationships, and rights in land had been just one of the expressions of those relations. The study of land owner-ship, therefore, is naturally of limited value in defining the social struc-ture of pre-colonial South Indian society, which was not necessarily stratified according to land ownership in the exclusive sense of the modern period.

What is attempted in this chapter is to present the pre-colonial society in terms of a 'share distribution system'. As will be shown, the produce of the village was divided into many shares in the shape of dues, state taxes, or allocations to cultivators. Those engaged in production were supported by these shares, which had been established by custom and were linked to the roles necessary for maintaining the local society (the term 'local society' will be defined later). A combination of roles with rights in the share of the produce, including those to tax-free land, was widely found not only in the south but also

---

[1] The wording of this chapter has been revised by Peter Robb.

in other parts of India at the start of British rule.[2] In South India the customary or inherited right was called *kani*, and a person with such rights was *kaniyatchikkaran* in Tamil. In the British revenue administration (and for convenience in this paper) the right was generally designated as *mirasi*, an Arabic term, and the holder was called *mirasidar*. *Mirasi* right was known to be transferable either by sale, mortgage, or inheritance.

Production was maintained by this share-distribution system, each share being linked to a certain assigned role. The mirasi system constituted the core of local society in the pre-colonial period—its economic structure, class structure, and other features. By detailed investigation into the system and by discerning its salient features, the structure of pre-colonial South Indian society may be clarified. As similar types of share distribution systems were observed in other parts of India, the findings of the present study will have wide applicability.[3]

The source used here is the Barnard report. In the 1760s, just after the Jagir (the area corresponding to the present Chengalpattu or Chingleput district, Tamil Nadu) was granted to the East India Company, Barnard, a British surveyor, collected detailed information on all the villages in the Jagir. With the assistance of village accountants and the village documents kept by them, he compiled statistics for over 2,000 villages more or less in the same format. Such records covering all the villages in one district in eighteenth-century Asia must be very rare. The contents included figures on many aspects of every village, such as land use, *inam* (tax-exemption), caste, shares in the produce, livestock, *poli-*

[2] Many of the collectors engaged in revenue administration at the initial stage of colonial rule noticed different types of tax exemption and allowances variously designated from area to area, though they did not necessarily locate them in the share distribution system discussed here. See, for instance, those reports contained in the Papers on Mirasi Rights.

[3] One attempt to understand the structure of local society in the precolonial period in terms of the *mirasi* (vatan) system has been made by H. Kotani on medieval Deccan. See, for instance, Kotani 1985. The present study, though based on completely different set of sources, owes much to his argument, with several important differences in understanding the nature of change occurring in the period concerned. As to the problems of Kotani's theory and the differences with the one presented in this paper, see Mizushima 1990a.

gars (in charge of law and order), landholders and crops.[4] Unfortunately, no descriptive account of the contents of the Barnard report is available in other sources, which poses some difficulty for the following analysis. The format or categorisation adopted in the record is here assumed to be significant, but it is not certain whether it was Barnard's invention or an exact copy of some original.[5] Nonetheless, though the report may have certain limitations, a careful statistical examination of its contents will give us more insights than any other records into the social structure of Indian society in the pre-colonial period. We will take up as a case study Salavaucum *pargana*, the records of which are comparatively more readable than those for other areas.[6]

Tax-free land, dues and the mirasi system

As explained above, the mirasi system was a share-distribution system by which production activities in local society were maintained. Any right took the shape of shares in the produce. These were of three different types. First was tax-free land, second were dues in kind, and third were shares for the state and the cultivators, all of which were distributed to the various recipients. Features of the mirasi system were clearly expressed in the allocation of these shares, details of which form the major portion of the Barnard report. We will examine them in order. Table 1 is the aggregated result of information showing the frequency of villages where the different items were recorded.

Land in Salavaucum pargana, where 72 villages were located (see

[4] The poligars' duty was to safeguard the villages under their jurisdiction. They usually had several talliars (*talaiyari* or watchmen) and prevented robbery or other crimes. Some of the bigger poligars had armed forces and played an important role in military operations. More details will be given below.

[5] Professor Y. Subbarayalu of the Tamil University kindly suggested to me that the original is in the shape of palm leaves and is kept in the Tamil University. I have had no chance to confirm this.

[6] The Barnard Report is kept in the Tamil Nadu Archives. It is bundled according to pargana, a pargana being an administrative unit smaller than a district. There are 26 volumes in total, excluding the duplicates. The record of Salavaucum pargana is titled 'Jagir—Barnard's Survey Accounts of Salavakam Vol.69'. Microfilming has been completed recently, and the data-processing is in progress. This chapter is the result of on-going computational analysis.

Maps 1 and 2, and Table 2), was categorised into three types in the report, that is, 'free-gift land', 'sirkar land', and other land. The first two can be called 'tax-free land' and 'state land' respectively (see Table 3).[7] The other land included woods and poor ground, and 'Yary' (*eri*: irrigation tank, or reservoir). These were basically the lands not available for cultivation, as indicated in Table 4. This type of land can be called 'land for public use', the type of land which was called 'poramboke' in colonial land administration. The total extent of Salavaucum pargana was 22,496 *kanis* (cawnies) or 29,694 acres, of which tax-free land occupied 10 per cent, state land 52 per cent, and the rest 38 per cent.

The point to be observed in this categorisation is that the 'sirkari' (state) land was the counterpart of tax-free land, and signified, therefore, taxable land, and not necessarily land in state ownership. The state had the right to levy taxes from land when it was brought under cultivation and some crop was produced. In the tax-free land, the exempted tax, or the state's share, was left to the holder. Even if the land was clearly demarcated, what was granted was not the land but the share of the produce to be levied by the state. Two broad categories of such land could be observed in the Barnard report, land 'belonging to the village establishment' and 'the property of strangers', both of which were then divided into 'old' and 'new' respectively. These were terms to distinguish the year of the grant of the tax-free land. If made before the reign of Nawab Sadatullah Khan (1710-32), the land was classified as 'old'. All the later grants were 'new'. The year of the grant, the grantor's and grantee's names were often recorded in the latter cases.

The usage of the terms 'village establishment' and 'stranger', needs careful consideration as they connote what was the 'village' and what 'local society' in the period. Analysis shows first that those receiving tax-free lands belonging to the village establishment did not always

---

[7] Terms used in the tables and figures, such as measures, caste names, personal names, and others, are reproduced from the original record in most cases, though some are partly standardised because the same person, caste, or place are very often spelled differently from page to page in the report. For instance, 'Brahmin' is spelled 'Braminy', 'Braman', or 'Bramin'. There are also many Tamil-style English usages in the report, which are also standardised in most cases.

reside in the village where the land was allocated. In Ahlapaucum village, for instance, the recipients of the irrigated 'old tax-free land belonging to the village establishment' were shown as 'pagodas' (temples), Vaithaverty Brahman, landholders, chief of the village, poligar, *kanakkuppillai* (village writer/accountant), shroff (money changer), artisans,[8] *eri* servant, cowkeeper, barber, and cornmeter (probably a measurer of land or grain). Of these, the Vaithaverty Brahman, poligar, kanakkuppillai, artisan, and barber were not to be found in the list of the households residing in Ahlapaucum. Therefore recipients of this tax-free land were not necessarily residents, but rather were those who in some way or other were involved with village activities. They were granted tax-free land for their services. A barber, for instance, is essential to village life in India, as he officiates over marriage and other rituals, and his wife often works as a midwife.[9] Thus, though no barber lived in Ahlapaucum, tax-free land was allocated to a barber from outside who provided the service. The sphere of such contacts, forming the basic unit of production activities and wider than a single village, is termed 'local society' in this paper. The spatial aspect will be discussed later in more detail.

A few more features of this type of tax-exemption can be indicated from total figures for the pargana. Table 5 shows the major recipients. Land in this category totalled around 10 per cent of the total of 14 per cent of the land fit for cultivation. Poligars (and talliars) occupied 31 per cent, kanakkuppillai 25 per cent, artisans 10 per cent, 'cornmeters' 7 per cent, landholders 7 per cent, chiefs of the villages 5 per cent, and shroffs 5 per cent. These seven recipients occupied around 90 per cent. Artisans, accountants, 'cornmeters', poligars, and shroffs were granted tax-free land in almost all the villages.

What is indicated by these figures is, first of all, that the grant was made not because of a contribution to the state, but because of a contribution to local society. This becomes clearer by examining the list of

---

[8] The artisans here most probably signify the Kammalan caste composed of five occupational sections, that is, goldsmith, brass-smith, carpenter, stone-mason (or sculptor), and blacksmith. Thurston 1909, vol.III, pp.106-25.

[9] Thurston 1909, vol.I, pp.32-41.

recipients as a whole. It included artisans, *panjangum* (astrologers), Brahmans, barbers, potmakers, snake-doctors, washermen, temples, landholders, and chiefs of the villagers. Most of these were apparently not directly involved in the state administration, but rather they performed essential roles in production and the local society. The major recipients were core members of that society—though the state was ensuring their co-operation by assuring their vested interests through tax-exemption. Secondly, this 'village establishment' was not necessarily contained spatially within any one 'village'. Though tax-free lands were located in almost all the villages, the recipients were not always resident. The land for the village establishment was located in the village, but could be allocated to any outsiders who performed the roles required. A further implication from the fact that poligars were included in this category, will be discussed later.

Another category of tax-free land was 'the free-gift land which was the property of strangers'. Tables 6 and 7 list the major 'strangers' who were granted tax-free land, old and new, in Salavaucum pargana. The 'old' occupied 110 kanis or 0.8 per cent and the 'new' occupied 101.875 kanis or 0.7 per cent of the cultivable land. Among the recipients of the 'old' grants, both the town kanakkuppillai (a revenue accountant probably based in Salavaucum town), and the Permal Temple of Wyouoor were found in twelve villages. The former was granted 18.605 kanis and the latter 18.5 kanis. Sydda Moostapah Fakeer (a Muslim priest)[10] of Salavaucum town had 18.5 kanis in three villages. Yawkoopshaw Fakeer of Seetaupoorum was granted the whole of Seetaupoorum (14 kanis). Other recipients received just a few kanis in one or two villages. Among the recipients of the 'new', Sydda Moostapah Fakeer and Sooltaun Fakeer, both of Salavaucum, predominated. The former was newly granted 41.5 kanis in ten villages, and the latter was granted 30 kanis in a single village. The high share of the Muslim priests in the new grants must have been the result of the Nawabi rule in the period. The rest were grants between one-half and 5 kanis of land in a village or two.

[10] The names of persons not otherwise known to the historical record are spelt in this chapter as they appear in the Barnard Report. [Note by eds.]

The list of the recipients of tax-free lands in this category indicates that the majority of grants were made personally to officials or priests of a higher level.[11] They were individuals or institutions that played some role at an intermediary level between the local society and the state. The reason why they were called 'strangers' is now evident. In contrast to those recipients of the tax-free land 'belonging to the village establishment', they were not only outsiders but people or institutions of much higher level or a level 'strange' to the villages concerned. We may conclude that the categorisation of tax-free lands into two types in the report was well based on notions prevalent in the period.

One last but nevertheless important issue regarding tax-free land remains to be noted. What was granted to the recipient of tax-free land was the share to be taken as land-tax by the state. The dues to be studied below as well as the share for the cultivators were still collected from this land. In this sense the ownership of tax-free land in this period had a completely different nature from the exclusive ownership of plots under the raiyatwari system in the later period, even though the tax-free land was clearly demarcated. It was ownership of shares in the produce like other types of shares in the mirasi system.

We will next investigate the nature of dues in the mirasi system, a subject which occupies a major portion of the report. Four categories were recorded. They were (1) 'dues paid before treading the corn', (2) 'dues paid before measuring the corn', (3) 'dues paid half by the sirkar (state) and half by the cultivators', and (4) 'dues paid by the sirkar alone'. The first two were paid gross, before the major division of the crop, and the second two were paid net from the two main shares allocated on the harvest floor. For convenience we will call them (1) bulk, (2) grain, (3) joint and (4) state dues. All were shares taken from the harvest as a whole: whether from the unwinnowed crop, or from the grain heap, or from the net share of the cultivators or the 'sirkar'.

A few sources in the period give accounts how the dues were dis-

[11] The recipients included among others the kanungo of the district, chiefs of the district such as Moodooramanaick, Vanagoovamoodelliar, and Vencatachela Brahmin, Sydda Esmall Fakeer of Salavaucum, and Ramia (Brahman) of Kanchipuram, a famous religious centre.

tributed. For instance, Ellis wrote that 'Mereis [dues or fees] of all descriptions are received in three ways, in the straw before threshing, in grain before measuring, from the government share after division…, the same officer sometimes receiving a merei once only, sometimes twice, and sometimes thrice'.[12] Buchanan, who travelled in South India at the beginning of the nineteenth century, vividly described the way of dividing the rice produce as follows:

In almost every village (Grama) the customs of the farmers, especially in dividing the crops, are different, The Shanaboga, or village accomptant, keeps a written account of these customs; which is referred to as being the law, or custom of the manor…. The corn, when cut down, is made up into burthens, as large as a man can carry on his head. From each of these is taken a bunch, equal in all to about $^{56}/_{80}$ parts of the seed down. These parts are divided thus: To the Nirgunty, or distributor of water 16 [seers], to the Toti, or watchman 16, to the Aduca, or beadle, called here Cauliga, 16, to the iron smith 8. Then from the heap is taken, by the Toti, or watchman, whatever sticks to the seals of mud, that he put on to prevent embezzlement, which may be about 3, by the Pujaries, or priests of the village gods 4, by vagrants of all religions and kinds, who, under pretence of dedicating themselves to God, live by begging 4, by the Gauda who rents the village, as his perquisite 8, by the government, as its perquisite, called Sadi 16, by the hereditary Gauda, or chief of the village, in order to defray the expense of the feast which is given to Ganesa, under the forms of a stake in the Cassia Fistula 16. The heap is then measured, and divided equally between the government, or renter, and the farmer; but a certain portion is left, which is divided as follows: From this portion twelve Seers for every Candaca [160 seers] in the heap are measured, of which the acccomptant takes one third, and the remainder goes to the renter…. From what remains there is taken, by the Panchanga, or astrologer 1 [seer], by the Cumbhara, or potmaker 1, by the Assaga, or washerman 1, by the Vasara-dava, or blacksmith and carpenter 1, by the measurer the sweepings about 8 [totally 12 seers]. It is evident, from the very unequal size of the heaps, and various rates of produce in different soils and seasons, that no exact calculation can be formed of the amount of these perquisites on the whole crop. If the heap contains 20 Candacas, and the produce be ten seeds, then they will amount to 17 per cent; of which the government gets $5^1/_2$ per cent; or altogether 47 per cent of the

---

[12] Replies from Mr. F.W. Ellis, Collector of Madras, to the Mirasi Questions, 30 May 1816, Papers on Mirasi Rights, p.178, footnote 6.

crop; from which is to be deducted the expense of the tanks.[13]

Though we do not have any descriptive account how the four types of dues recorded in the report were distributed, it must have been as follows. When the crop was harvested, bulk and grain dues were taken aside and were paid to the respective recipients. Then the rest of the produce was measured and divided between the state and the cultivators. After this, joint and state dues were paid.

The first notable aspect to be observed about the format of the Barnard report is that the net produce, after measurement and distribution of shares calculated gross, was equated with the total harvest. The shares for the state and the cultivators and the dues which were drawn from them always amounted to 100 per cent. This fact implies that the amount of dues recorded in the report as *kalam*, *marakkal*, and *measure* (k.m.m.)[14] was not an absolute total but a proportion of the produce. Another implication is that 'measuring the corn' marked the start of the state's intervention. Shares which were paid before this main measuring, were in this sense beyond the reach of the state. This aspect will be studied later. The second important point to be observed is that dues, and the state's and the cultivators' shares, were treated in the same way in the report, that is, as a proportion of the produce. Even though the state and the cultivators were by far the largest beneficiaries, yet all rights in the harvest—tax, rent, dues, or whatever—took the shape of shares.

The study and the comparison of the categories of dues gives us some more information about the nature of the mirasi system. We will first examine the dues paid in bulk (before the treading of the corn). In Salavaucum pargana there were 57 types of recipient of such bulk dues. The villages with recipients, the range of dues, and the average share for the major recipients are indicated in Table 8. Artisans, barbers, kanak-kuppillai, shroffs, and washermen received such payments in almost all

[13] Buchanan, I, pp.299-300.

[14] These units of measurement will subsequently be referred to as 'k.m.m.' or in figures thus (for example): 1-10-5 kalam, meaning 1 kalam, 10 marakkal and 5 measures. One kalam was equal to 12 marakkal, and one marakkal to 8 measures. [Note by eds.]

the 72 villages. They were followed by snake-doctors (curers of snake bites), 'cornmeters', Panjangum Brahmans, and poligars with talliars. The amount differed greatly from village to village and from recipient to recipient. On average poligars received the highest share of one kalam per hundred (that is, one per cent of the gross produce), followed by artisans with 8 marakkal, kanakkuppillai with 6, shroffs with 4, and 'cornmeters' with 3. The total amount in the respective villages ranged from 1-06-0 to 7-02-3 kalam. On average it was around 4 per cent of the gross produce.

There were 73 types of recipient of gross dues paid from the grain heap, before the corn was allocated between the cultivator and the state. The details are indicated in Table 9. In most villages the recipients were landholders, barbers, Panjangum Brahmans, 'cornmeters', and washermen. They were followed by cultivators' servants, snake-doctors, shroffs, kanakkuppillai, the O(A)mmun Padari temple, artisans, Valloovuns (priests for the Pariah, Pallas and other out-castes),[15] and village temples. Among these, the cultivators' servants received the highest share of 2-6-0 kalam in every hundred (2.5 per cent) of the gross produce. The rest received less than half a kalam each. The total amount of these dues ranged from 1-06-0 to 8-08-2 kalam. On average it was around 5 per cent of the gross produce.

The first point to be noticed is that the recipients of these two dues were all at the 'village' level. They contrasted with the recipients of joint dues paid from the net shares of the state and the cultivators; as will be shown, most of these were intermediary between the local society and the state. Many of the recipients of dues from the gross product also held tax-free land 'belonging to the village establishment'—a fact which explains why their share pre-empted the state's. We must remember, however, that the recipients did not necessarily live in the village

[15] There are three main Brahman categories, of unequal rank: those who served as priests to non-Brahmans, temple priests, and learned priests; Brahmans may also be divided by occupation rather than sub-caste into 'vaidic' and 'laukic', the orthodox who served as priests and the lay who were landholders or engaged in other secular occupations. See for example André Béteille, *Caste, Class and Power* (Berkeley, 1971), p.15 and ch.3. [Note by eds.]

where the dues were paid. Each unit of production extended over a wider area than a single village, and the dues related to roles essential to production. Secondly, it seems that, by effectively exempting these payments from taxation, the state admitted that the arrangements of the village establishment were beyond its reach: the right of tax-exemption was a customary right with which the state could never interfere, and which, on the contrary, it confirmed. Such vested interests encouraged the collaboration of the local society with the state.

Thirdly, the timing of the payments seems to have been related somewhat to the recipients' roles in production. The first set of dues were paid, as said, in bulk, in both grain and straw; the second in grain only. Most of those who received the first received the second as well, though some only the second. For example, poligars received grain and straw in 53 villages, but grain only in 21; this may reflect the fact that their main role was to protect the standing crop from thieves. Cultivators' servants received both grain and straw in one village only, but grain shares in 62 villages. This may be because their main role was in harvesting, threshing, and carrying the crop before its measurement. Landholders received grain and straw in only five villages but grain dues in almost all—perhaps because they were not necessarily involved in agricultural operations.

After these dues were deducted, amounting to 9 per cent of the gross produce, the remainder of the harvest was measured, and two other types of dues were deducted, as said, either half from the state's share and half from the cultivators', or wholly from the state's share. In Salavaucum pargana there were 22 categories of recipient of the first of these, which was by far the largest set of dues. It ranged between 9-10-6 and 15-10-1 kalam, and on average came to around 13 per cent of the net or 12 per cent of the gross produce. The major recipients are indicated in Table 10. In more than 50 villages they included the kanungo and the deshmukh of the district, the 'Dovetraw' grant (see below), the Kanchipuram (Conjiverum) Temple, Mullah Saib of Salavaucum, the Paulia Pandarum (non-Brahman temple priest) of Kanchipuram, and cultivators' servants. The *eri* (reservoir) was a recipient in 27 villages. Other categories appeared in fewer than four villages. Cultivators' servants received the highest share at around 4-9-1 kalam or 4.8 per cent of the net produce. Next came 'Dovetraw', a two (*do*) per cent share originally granted to Toorelmull who assisted Nawab Sadatullah Khan in negotiations with the Marathas.[16]

First to be noticed is the contrast in the composition of recipients of these payments and those whose shares were calculated gross. With the exception of the cultivators' servants, almost all were from outside the village. They were high officials of the state, highly-esteemed priests or religious institutions, either intermediary between local society and the state or without any place in the village concerned. They were very similar to those granted tax-free lands as 'strangers'. Many of them seemed to be closely attached to the state by their administrative and religious contribution, but at the same time they were indispensable for the local society, by connecting it to the state or to centres of worship. Their dual role is mirrored in their being paid half by the state and half by the cultivators. The reason why cultivators' servants were included in this category needs some interpretation. Why did the state grant dues to those not supposed to play any role for the state? One interpretation may be that the state utilised these servants for public works such as construction or transporting ammunition. The majority were Pariahs, Pallas (agriculturists of low rank), or other lower castes,[17] who made up around 30 per cent of the total population in the period.[18] Perhaps the

---

[16] Report of Mr. Place, 6 October 1795, extract from Chingleput Districts Records, 15, followed by Saismooc of the district (1.11.5), Conigo of the district (0.11.7), Mullah Saib of Salavaucum (0.11.7), and Conjiverum pagoda (0.11.7).

[17] According to Ellis, most Pariahs and Pallas were slaves attached to the lands of the Vellala (agricultural caste) and the Palli were generally serfs on the lands of the Brahman. (Minute of the Board of Revenue, on the different modes of Land Revenue Settlement, as existing in different Districts, 5 January 1818, Papers on Mirasi Rights, 368.)

[18] The total number of households and the number of casts in the 72 villages in Salavaucum Pargana were 1,894 and 67 respectively. Salavaucum Pargana had the largest number of households (192). All others had less than a hundred, including five uninhabited villages. Of the 67 castes, Pariah had 286 households (15 per cent), followed by Pallis (211 or 11 per cent), Vistnoo Brahman (143 or 8 per cent), Puccolum Vellala (agriculturist of high rank, 109 households or 6 per cent), cowkeeper (109 or 6 per cent), Tooliva Vellala (89 or 5 per cent), Reddi (agriculturist of high rank, 79 or 4

state, by paying these people, sought to exercise control over them.

Another feature to be noticed was the uniformity of the amount of dues of the kind compared with other types (see Table 10). This uniformity may be explained first by the fact that they were allocated to those who did not have any base in the village concerned. The shares, therefore, were not affected by local conditions except in the case of the '*eri*' fund, for irrigation, which naturally varied a lot according to the size or condition of the tank. One might argue too that the payments were uniform because, as they were made jointly by the state and local society, they resulted from a competition between them, which produced standard rates.

The dues paid by the state alone had some different features. There were very few recipients, even though in the 72 villages in Salavaucum there were 15 types in all (see Table 11). Only artisans, kanakkuppillai, shroffs, poligars, and 'cornmeters', occurred in more than 50 villages, and the others in fewer than seven. The recipients were predominantly drawn from the local society. Though some from beyond the village level, such as the district kanungo and deshmukh, the Kanchipuram temple, or 'Dovetraw', were also included in the list, they were found in only the same two villages. In short, such payments were exceptional.

The amount of these state dues differed greatly from village to village, ranging from 2-4-4 to 9-10-6 kalam (on average around 4.5 per cent of the net or 4 per cent of the gross produce). The amount should have been more uniform, if the payments being defrayed by the state were under state control. In what sense were these payments for contributions to the state? No doubt the kanakkuppillai, shroffs, poligars, and 'cornmeters' could be said to receive the dues because of their contribution to administration.[19] Even artisans (carpenters, smiths) may have

---

per cent) and others.

[19] The Tiruvendipuram Report of 1775 gives the following account. Kanakapillai was a village accountant to keep revenue records. The main duty of the shroff (money-changer) was to sort and examine the money paid by the husbandmen (cultivators) for their rent and to keep it for payment to the revenue collector. 'Cornmeters' surveyed land and measured produce for assessment. The poligar was the head of the talliar, or village watchmen. He kept law and order and safeguarded the crop. Tiruvendipuram was located near Pondicherry, and consisted of 32 villages. (Reports and Accounts of the Old

been included in order to ally them to the state as core members of the local society. Artisans, for instance, were not only skilled people but the leaders of left-hand factions in the right-hand versus left-hand conflict. Yet, judging from the fact that the same people received harvest shares of other kinds, and tax-free land in almost all villages, we may conclude that the variation in the rates of deduction from the 'state's share' was due to local conditions.

To sum up, the average proportion of the gross produce in Salavaucum pargana paid out in dues were 4, 5, 12 and 4 per cent respectively for the four kinds we have considered. In total the dues amounted to about one-quarter of the gross produce.[20]

The last dimension of the mirasi system was the division of produce between the state and the cultivators. The produce was divided in different proportions. In principle the cultivators' share was higher (1) when the cultivators were not village residents, (2) when the cultivators were not landholders, (3) when the land was not irrigated or used for paddy (it was *cumbu* or unirrigated), (4) when it was watered by lift irrigation (by 'picotash', a less efficient see-saw method) rather than from reservoirs and channels, and (5) when irrigation was available only for short peri-

---

Farm of Tiruvendipuram, Selections from the Records of the South Arcot District, No.IV, printed at the Collectorate Press in 1888.)

[20] Place reported in 1795 about the proportion of fees in the three parganas of Tripassore, Carangooly and Conjeveram. Swoduntra Dittum, or fees collected before the produce was measured, amounted to 12 per cent of the gross produce. Purdie marah and Aurdie marah, both of which were collected from the Teerva (assessed) produce, was from 9 46/64 to 12 53/64 per cent in the former and 2.5 per cent in the latter. Adding the share from the Mauniam or tax-free land, the total deductions comprised from 30 25/64 to 34 5/8 per cent of the gross produce in the three parganas. (A letter from Mr. Place, Collecture of the Jaghire to the Board of Revenue, 6 October 1795, Papers on Mirasi Rights, pp.19-24.) Actually the proportion of these dues in the total produce was variable from place to place and from time to time. For instance, Hodgson reported from Dindigul that the share was stated to have been 6.25 in every 100 of the gross produce prior to the acquisition of Dindigul, which was estimated at 400 in 100 in 1794, and was regulated at about 12 in 100 in later years. (Mr. Hodgson's Report on the Province of Dindigul, 28 March 1808, Fifth Report, III, p.552.) Wallace reported from Tanjore that Sotuntrums (dues) and Mauniams (tax-exemption) amounted to about 12 per cent on the gross produce. (Mr. Wallace's Report on Tanjore, 8 September 1805, Papers on Mirasi Rights, p.100.)

ods. In some villages there were as many as ten different proportions, but, in short, the share for the cultivators was higher when more labour was needed or a lower yield was expected.

The state land, both cultivated and uncultivated, occupied 52 per cent of the total area. The proportion of cultivated land in the state land was 66 per cent, from which the state collected its share as land-tax.[21] The cultivators' share in Salavaucum fell between 63-10-5 and 34-4-0 kalam, and that of the state between 51-6-3 and 20-1-6. The average of the gross produce was around 43 per cent for the cultivators and 32 per cent for the state. Calculating from the extent of tax-free land and the state's share of dues paid out, the final proportion for recipients of tax-free land comes to 7 per cent and that of the state to 25 per cent.

The allocation of produce between the state and the cultivators indicates the following points. Firstly, there are variations which again suggest a delicate balance between the two. Secondly, the apparent relationship between shares and local conditions confirms that produce was at the heart of revenue administration. Tax was imposed not upon land but upon the produce from cultivated land. Thirdly, the share for the state was very high, at nearly a quarter of the gross produce. Finally, the working of the system of harvest shares indicates the significant involvement of the state. It did its best to maximise its own share by intervening in the local society, and at the same time made compromises to gain collaboration by affirming tax-free land or dues. The conclusion is that we cannot imagine the state to be confined within a ritualistic sphere in the period under study.[22]

This relationship between the state and local society leads us to discern two highly-competitive forces. We may term one 'community formation', the attempt to protect and maximise the interests of the local society, and we may call the other 'state formation', a counter-force to enlarge the state's interests. Each of these sought dominance over the other and to reach a maximum of autonomy. We can clearly

[21] The proportion of the irrigated land in the state land was as high as 56 per cent in the area. The taluk maps prepared in the 1960s indicate at least one or two reservoirs in every village. Besides the Cheyyar and the Palar rivers run in the north and east of the area.

[22] See B. Stein, *Peasant and State in Medieval South India* (Delhi 1980).

understand the nature of the dues, tax-free land, state's and cultivators' shares so far examined, or the mirasi system itself, if we characterise this competition. The two elements competed to enlarge their own shares, and thus the various types of dues and the extent of tax-free land; the number of villages where dues or tax-free land were granted, and the range of the beneficiaries, could all be considered as expressions of their rivalry. We can suppose that the state and the local society had an interest in accommodating each other, but also that there must have been serious conflicts between them to finalise the shares, as they were observed in the initial period of colonial rule. The mirasi system was formed and maintained in the course of such conflicts and compromises.

Secondly, the mirasi system was highly flexible. The right to the share, though customarily established and frequently transacted, was linked to the role performed and not necessarily to the recipient. Hence the shares were preserved even when the personnel changed. Vacancies could be easily filled by a newcomer, which was appropriate given people's high mobility due to the precarious nature of Indian agriculture and to nomadic herding. Newly-created roles could also be absorbed into the system by the allocation of a new share. Moreover, the activities covered by the system were social and religious as well as economic, as was indicated by the inclusion of temples and Brahmans among the recipients of various dues.[23] Finally, any change in the power balance between the state and the communal formations, or in the class relations among the members of the local society, could be accommodated by an adjustment of internal mechanisms (shares, tax-free land, and their recipients), so that the system itself would never break down. Though it is not yet ascertained when mirasi was established as a system in South India,[24] its flexibility surely allowed it to survive over

[23] The role of religious institutions such as temples and maths is considered important in generating identity for any local society. The issue will be undertaken on some other occasion.

[24] James Heitzman investigated the usage of the term 'kani' or 'cawny', the Tamil equivalent of 'mirasi', in the Chola inscriptions, and argued that the term occurs within the following three main contexts. They are: (1) the situation in which property or possession entailed performance of a specified duty within a village or a temple; (2) the enjoyment of properties or prerogatives in connection with membership in a temple staff; and (3) the

a very long period.

A few more important issues remain in the analysis of the mirasi system. In the report almost all the villages have some account of 'landholders' and the 'shares' held by them (see Table 12 for the number of villages held by the respective castes). For instance, in the case of Ahlapaucum village, the report recorded 'Landholders shares Vistnoo Braminies 8 fixed'. Other villages were also held in several 'shares', the biggest as many as 80. The 'share' in this usage is the equivalent of '*karai*' or '*pangu*', that is, with regard to the ownership of the village. In the colonial records such owners as were entitled to *karai* or *pangu* were called *kaniyatchikkaran*, and often mirasidar. 'Landholder' in the report therefore signifies mirasidar.[25]

How can we, then, locate the landholders in the mirasi system? Strangely enough, the Barnard report does not have any information proving their superior position in the village. For instance, the Shri Vaishnava Brahmans in Ahlapaucum, whose households numbered 31, held the village in eight fixed shares. If they were 'owners' of the village, there should have been evidence of some privileged right, namely the landlord's share in the produce. It is true the landholder in Ahlapaucum received two kanis of tax-free land belonging to the village establishment, and six marakkals (half a kalam) in every hundred kalams of the gross produce before threshing. Such benefits, however, would be negligible once divided among 31 households. But no other privileges were recorded in the report. Interestingly enough, Place, the collector of the Jagir at the end of the eighteenth century, did not record any figures for the mirasidars' rent as landlords. Though he acknowledged the mirasidar's hereditary right, and considered it sequestable for neglect of 'duties'—cultivating lands, paying tax, obeying the state—yet he noted:

Were I called upon the define the term meerassee, and its properties, I think it bears exact analogy to a fee. I would call meerassee a freehold estate of

---

property in land. According to him, the term had never been used till AD 985, but it came to be used far more frequently after AD 1070 (Heitzman, 1987, pp.54-7).

[25] Though the term 'mirasidar' could be applied to anyone having hereditary right, it was mostly used to designate the kaniyatchikkaran who owned the village.

inheritance, and a meerassadar a tenant in fee simple, holding of a superior lord, on condition of rendering him service. His lord is the Sirkar, his estate the usufructuary right of the soil, and the service he owes, a renter of a stated portion of the produce of his labour. [26]

By contrast Ellis, who made a fairly long report about the mirasi right as a collector of Madras, clearly noticed the landlord's rent. Indeed the object of his report was to reverse Place's view of mirasi rights.[27] Unfortunately for Ellis, he could not find any information about the rates of rent taken by the landlord.[28] Not until 1820 were rents in Chengalpattu recorded with concrete figures. Smalley noted that the landlord rent in the district varied from less than one per cent to 11 per cent of the produce, and that the average was about 3.5 per cent.[29]

The existence of mirasidars and their landlord's share was, however, not a rarity in South India. Several accounts of the landlord's share were recorded by the administrators of different districts, and are found in the early colonial records. By far the highest share taken by the mirasidars was in Tanjavur (Tanjore), the stronghold of the mirasidars. According to Harris, the Collector, mirasidars received nearly half of the 'coodee-wayryn' (cultivators' share).[30] The stronger position of the mirasidars in Tanjavur and their higher share seemed to reflect their more direct

[26] Extract from Mr. Place's Final Report on the Jagir, 6 June 1799, Papers on Mirasi Rights, pp.53-4.

[27] This is most clearly expressed in his following note: 'Mr. Place, whose authority on all subjects connected with Mirasi must be admitted to be great, holds exactly the reverse of what is here stated...[this view is] totally inapplicable to Mirasi, as it originated, and as it exists'. Replies from mr. F.W. Ellis, Collector of Madras, to the Mirasi Questions, 30 May 1816, Papers on Mirasi Rights, p.341, fn.4.

[28] Replies from Mr. F.W. Ellis, Collector of Madras, to the Mirasi in Chingleput

[29] From Mr. Smalley, Collector of Chingleput, to the Board of Revenue, 4 November 1820, on the Introduction of a Ryotwar Settlement into his District, Papers on Mirasi Rights, p.398.

[30] Harris noted that under the Maratha management or pre-British period, the 'coodee-waurum' was 40 per cent and the miraisidars had 20 per cent, 'and in some places even 25 of it. At present the coodee-waurum varies between 50 and 60 per cent, and their receive from 23 1/2 to 26 1/2 of it', whereas 'in other parts of India' it was 'never more, but often less, than 5 per cent of the net produce'. (Report of Mr. Harriss, to Committee at Tanjore, 9 May 1804, Papers on Mirasi Rights, p.87.)

control over the tenants. According to Wallace, when Collector, the mirasidars 'superintend and direct the labours of the poragoodies [tenants] in all the particulars of rural economy.... In cases...[when] they are unable to supply the agricultural stock, advances are made to them either in grain or money by the meerassadars....'[31] Alternatively, 'The far greater mass of them [the mirasidars], till their lands by the means of hired labourers, or by a class of people termed Pullers, who are of the lowest caste, and who may be considered the slaves of the soil'.[32] These remarks seem to imply a kind of individual control over the cultivators. They do not, however, deny the existence of the share distribution system discussed here. Wallace also gave details of 'sotuntrums' (dues) and 'mauniams' (tax-free land), of three classes—for agricultural purposes, in support of certain village officers, and for charitable and religious purposes or the maintenance of the police. They amounted to about 12 per cent of the gross produce and were borne jointly by government and the mirasidars.[33]

Compared with Tanjavur, the landlords of other areas had a lower share. Hodgson reported from Tirunelveli (Tinnevelly) that the share for the mirasidar was about 13.5 per cent of the gross produce.[34] J. Cotton quoted, for the same district, the figure of $4^7/_8$ per cent for the land of 'swamy bogum' tenure, by which the mirasidars received the landlord's share from the tenants who bore all the costs including the land tax, and a figure of around 10 to 12 per cent on other types of land.[35] R. Peter reported that the average annual value of mirasi right was about one-tenth of the produce in Madura.[36] Another important source of the period, the Tiruvendipuram report of 1775, recorded the existence of

[31] Report of Mr. Wallace on the Settlement of Tanjore, for Fusli 1214, 1 May 1805, Papers on Mirasi Rights, pp.96-7.

[32] Mr. Wallace's Report on Tanjore, 8 September 1805, Papers on Mirasi Rights, p.99.

[33] Ibid., p.100-5.

[34] Report of Mr. Hodgson on the Revenues etc. of the Province of Tinevelly, 24 September 1807, Fifth Report, III, p.346.

[35] Replies from Mr. J. Cotton at Tinnevelly to the Mirasi Questions, 8 May 1817, Papers on Mirasi Rights, p.162.

[36] Replies from Mr. R. Peter at Madura to the Mirasi Questions, 10 January 1815, Papers on Mirasi Rights, p.159.

landlord's rent in the area near Pondicherry. It amounted to 10 per cent of the produce or 10 per cent of the tax from the land cultivated by the non-landlords.[37]

Thomas Munro, chief protagonist of the raiyatwari system, had a slightly different view of the landlord's right. He did not deny altogether the mirasidar's right to claim rent from the tenant, when the mirasidar paid the public revenue and made terms with the tenants on his own accord. In his view, mirasidars, however, had no claim to rent on land for which they neither paid the public revenue nor found a tenant (because the government provided one). Secondly, Munro thought the mirasi right worth little and that it was seldom saleable except in a few districts. [38]

Why, then, were there no figures about landlord rent in the Barnard report? The first possible reason is simple, that is, that landlord rent did not exist. This explanation is, however, hard to sustain. For instance, the Ahlapaucum Shri Vaishnava Brahmans, who were by no means engaged in cultivation, would not have been able to maintain themselves without receiving rent. A second and more plausible explanation is that the landlord rent was not recorded in the village accounts Barnard depended on, as landlord-tenant relations were not directly related with the revenue administration. This is implied by the fact that in 39 villages in Salavaucum the number of shares was recorded as 'unknown' in the report, as were the residences of the landholders in 13 villages. Barnard might have attempted to obtain information about these rights in vain. He did collect much information which was not about revenue, and thus a third possible explanation is related to this one—that Barnard did not expect to find landlord rent due to the notion prevalent among eighteenth-century Europeans that the state was the sole owner of the land. This notion is clearly expressed by the following statement: 'It is the first feature in all the Governments of India that the Sovereign, whether he be a Mussulman or a Hindoo, is lord of the soil'.[39] The

[37] Tiruvendipuram Report, p.10.
[38] Sir Thomas Munro's Minute on the state of the country, and condition of the people, 31 December 1824, Papers on Mirasi Rights, pp.435, 444.
[39] Extract from the Minutes of Consultations, 8 January 1796, Papers on Mirasi Rights, p.26.

categorisation of all the land as 'state land', except land reserved for public purposes and the tax-free land, makes this explanation seem plausible.

Though further verification is needed, it can be assumed that the landholders received a considerable share in the produce as landlord rent, and that this share was then divided among them in proportion to their 'ownership' rights. This leads us to discern a different type of produce-distribution, namely, a higher or secondary flow from the initial recipients to the secondary recipients in the mirasi system. Included in this category might be not only the rent paid by the tenant to the landholders, or by the cultivators to labourers, but also the payments from a poligar to his followers or from a temple to its servants. There is evidence of customary wages for labour;[40] but further information on such secondary distribution cannot be discerned from the Barnard report.[41]

Another issue to be considered in this regard is the position of the landholders in the mirasi system and in the formation of the community. In the Chola period we know, according to Subbarayalu, that *nadu*

---

[40] The following accounts indicate that daily wages was paid to the agricultural labourers: '...in every village a custom prevails..., whereby the servants [of the husbandmen] during a stated number of months in the year, are to receive a further payment in grain from their masters, which continues or ceases in proportion to the extent of their fees, and in the end equalised the whole to about 2 1/2 cullums or 105 pucca seers per month' (a letter from Mr. Place, Collector of the Jaghire to the Board of Revenue, 6 October 1795, Papers on Mirasi Rights, p.20). 'The labouring servants are for the most part Parias...; they receive wages, partly in money, and partly in those fees...' (Place's Final Report on the Jagir, 6 June 1799, Papers on Mirasi Rights, p.47). The following account indicates the prevalence of daily wage for the works other than cultivation: 'The paragoody warum [the labourers' share under mirasidar] to the paragoodies, and the allowance to the slaves, are granted for the labour of cultivating. In every other work [than cultivation], the paragoodies and slaves fare alike, and are considered as coolies. When employed on the circar maramut [original note: repairs performed, at the expense of the circar, to the rivers and great channels], they receive from the circar; and when employed on the coodeemaramut [original note: repairs performed, at the expense of the tenantry, to the small channels, and to the banks of the paddy fields], they receive from the meerassadars, daily hire, at fixed rates' (Report of Mr. Harriss, to Committee at Tanjore, 9 May 1804, Fifth Report, III, p.341).

[41] This question will be examined on another occasion using other sources.

were the groupings of agricultural settlements formed by natural factors conducive to agriculture. Each was basically a cohesive group of agricultural people tied together by marriage and blood relationships. Members of the group met together in a local assembly, also called the nadu assembly. Among them, the *nattar* occupied the dominant position, being the representatives of the villages of agriculturists, the prime landholders, and the chief spokesmen of the people in the region.[42] If we accept this view (verification of which is beyond the scope of the present study) then, in its ideal or original form the nadu may be presumed to be the spatial unit and nattar its leading figures. These nattar must have had some caste or kinship relation among themselves in the respective local societies. Though originally just the representatives of their community, they must have taken a dominant position in the locality as time went on. As more and more communities with different interests were incorporated in the respective localities and formed a heteregeneous society, the mirasi system as observed in the eighteenth century must have gradually taken shape. Landholders or mirasidars, their dominance extending beyond a single village, somehow or other inherited the status of the nattar. They too represented the local society in negotiations with the state while being co-opted by the state to represent its interests.[43] The landholders maintained the cohesion of the local society by holding the villages in it either jointly or individually; their role depended on the communal identity among themselves.

The last issue to consider about the mirasi system is state formation. It is not easy to define what the state was and what its components were in the period of political turmoil in the eighteenth century. Not a small

[42] Subbarayalu 1973, pp.33-4, 39-40.

[43] The following remark by B. Sancaraya can be clearly understood in this context: 'The Mirasudar looks on himself as entitled to direct the affairs of the village, to stand forward on all occasions when the affairs of the Sircar are in discussion, and to receive any Tasrif [original note: honorary presents at the time for forming the annual settlements] given by it, and the pre-eminence thus claimed is allowed him by others' (translation of Answers to the Questions enclosed in mr. Secretary Hill's letter to the Board of Revenue, 2 August 1814, by B. Sancaraya, late Sheristadar to the Collector of Madras, Papers on Mirasi Rights, p.227.

number of local powers, including Europeans, struggled hard to gain control in different parts of South India. Three factors should be considered in this regard. First was, as said, that as much as one-quarter of the gross produce was set aside for the state. Second was the payment of local dues out of this state share through the mirasi system. Third was the existence of supra-local administrators such as kanungos, deshmukhs or town accountants, and of local state servants such as village accountants, measurers and so on, among the recipients of shares under the mirasi system.[44] The second point implies the deep involvement of the state with the system, and the third point implies the existence of state revenue-collection machinery. The first point, on the other hand, implies a very firm notion of a harvest share for the state: accordingly, though the present paper does not discuss these functions, the state must have played an important role in public works, military operations and so on. Moreover, from the viewpoint of the distribution system, what as important was that the state's share was always reserved regardless of who held state power. Those who wished to seize power tried hard to usurp this share by acceding to the throne as Nawab or, more often, by resorting to coercive measures. In the political situation in the eighteenth century there was a frequent shift of overlordship from year to year. In a sense whoever succeeded in controlling the state's share in the produce by any means could be called the holder of state power. It was immaterial in the mirasi system who performed an assigned role; it mattered only that the expected role was fulfilled. The highly flexible nature of the mirasi system could be observed even here.

Emergence of a new formation and the decline of the mirasi system

It has been shown so far that the mirasi system was a distribution system in which shares took several forms, were established by custom, and were linked with specific roles necessary for maintaining a local society. It was also indicated that the system was the product of conflict and compromise between the state and communities in the sphere of a

---

[44] As to the roles of kanungo and deshmukh, see letter from Mr. Place, Collector of the Jaghire to the Board of Revenue, 6 October 1795, Papers on Mirasi Rights, pp.21-2.

local society. In the period under study, however, there occurred several noticeable changes leading to the decline of the mirasi system. The changes were multiple, and several approaches are needed to clarify them. First we will take up the collapse of the cohesion of local society. As said, the primary unit of the mirasi system was not a single village but an area comprising several villages. This can be verified by the simple fact that the majority of villages of the period had an insufficient number of castes to maintain the village reproduction cycle. In Salavaucum pargana, for instance, the number of castes was largest in Salavaucum village where members of 33 different castes resided. Four other villages had more than 20 different castes. The others, however, had fewer than ten. Table 1, which indicates the number of villages in which each of the castes resided, and the number of villages in which several types of dues or tax-free lands were granted, clearly shows that many of the villages lacked essential functional castes, even though dues or other shares were reserved for them. Some of the activities essential to village life were performed by the people from outside, which implies that the village was not self-contained but incorporated in a wider societal unit. A village was, thus, nothing but a part of the local society and could survive only as a part of a larger whole. This raises the question of which territorial unit among those prevalent in the period actually corresponded to this 'local society'. The Barnard report used three territorial categories—village, *magan(-am)* and pargana. Salavaucum pargana consisted of 72 villages which were divided into eight magan. They were called Salavaucum, Cuddungherry, Sautanunjerry, Culliapettah, Perramatoor, Mungalum, Vellapootoor, and Undavaucum, names which corresponded with those of the principal villages. These had comparatively large populations and were often called 'cusbah' (*qasbah*) or town in other parts of the report. It seems they had a town-like nature. Though no information is available as to how each magan boundary was demarcated, a magan seemed to be a cluster of villages in the centre of which a principal village was located. We might take the magan to be the basic territorial unit within which the mirasi system functioned.[45] Investigation into several other aspects

[45] There is some evidence indicating a magan to be the basic unit for ad-

of spatial distribution, however, does not confirm this and rather indicates an inconsistency between the magan and the local society. We will examine the spatial aspect of poligars' jurisdiction of landholding castes, and of service.

There is an argument that views the Vijayanagar state as a feudal one in which the *nayaka* played a key role and ruled a territory called *nayakattanam*.[46] Poligars in the period concerned are also known to have taken overlordship in some parts of South India as the nayaka had in the previous period. The first point to be clarified is, therefore, the relation between the nayaka in the previous period and the poligar in the period with which we are concerned.

Most of the poligars had the title 'Naick'. All but one of 53 did so, for instance, in Salavaucum pargana.[47] So far as names are concerned, a close relationship is suggested between the nayaka in the previous period and the poligars in the eighteenth century. A second crucial point to consider is how the status and power of the poligars compared with those of nayakas. The poligars were thought to belong to the village establishment in the Barnard report. Their tax-free lands or dues were not classified in the categories for 'strangers' or for those in the intermedi-

ministration. For instance, the Committee of Revenue proposed in 1775 to set up local courts in every magan. A few other sources of the period also referred to magan as the basic local unit (letter from the Committee of Revenue to the President and Governor and Council of Fort St. George, 28 July 1775, in Madras Revenue Proceedings, 12 July 1775). See Mizushima 1986, III-2 and Appendix 14, for the discussion on magan and the list of magans with constituting village names in the Jagir. For the argument regarding the geographical units in South India, see Stein 1977 and 1982, and Subbarayalu, 1973. As to the views and the problems concerning the demarcation of region, see Saberwar, 1971.

[46] Karashima, 1992.

[47] As the overlapping cases cannot be ascertained by names alone, the total number of poligars could be slightly few, most probably 49. Permal Pagoda of Wyouoor is recorded as the poligar of Wyouoor, a shrotrium village given to the same Pagoda. 'Moodrial', the caste name, was used along with 'Naick' in 15 of the cases. Another title, Puttrawar, was used in the same way by ten poligars. Of 120 poligars appearing in the villages of Salavaucum, Parumbauk, and Saut Magan parganas in the report, 106 had the same title. Of 177 poligars in the Jagir recorded in the Permanent Settlement Records prepared in 1801, 150 of them had the name 'Naick' (Permanent Settlement Records, vol.26, Privilege of Poligar, in a letter from mr. Greenway, 30 October 1801).

ary level. Instead poligars received them apparently as members of the 'village establishment'. They occupied nearly 30 per cent of the total of the tax-free land belonging to the village establishment and were granted land in almost all the villages. They received the largest proportion of the bulk dues. They were given grain dues in 53 villages, joint dues in 20, and state dues in almost all the villages. These figures clearly indicate their position as one of the village establishment. According to the Tiruvendipuram Report,

The duty of the Polygar is now, and is said immemorially to have been to appoint Taliars to watch in 26 villages of the Farm both by night and by day at his own expense, for to prevent any of the Inhabitants as well as strangers from being robbed; and in case any person is robbed within the precincts of the 26 villages, he is to make good his loss to him, unless he apprehends the thief; one month before the paddy and small grain is fit for cutting in these villages, he is to provide Taliars to watch it, as also after it is cut and lyes in the fields; and if any of it is stolen he is to make it good to the cultivator. He is to take care of all thieves that are apprehended in these villages, he is to prevent the husbandmen from quitting the villages themselves, or from driving away their working cattle, and from carrying away their instruments of husbandry. In case an Invasion is apprehended from a Foreign Enemy, he is to furnish if ordered by the Amuldar or Renter a body of Peons or Subbendy to protect the whole Farm, for which charge of Subbendy he is paid besides his usual privileges.[48]

It is indicated from this account that the role of the poligar in Tiruvendipuram was basically to maintain law and order in the 26 villages under his jurisdiction. He performed a role in the local society as part thereof. It is also observed that he was placed under the amuldar (revenue collector) or revenue renter even though he had some military force. Viewed from this aspect the poligar did not seem to be a kind of feudal lord but a leader of peons.

A few other aspects should be studied to assess the poligars' status in the society. The following indicates the caste composition of the poligars in the three parganas of Salacaucum, Parumbauk, and Saut Magan as recorded in the Barnard Report. Of 120 in all 56 were recorded as Naick, 34 as Moodrian Naick, 10 as Rajah, 9 as Puttrawar Naick, 3 as

---

[48] Tiruvendipuram Report, p.4.

Palli Naick, one as Moodrian, one as the Perumal Temple of Wyou-noor, and one as Pillai. Five others were unidentified. Of these poligars the caste origin of Naick or Rajah, which comprised the majority, is hard to ascertain. What we do know is that a few Tamil castes like Moodrian and Palli occupied an important portion. Neither Moodrian nor Palli was high in the caste hierarchy of Tamil society compared with Vellala, Pillai, and other castes. For instance, there were no Moodrian landholders in the Barnard report. Though there were several Palli landholders in Salavaucum, both Place and Ellis considered Pallis to be a servant class under Brahman mirasidars.[49] Except for one Pillai, we cannot find any other high Tamil castes among the poligars.[50] We also have another record prepared by Place for the year 1792-93. It showed 80 Naick, 67 Naidoo, 20 Rajah, 3 Pillai, 1 Reddi and 10 unidentified, a total of 181 poligars.[51] Here the majority were of Telugu origin (Naidoo or Reddi). A few of the influential local agricultural castes were also on the list.

The last point to clarify is the spatial aspect of poligars' jurisdiction. The number of villages under each poligar in the three parganas of Salavaucum, Parumbauk, and Saut Magan was recorded in the Barnard report. Apparently most of the poligars (80 out of 120) had just one village under their individual jurisdiction.[52] The same findings can be obtained from another record on the poligars in Chengalpattu prepared by Greenway in 1801.[53] Nearly half of the poligars had fewer than two villages under their individual jurisdiction, and three-quarters had fewer

[49] Minute of the Board of Revenue, on the different modes of Land Revenue Settlement, as existing in different Districts, 5 January 1818, Papers on Mirasi Rights, p.368; Mr. Place's Final Report on the Jagir, 6 June 1799, Papers on Mirasi Rights, p.47.

[50] Professor Y. Subbarayalu suggested to me that Puttrawar could be the title in this period (personal communication). If so, there is a possibility of finding some high castes among the poligars. This point needs further verification.

[51] Permanent Settlement Records, vol.44, Account of the Privileges enjoyed by Poligars of the Jagir for Fusly 1202.

[52] Another 18 had two villages, and six had three. There were four with four villages, five with five, and another five who had between six and nine. In addition one had 12 villages, and one had 21.

[53] Chingleput Permanent Settlement Records, vol.26, a letter from Mr. Greenway, 30 October 1801, in Board's Consultation, 31 March 1802.

than ten.[54] Clearly the jurisdiction of the average poligar was rather
small. The Barnard report suggests, however, another spatial aspect. In
Salavaucum there were 33 poligar households, spread over 20 villages.
But there were villages where several poligars resided, and from which
the poligars extended their jurisdiction. We will call such a village a
'poligar-village'. Yeddamitchy, for instance, had eleven poligars,[55]
whose jurisdiction spread over 19 villages. Next came Comarravaddy
whose poligars covered nine villages. Other poligar-villages covered
three to six villages each. It should not be thought however that the
poligar-villages created a kind of 'circle of villages' which might be
related to the magan: such circles had little to do with the magan boundary. For instance, the villages under the Yeddamitchy poligars were
scattered across four magan (shown in Table 2 as 1, 2, 5 and 6), and of
the four all except one (number 1) had some villages with poligars who
were not from Yeddamitchy. The reason for the distribution of villages
between poligars seems to have been the proximity of the poligars'
place of domicile.

To sum up, the poligars performed the role of safeguarding the local
society as part of the village establishment, and originated not from the
influential agricultural castes but mostly either from low castes or Telugu castes. They were in many cases village watchmen, even though a
few of them had fairly considerable military power similar to that of the
nayaka in the previous period, and though they sometimes offered military service to the state in some parts of South India. It is not possible
to connect the influential nayakas in the Vijayanagar period directly
with the poligars with that very limited number of villages in the eighteenth century, at least in the area studied.[56]

---

[54] The figures were that, of 177 poligars, 88 had 0 to 2 villages, 26 3 to
5, 19 6 to 10, 23 11 to 20, 12 21 to 50, and five 51-100, while one had
128, one 131 and one 388.
[55] The number could be smaller as the actual identification is hardly possible by poligar names alone.
[56] Generally speaking, soldiers in the eighteenth century were paid soldiers. They very often changed their loyalty due to the arrears of wage payment. We can find many accounts of such cases in the private diary of
Ananda Ragga Pillai. It may be noted, therefore, that there is no necessity
for us to take the military contribution of poligars to the state as the ground

We will next examine the spatial distribution of village ownership. In the previous section it was suggested that the landholders or mirasidars represented the local society. Can we, then, discern any communal tie among the landholders? We will investigate the spatial distribution of the landholding castes, as well as the villages held by each of them, in relation to the boundary of the magan.

The largest landholding caste in Salavaucum pargana was that of the 'Gentoo' Brahmans, who held 24 villages while being resident in 17 (see Table 12). The Shri Vaishnava and the Shiva Brahman castes held 18 and four villages respectively.[57] In total Brahman castes held 46 villages. Next came the Vellalas, who held 17 villages; Tooliva Vellalas held ten villages, Puccolum Vellalas six, and Conidighetty Vellalas one. Among others, Pallis held nine villages, kanakkuppillai two, 'cowkeeper' one, and Raja one. The temples at Eishverum, Permal , and Seetanunjerry held three, three, and one respectively. The holders of the remaining three villages were not known to Barnard.

A few of these castes held villages in concentrated localities. For instance, Shiva Brahmans held villages in the northern part of Salavaucum pargana, and Puccolum Vellalas in the north-eastern part. On the other hand, the Puccolum Vellala, Reddi, the Tooliva Vellala castes each show separated and concentrated areas of habitation. The distribution of the castes, however, did not correspond either with the landholding area or with any magan.

What is important in this regard is that there were many instances of villages held by absentee landholders. Of the 72 villages in Salavaucum, 28 were held by those living somewhere else (with two overlaps). Among these villages, 25 contained no members of the landholders' caste. The one exception were the Puccolum Vellala who dominated one magan (number 4). Villages held by other castes were widely scattered.

---

in placing poligars as state agents.

[57] See Béteille, *Caste*, pp. 67-72, for an explanation of the differences between Shri Vaishnava Brahmans (followers of Ramanujan and worshippers of Vishnu) and, on the other hand, Smartha Brahmans (followers of Shankaracharya and *smrti*), sometimes called Shaivites or worshippers of Shiva, and including priests in Shiva temples, called Kurukkal. [Note by eds.]

One might say that they were like a mosaic. Thus the cohesion of local society cannot be observed in terms of landholding.

Lastly we will examine the areas of service of some of the so-called 'village' functionaries. The point will be whether every magan had them in its villages. First were the artisans. There were 27 carpenters residing in 21 villages. All the eight magans in Salavaucum pargana had from one to four villages where carpenters lived. Sixteen households belonging to ironsmiths were scattered among 14 villages. Dues and tax-free land were granted to the artisans in almost all the villages. One magan (number 6), however, had no ironsmith in any of the villages.

Sixteen barbers were living in 13 villages. Salavaucum and Annahdoor villages had three and two households respectively, others one. They received gross dues (bulk and grain) in almost all the villages. No barber, however, was found in one magan (number 2), though the tax-free land belonging to the village establishment was allocated to them in two villages where no barbers resided.

There were 13 households of potters in eleven villages. One magan (number 6) did not have any potters while other magans had one or two. Bulk and grain dues were granted in half the villages, but only a few in two of the magans (numbers 6 and 7). Tax-free land was granted in nine villages in which potters did not necessarily reside.

There were 24 households of washermen in 24 villages.[58] All magans had from two to five washermen respectively, except one magan (number 2) where no washermen lived. Bulk and grain dues were granted in almost all the villages. There were seven villages where tax-free land was granted to washermen, but their location did not always correspond to their places of residence.

There were 22 households of 'snake doctors' in 15 villages. All the magan, except one (number 6), had from one to five. Bulk and grain dues were granted to them in almost all the villages including magan 6. Their tax-free lands were found in seven villages, though the lands were not always located in the villages where they lived.

[58] There was another village where some washermen lived. But the number of the households is unreadable due to the damage of the original record.

There were only two shoemakers, living in Salavaucum and Pan-nioor. Salavaucum was the largest and Pannioor was the second largest village in the pargana. In this sense shoemakers should not be called 'village' functionaries. Neither dues nor tax-free land for them were recorded in the report. Usually Sakkili is the caste name for shoemakers in South India, but their name does not appear either. The two shoe-makers must have been engaged in shoemaking in town bazaars.

There were six astrologers (Panjangum Brahmans) living in four vil-lages. None was found in six magans, but bulk and grain dues were granted to them in most of the villages. They were given tax-free land in twelve villages, but hardly any connection can be observed with their places of residence.

Sixty-six households of kanakkuppillai lived in 27 villages and in all the magans. They were granted bulk, grain and state dues, and tax-free land belonging to the village establishment in almost all the villages in the pargana. There were also 35 shroff households in 21 villages. In other features their position was the same as that of the kanakkuppillai.

The above examination indicates, first of all, the inability of any single village to carry on production activities independently, due to the lack of essential functionaries. Even if some shares were reserved for the functionaries, many villages were found to be without any resident recipients. The location of the tax-free land did not always correspond to their holders' places of residence, either. Thus a certain number of vil-lages were covered by one functionary. In other words, he was not con-fined to his village of residence, but worked among a certain number of villages. On the other hand, however, a positive correlation between this service area and the magan could not be clearly observed. There were some magans without essential service castes.

It seems very doubtful therefore that the magan was the basic spatial unit of the mirasi system, given what can be discerned of the distribu-tion of poligars, landholding and service castes. How then was the 'local society' composed? Or, rather, did the local society still exist? The answer seems to be in the negative. It is by now quite clear that by the early nineteenth century there was no longer any role either for the caste-unity which had provided the base for the nadu in the Chola period, or for the nayaka rule which once demarcated the boundary of the territory called nayakattanam in the Vijayanagar period. Though features of the share distribution system could still be observed, sustaining production over an area greater than the village, yet it seems that they were just remnants, and that the classical type of local society no longer existed. Some process must have been destroying the integration of the local society.

Our next task is to identify the main forces which caused this change in so far as they are revealed in the Barnard report. It will be shown that the cause was no less than the emergence of a new form of society which we will call individualistic. Its chief actor was the village leader who utilised his village(s) as a power base. Due to the emergence of this individual formation, led by the village leaders, local society was being undermined from two sides, both within the mirasi system and outside it. We will investigate the former first.

The new individualism was expressed in a change in the nature of landholders and the emergence of village leaders from among the landholders. It was previously assumed that the landholders were, following nattar of the Chola period, the main co-ordinators of community life. They had some communal ties among themselves extending over several villages, even if they held ownership shares individually.[59] However, these communal ties had become fairly weak. We saw this in the mosaic of village ownership by castes, and the general lack of communal unity in any locality. Ownership shares had very often passed to those outside the local society including merchants. The village leaders, who led the shift to individualism, grew out of landholders who had lost their communal ties and ceased to identity with the local society to which they once belonged. Instead of being dependent upon a wider area, they utilised villages to support their independent power.

Such change can be discerned in the Barnard report too. The landholders, who were allocated bulk and grain dues in almost all the villages, acquired over 144 kani of tax-free land belonging to the village estab-

---

[59] An individual type of landholding, for instance, can even be included in the category of communal landholding if the individual landholders had some communally exclusive tie with other landholders.

lishment in 34 villages. The fact that they were granted in half of the villages indicates that such grants had originated comparatively recently in comparison with those given to 'village establishments' such as artisans, shroffs, kanakkuppillai, 'cornmeters', or poligars, who were granted the same type of tax-free land in almost all the villages. The recent origin of the grant is further indicated by other evidence. Of the 16 kani of 'new' tax-free land belonging to the village establishment allocated to nine recipients, landholders occupied five kani in three villages. Such new grants made to the village leaders can be interpreted as an expression of the change in the landholders' position. Formerly the landholders occupied, as the owners of the village, a higher position in the village structure; they held the village as a result of their status. Rather than expressing it by acquiring tax-free land, they had chosen to hold the leading position while demanding rent from the under-raiyats. But, as the nature of the landholders changed from being buried in a communal tie to being based on personal power, they must have begun to seek the privilege of tax-exemption too. The state naturally responded to their request in order to incorporate them and associate them with state power. In this regard we might refer to an argument of Eric Stokes. In his comparison between the 'office' and the 'landed' mirasi in the raiyatwari tenure in South India and in the joint landlord tenures in North India, Stokes seems to suggest that two choices were available to the village elite: to seek grants of tax-free land from the state, or to impose rent upon the tenants. He then regards the regional differences as resulting from different decisions, creating two types of mirasi system.[60] By contrast, it is asserted here that the landholders started seeking tax-free land ('office mirasi' in Stokes' usage) as they began to

---

[60] Stokes, 1977, pp.55-7. He states that the landed mirasi was found where the system of village officers supported by office mirasi was weakest. Though this paper does not go further into this issue, it should be noted that even Tanjore villages had many types of office mirasi (see several types of Inam in Gough, 1981, pp.38-44) even though the landlord's rent occupied a very high proportion. These office mirasi tended to be buried unnoticed in the statistics as they were often located in Inam village. Anyhow it may suffice here to say the structure of the mirasi system did not fundamentally differ between Tanjore and Chingleput even though the proportional shares might greatly vary.

separate themselves from the wider community and its collective arrangements.

Similar evidence may be obtained about the 'chief of the village' in the Barnard report. In total, in 25 villages, the village chief was granted over 97 per cent of the 'old' tax-free land belonging to the village establishment. Of the nine recipients newly granted 16 kani of such tax-free land, one village head obtained seven kani in one village. Both these figures again indicate the recent origin of the 'chief of the village'. There is an important difference between this chief and the landholders in the mirasi system. Landholders received either bulk or grain dues in almost all the villages, whereas the village heads received no dues in any. As said, these gross dues were usually allocated to the 'village establishment'. Given their traditionally-established position in the local society, it was natural that the landholders should be allocated them. On the other hand, it was very remarkable that the village heads should have had no dues in any of the villages, for some dues were usually allocated whenever tax-free land was reserved. This difference once more indicates that the position of the chief of the village was, unlike that of landholders, of very recent origin. Indeed the non-existence of the office of patel or village headman in Tondaimandalam (including Chengelpattu) was well known at the beginning of colonial rule.[61] It was only in 1816 that the appointment of village headman in all the villages throughout the Tamil country was administratively enforced.[62]

The individualism brought about by these village leaders was incompatible with the mirasi system. It gradually broke down, as villages and their leaders strengthened a position separate from the rest of the society. The local society accordingly began to lose its cohesion. The newly-emerging village leaders, by increasing their dues and tax-free land and by usurping other shares, broke down local society into smaller autonomous spheres, usually one or more villages. Hence the

[61] Minute of the Board of Revenue, on the different modes of Land Revenue Settlement, as existing in different Districts, 5 January 1818, Papers on Mirasi Rights, p.377.
[62] Replies from Mr. F.W. Ellis, Collector of Madras, to the Mirasi Questions, 30 May 1816, Appendix, Papers on Mirasi Rights, p.254, fn.25.

mirasi system was on the verge of collapse in the eighteenth century, as village leaders emerged as the dominant force transforming South Indian society.

It may be noted here that there was a difference between the 'quasi-individualism' in the earlier period and the individual formation discussed here. For instance, many individuals or institutions, mostly government high officials or religious authorities, were included in the list of recipients of joint dues or of the tax-free land allocated to outsiders. These individuals received grants because of their contribution to the state or to the local society. The grant was, therefore, backed by the state and the community, without which individuals would soon have lost their rights. On the other hand, later village leaders maintained their status and rights not by the support of the state or the community but on their own. Even if, in both cases, privileges were granted and appropriated individually, in each the power base was completely different.

The role of the emerging village leaders and their relationship with the mirasi system can be clarified by the list of collections and disbursements by the *gramattan* (patta 'monigars' or village headmen) contained in a Collector's letter from South Arcot.[63] The letter contains the accounts concerning the 31 gramattan. We will take up one of them, the gramattan of Acolagramum village in Tindivanam. Table 13 gives the accounts. The gramattan's 'unauthorised collection' amounted to 45-2-65 pagoda.[64] He disbursed the amount for various purposes. For instance, he paid 16-23-58 pagoda to the Mahatady peons who came to collective revenue; 1-22-40 pagoda to the peons of Vatavalum Bundary (probably a poligar) who came to apprehend thieves; 24 pagoda to repair a tank; 2 pagoda to a temple Brahman; and one pagoda to a carpenter. The accounts of other villages recorded in the same letter also indicate

---

[63] Letter from the principal collector in the Southern Division of Arcot, 15 December 1805, Madras Board of Revenue Proceedings, 2 January 1806. [A *gramam* was usually a village dominated by Brahman landlords; note by eds.]

[64] In the amount shown, this means 45 pagoda (the gold currency), 2 *fanum* and 65 cash. Subsequent references to this currency will be given in the same form. [Note by eds.]

various items for which the gramattans provided money. They include expenses for festivals, rituals, temples, road and tank repairs, the fees to Brahmans, Muslim priests, kanakkuppillai, watchmen, pilgrims, surveyors, carpenters, ironsmiths, dancing girls and actors, and the bribes to revenue collectors. The village leaders had thus become the sponsors of those engaged in various activities in the village. The village leaders took over most of the activities the mirasi system had previously performed. For instance, the gramattans paid the kanakkuppillai (or accountant) 'above his pay', or embezzled the payment due to artisans, and so on. And, as was clearly indicated by their support of the village festivals, Brahmans, and temples, they took the lead not only in secular but also in religious activities. Though this study does not discuss the nature of leadership in general, it may be remarked that such support for religion was always an essential qualification in India.

The second noticeable aspect is that the amount of such 'unauthorised collection', which meant the collection done by the gramattan privately without state sanction, amounted in some villages to as much as 50 per cent of the state tax from the village. Such status and influence as possessed by the village heads must have forced the state to allow them tax exemption to win their co-operation. Lastly, the collection was not only 'unauthorised' by the state, but also independent of the mirasi system. In other words, it went on outside the sphere of both the state and the communal formations. Such village leaders inflicted critical damage on the mirasi system. They behaved independently, and completely differently from the nattar, who, in Chola times, endeavoured to protect the local society from state intervention. As a result local society, in the sense the term has been used here, was in the process of collapse in the eighteenth century, and villages under their leaders had steadily become the basic unit. We will examine an example of such a village head in the next section.

Another major cause for the collapse of local society was the development of commercial activities outside the mirasi system. A detailed study is not attempted here; but commerce seems to have caused a crisis, particularly due to the drastic increase in the trade in cotton piece-goods associated with European commercial power from the late seventeenth century. That the mirasi system had to give way is again verified by the Barnard report.

Table 1, which is the aggregated result of grants in dues and tax-free land to the different castes in Salavaucum pargana, indicates that the traders such as Chetti, 'Caumities', *chunam*-sellers, toddy-sellers, and oil-sellers, or the 'non-village' artisans such as weavers, cotton-refiners, or woodcutters, were granted hardly any dues or tax-free land in any of the villages. This signifies that their position in the local society differed from that of 'village' service castes such as barbers, artisans, washermen, or astrologers—that is, their activities were performed outside the sphere of the mirasi system or of the local society even if they resided within it. Their relations with the local society differed completely in the sense that the village functionaries were totally supported by or incorporated in the local society, whereas the non-village artisans were only partly incorporated in it, making only nominal payments to the system (more accurately to its representatives, as will be discussed below). Their relations with the local society was partial and the major part of their wealth was accumulated outside the mirasi system.

The state and the local communities attempted to incorporate the developing commercial activities too. The former imposed customs duty on the passage of goods, collecting fees for the local society at the same time. The Tiruvendipuram report has information on this issue. In the Tiruvendipuram jurisdiction there were eight customs offices in the 1760s, when the area was surveyed by the East India Company. Different rates were imposed according to the nature of the goods or the status of the traders. The rates of duty and fees imposed on the goods in the six customs offices in the area are indicated in Table 14.[65] Duty collected there was paid to the state, whereas the fees collected by the revenue-farmer's kanakkuppillai were distributed among temples, nattars, the head kanakkuppillai, Brahmans, a fakir (Muslim priest), and a *pandarum* (non-Brahman priest), as indicated in Table 15. The poligar received one-eighth of the total customs duties.

Two points should be observed from these tables. The first is that the

[65] The offices (in the original spellings) were at Sharady, Ramaporam, Padrycopang, Comerapuram, Toutaput, and Cuddalore riverside.

relations of the state and the community with commercial activities were symbolically expressed in the form of customs and fees. Secondly, these relations were only partial. More concretely, the recipients of the fees were not the local society in general but temples and certain leading persons. The fee-payers related to the local society only through these recipients. By contrast, a village functionary—a carpenter, for instance —performed his role in exchange for dues collected from the whole produce in the local society, and so formed a relationship, in this sense, with the local society as a whole. The local society failed to incorporate those engaged in commerce and industry by granting dues or tax-free land. In other words, it failed to incorporate them in the mirasi system, which could not adjust itself to the development of commerce. The accumulation of wealth took a different form in commerce and in the mirasi system.

The state, on the other hand, seems to have attempted comparatively more direct relations with commerce and industry. It imposed taxes directly upon producers such as weavers and dyers. Tax was also levied from the shops for betel-nut and tobacco, on oil, toddy and arrack and salt, and on carts, fishermen, and the coolie houses in Tiruvendipuram. Some additional fees were collected along with the state tax, which were taken by the poligar or watchman in most cases. In the Jagir, for Sattavaid, there is a list of the state taxes imposed upon those engaged in commerce and industry (see Table 16). But it was the state and not the local society which tried to control commercial and industrial activities, and the amount of revenue collected from trade was fairly negligible compared with that from the land. Like the local society, the state was only partially successful in attempting to control commerce.

Those engaged in commerce and industry, who had grown outside the sphere of the mirasi system or local society, began to leave the local society. Their destination was urban centres, especially those coastal cities developing rapidly in the eighteenth century due to the increase of foreign trade encouraged by the European East India Companies.[66] It is suggested here that the village leaders participated in the increasing ru-

[66] I have made a study of this aspect elsewhere; see Mizushima, 1991, ch.3.

ral-urban grain-trade by utilising their control over production activities. It is highly possible that they accumulated wealth through trade and strengthened their economic position and independence. All these factors helped to generate the new structure of wealth-accumulation, one independent of the outdated mirasi system.

The local society was thus steadily collapsing, both internally and externally, and individualism had become the dominant form. The village leaders took political leadership in the period as well, and threw South Indian society into the turmoil that finally led to colonisation. The political unrest which went along with such structural change, led to the emergence of village heads as leaders of much bigger territories consisting of a number of villages. No matter how large their territory might be, however, their rule had nothing to do with the communal formation or the mirasi system.[67] They simply utilised villages as their power base to separate themselves from the pre-existent system and assert their individual power. To sum up, the mirasi system, which had long functioned to reproduce the local society, was being deformed from within by the emerging village leaders and from outside by developing commercial activities. In this process the local society came to the verge of breaking down into villages controlled by the village leaders. Local society in eighteenth-century South India was thus the remnant of a lost cohesion, which persisted only notionally in the collection of fees in the customs office by nattar and others who had represented the local society.

We will now examine the significance of such a change by comparing it with the developments after colonial rule.

---

[67] A name list of renters in the village settlement of South Arcot in the first year of colonial rule or 1801 (Board of Revenue Miscellaneous Records, vol.12, Jummabundy of Each Village for Fusli 1211, Board's Consultations, 13 July 1802) gives some interesting information. The analysis of the renters' names is not yet completed, but it seems that around one-third of the renters rented a single village, though there were also a few renters who rented more than a hundred villages. Larger-scale renters were in some cases those having 'nayaka' title in their names. They were probably the former poligars. The majority were, however, from socially high-ranking castes like Brahman, Mudali, Reddi, Pillai, and Row. More detailed analysis of the record will be done on some other occasion.

South Indian society since colonial rule

South India, which the English East India Company began to rule in the late eighteenth century, was in transition and political turmoil due mainly to the structural change described above. Among other factors, the land policy enforced by the colonial government had a critical impact upon society. In this section the developments of the last two centuries are briefly summarised to clarify the significance of pre-colonial changes.

At the initial stage of colonial rule, the revenue administration of most of the areas which came under the Company was left in charge of the revenue farmers, so that the Company did not pay much attention to the internal structure of the society. The first of its revenue systems, the zamindari system, also allocated the ownership of large areas to the zamindars, so that the Company did not need to know the condition of each farm so long as the revenue was secured. At the end of the eighteenth century, however, the British started revenue administration on their own. Soon they encountered great difficulties in deciding to whom land ownership should be granted. Facing the Board of Revenue in Madras were two choices, other than the zamindari system (or its equivalent in smaller scale, the mittahdari system). The first was to make the village the primary revenue unit and to give landownership to the landholding body of the village. This was called village settlement and was supported mainly by the administrators (Collectors) in charge of the richer coastal districts. The second was to make the land plot the primary unit and to give its ownership to the respective raiyats. This was called the raiyatwari settlement, and was supported mainly by Collectors like Thomas Munro who worked in the drier inner districts. In a sense, the revenue officers all had the same view about the influential position of the village leaders, even though they differed in their opinion as to whom land-ownership should be granted. The supporters of the raiyatwari settlement aimed to eliminate the power and influence which leaders exercised over large areas, by making settlement directly with the 'inferior patels or other substantial farmers' in the villages.[68]

[68] Murton, 1973, p.167.

On the other hand, the supporters of the village settlement intended to collect revenue by utilising the village leaders' control over the village.[69] After a period of trial and error in the first decade of the nineteenth century, the final decision to introduce the raiyatwari system came from London, and was enforced accordingly.

Under the raiyatwari settlement, the village land was divided into hundreds of plots, each demarcated, assessed, and granted to a raiyat who took responsibility for paying the tax. Making the land plot the primary unit of revenue administration had a critical impact upon the later development of South Indian society, as well as upon the mirasi system. First of all, the raiyatwari settlement gave a final blow to the system by abolishing dues and by separating tax-free land from the reproduction structure of local society. Dues allotted to various recipients were amalgamated into the land assessment. Differences in land tax according to the cultivators' status was abolished in the late nineteenth century. Though considerable lands were left unassessed or favourably assessed, they were separated from the production system of local society, and played a very limited role. Both the mirasi system—which had until then somehow persisted as the production system of the local society—and the communal formation which had worked as its core, finally ceased to play their historical roles. Those previously supported by the mirasi system had to enter into individual relationships with particular households.[70] The second and more drastic result was a shift in

[69] For detail, see Mizushima, 1986, ch.II.

[70] From this viewpoint the jajmani system can be characterised as a disguised system of the old mirasi system. The individual service relationship among the villagers under the jajmani system was the one newly formed after the collapse of the mirasi system. Even though the same allowance (dues) or the service-exchange relation would remain, the whole context differs in the relation of the recipients with the local society, which no longer exists. The same interpretation can be applied to the tax-free land called *inam*. It seems as much as one-third of the total cultivated area in the Madras Presidency was exempted from the full rate of assessment (Stein, 1977, p.68). Though further study is essential to clarify the nature of Inam tenure in the colonial period, there is the possibility of interpreting the grant of extensive inam as strengthening the individual utilisation of landed plot separated from the reproduction activities once observed in the local society. I would like to advance this argument on a future occasion. Inam-holding was finally abolished by the Madras Inam Estate (Abolition and Con-

the basis of individual rights. Formerly the village leaders utilised villages as their power base, which often gave cohesion to the villages. In the raiyatwari system, however, individual rights were centred upon a land plot. This gradually led to the breakdown of villages as a collectivity.[71]

By taking up Reddimangudi village in Tiruchirapalli district as a case study, we will trace the process of collapse of the village society.[72] In Reddimangudi the Reddiyar (Telugu agriculturist) caste owned 78 per cent of the occupied land (excluding inam land) in 1864, the first year for which a detailed village record of landholding is available (see Table 17). By the time a village study was conducted by the author in 1982-83, the Reddiyar share had fallen as low as 27 per cent. Other castes, which possessed hardly any land in 1864, had acquired large amounts during the period. The share of the Gounder (shepherd) caste in the year 1864, 1898, 1924 and 1982 was 0, 2, 12 and 18 per cent respectively. That of the Muthraja (Muttiriyan, a Tamil agriculturist caste) was 8, 17, 16 and 22 per cent. The same was true of the 'untouchable' castes like Pariah, Pallan, Sakkili, or Christian Pariah, whose share had increased gradually from one to 19 per cent over the last hundred years or so. Similar alterations in the landholding ratio were also observed in the neighbourhood of Reddimangudi.[73]

The process underlining the change in the landholding structure can be analysed in the following way. The fact that Reddiyars almost monopolised the village in 1864, despite the availability of a large amount of

---

version into Raiyotwari) Act in 1956.

[71] At the initial stage of revenue administration, the colonial government noticed the possible effect of raiyatwari settlement. The Committee at Tanjore, for instance, reported as follows in 1807: 'Land in India is seldom a separate farm. All land belongs to some village or another, whether it be cultivated or waste...there is then always, in the arable and cultivated land generally, a community of interests...; all labour for village works of general utility; all contribution for religious ceremonies; all the pay and labour of the village artisans and officers, are regulated by this communion of interest. A ryotwar rent may separate the villager's stock from that usually clubbed for public rent' (Report from the Committee at Tanjore, 22 February 1807, Fifth Report, III, pp.520-1).

[72] For detail, see Mizushima, 1983.

[73] Ibid.

cultivable land for other castes, implies that those who were regarded as raiyats and granted landownership under the raiyatwari settlement were not always direct cultivators but often members of the community to which the village leader belonged. Among the early records we can trace the protest movement against the high government tax in the area concerned. In April 1805, the *tahsildari* office was attacked by the inhabitants of Rutaiyur (an area in the north-western part of Tiruchira-palli), and several guards were injured. The leaders of the riot then ran off to Salem, a neighbouring district, but later asked the government's permission to return home. As one of the leaders was refused permis-sion, some of the leading people in the area petitioned for his return. Among the petitioners was found the name 'Mangooly [Reddimangudi] Moodoo Reddy'. The disturbance continued even after the incident, and some leading Reddiyars went to Madras to demand a tax reduction directly from the authorities. Wallace, the Collector, tried hard to pacify the area, and wrote to the Board of Revenue to request the arrest of the people who had gone to Madras. In the list of the warrant, we can again find the name 'Mootoo Reddy of Maungoody'.[74] Though it cannot be verified whether Reddimangudi was under his sole control or not, we may guess that some leading figures had emerged among the Reddiyars there. The monopoly of landholding for one community, at this early stage of the colonial period, implies that the leadership in the village had been too powerful to be ignored in the settlement, even though such a landholding structure was quite contrary to the original intention of the raiyatwari settlement, in which the state and the raiyat were intended to have a direct interest in each plot of land.

However, though it took a long time, the raiyatwari settlement grad-ually had the effect intended. Individual ownership based on separate land plots began to take root in South Indian society, and started break-ing down the village just as the individual rights of control in villages had previously broken down the wider local society. In this process the

[74] Letter from Kinlock to the Board of Revenue, 27 April 1805, Board of Revenue Proceedings, 2 May 1805, letter from the Principal Collector in Tanjour and Trichinopoly, 20 July 1805, Board of Revenue Proceedings, 29 July 1805. For detail, see Mizushima, 1986, pp.107-9.

ruling village elite's share in landholding dropped drastically. Three factors can be considered to have caused the change in the case of Reddimangudi. First was the loss of unity among the Reddiyars caused by their members' migration. A survey of all the genealogies in the village indicates the movement of considerable numbers of villagers to other parts of India, or to foreign countries. A number also left through marriage. By the time of survey, many families no longer had any descendants in the village, and the Reddiyars were no exception. The symbolic case was the disappearance of three among the four kaniyatchikkaran (mirasidar) families said to have owned Reddimangudi in shares. Two of them had disappeared without leaving any descendants, and one had emigrated to some other village. Only one kaniyatchikaran family still lived in Reddimangudi.[75] Beyond doubt, such migrations of the ruling caste members destroyed the unity among them, and led to the weakening of their monopoly over landholding. Secondly, the shift of the main irrigational facility from reservoirs to wells had changed the land-use pattern in the village.[76] Reservoir-maintenance and operation had necessitated unity among users (the majority were Reddiyars), but a well offered an individual use of water, and reduced the need for co-operation. The third factor was related to the second. Due to the increased importance of irrigated land, the exclusion of other castes from the use of reserved cultivable land lost its importance for the Reddiyars. The availability of water rather than the extent of landholding attracted the attention of the Reddiyars. A study of the ownership of newly-reclaimed land in Reddimangudi indicates that castes other than Reddiyars acquired a considerable extent of what had formerly been categorised as cultivable waste.[77]

The disappearance of the basis for unity among the ruling village community and the resulting decline of their landholding, occurred in association with a process in which the basic production unit shifted from the village to the land plot, and hence to individual ownership and

[75] The kaniyatchikaran's status was still observed symbolically in his leading role in the rituals of the village festival.
[76] The number of wells recorded in the village map of 1892 was only 23, but increased to 161 in 1982.
[77] Mizushima, 1983.

effort. Village leaders with their followers of the same caste, who once invaded the wider local society and alienated the villages from it, lost out due to the incursion of other raiyats who alienated plots of land from the village. Individualist property rights were thus focused upon the raiyats rather than the ruling community of the village. Another important change is that individuals came directly to confront the state. They had to find new means of coping with colonial rule. Though nationalism seemed to have offered itself as a new possibility, Indian villagers today are still desperately in need of finding out what model of society should be established, and in what units they should focus their efforts.

This paper is an attempt to describe the mirasi system, and to trace briefly the developments which occurred in South Indian society from the eighteenth century onwards, by using the Barnard report and some other sources, including information obtained through the author's field studies. The main points discussed can be summarised as follows. Share distribution, which can be termed the mirasi system, created and maintained a 'local society'. The system had been formed from the competitive power-balance between the state and the community, and had been based upon a wider regional sphere than a single village. The power-balance between the two formations was expressed as the various types of tax-free land and produce shares. The system could absorb changes in the power-balance by adjusting the proportion of the shares. In the later pre-colonial period, however, it was seriously challenged by two factors irrelevant to it—that is, by the emergence of individual rights in the hands of the village leaders, within the system, and secondly, by the development of commercial activities outside the system. The system could not retard the progress of these forces by adjusting the share-balance, and thus was beginning to collapse in the eighteenth century. The village leaders gradually established their independence in both economic and political fields by participating in commerce, and by taking over various activities once performed by the system. The emergence of individualist village leaders threw South Indian society into political turmoil and finally led to colonisation.

Under British rule the raiyatwari settlement ensured that the mirasi system would be demolished by doing away with many vested interests and by confining the unit of production to the individual land plot, thus bypassing the village. With the disintegration of the villages, the village leaders, who had emerged as bearers of individual rights, lost out. The main power-struggle was now between the state and the raiyats. A model of society based on individual rights came to be centred on the raiyats and the plot of land. Finally, this deformed South Indian society so that it became highly disintegrated. Its people need a new model of society to cope with the problems which resulted.

In this paper the alternative bases of society in the state, community, or individual have been discussed in rather an unconventional way. Each type is assumed to have had a specific territorial base and a representative or controlling figure. As territorial units we had, in turn, the state, a local region wider than a village, the village before the raiyatwari settlement, and subsequently the land plot. The representative figures were, in turn, the state, the nattars in the Chola period, the mirasidars in the later period, the individual village heads, and finally, after the raiyatwari settlement, the raiyats. These differences expressed the character of each of the *formations* of society.

On the other hand, 'local society' is also considered to be the sphere in which these variations work among each other. It is a sphere of particular human organisations and environments, where a man is born, mixes with others, and inherits and creates a world view, culture or tradition. The sphere of such local society expands or shrinks in the different stages of historical development. In South India, in the original form, or I should say the ideal type, the sphere of the local society was identical with the sphere of the community. Local society had been repeatedly reproduced by the mirasi system, which was created and maintained in the competitive relationship between the state and the community. In the eighteenth century, however, local society was breaking down into villages due to the emergence of village leaders, the bearers of an individualist basis for society. South India was colonised midway through this historical process, and the raiyatwari settlement, which totally differed from the mirasi system, was enforced. As the raiyatwari settlement was based on a tiny land plot and many of the institutions connected with the mirasi system were abolished, local society was destined to disappear from the scene in South India.

Table 1: *Distribution of castes and the number of villages where dues and tax-free and were granted in Salavaucum Pargana*

Recipients	Number of			Tax-free Land				Dues			
	H	V	L	a	b	c	d	e	f	g	h
Agriculturists											
Conidighetty Velalla	2	2	1								
Puccolum Vellala	109	18	6								
Tooliva Vellala	89	14	10								
Paly	211	41	9								
Reddy	79	13									
Commavar	58	9									
Moodrial	3	3						26	23		
Landholders, village chiefs											
Landholders				34	3			5	70		1
Chiefs				25	1						
District chiefs (individuals)											
Conicoply						1					
Chiefs							1				
Moodoorama Naick					1						
Vanagoova Moodellar					1						
Vencatachel Brahman					1						
Agricultural labourers											
Pariah	286	47									
Cowkeeper	109	35	1	2				21	28		
Cultivators' servants							1	62	67		
Village servant								2	1		
Village officials											
Shroff	35	21		69				69	60		70
Conicoply	66	27	2	71				70	57		70
Conigo Braminy	3	1									
Mautot peon	10	1									
Government high officials											
Daismoc of district									66	2	
Town conicoply				1		12	2				
Conigo of district						1			69	2	
Dovetraw										68	2
Service Castes											
Artificers				70				71	53		71
Carpenter	27	21									
Ironsmith	16	14									
Thauthun (goldsmith)	3	1									
Barber	16	13		17				71	68		7
Potmaker	13	11		9				35	32		
Shoemaker	2	2									

Table continues

Table 1 continued

Recipients	Number of			Tax-free Land				Dues			
	H	V	L	a	b	c	d	e	f	g	h
Snake doctor	22	15		7				63	61		
Washerman	24	25		5	1			69	67		1
Panjangum Brahman	3	3	13	1			55	68			
Panjangum	3	1									
Poligars											
Poligar /Talliar	34	21		70				58	24	1	71
Cavare	30	13									
Puttrawar	1	1									
Rajah	1	1	1								
Other service castes											
Yary servant	6	4		2				42	22		
Dancing-girl	26	4		4				3	7		
Malabar schoolmaster	1		1				10	14			
Schoolmaster								1	4		
Fluteman	2	2		1				2	4		
Tomtomman	1	1		2				7	11		
Waterwoman	2	2						7	7		
Cootaudy	1	1		2				2	5		
Brahmans											
Gentoo	46	17	24								
Siva	54	15	4								
Vistnoo	143	20	18								
Vaithaverty				11	1			6	8		
Brahman woman									1		
Village doctor							1				
Paitha				2							
Brahmans (individuals)											
A. S. Brahman, Conjiveram[1]				1							
Groocul, Conjiveram					1						
Mauhaaveraputher Brahman								1			
Ramacharia, Conjiveram					1						
Annavia Brahman (131)					4						
Ramia Brahman (1)								3			
Ramia Brahman (35)					3						
R. Brahman (57)[2]				2							
S. Brahman, Carangoly[3]				1							
T. Brahman, Conjiveram[4]				1							
Muslim											
Moorman	23	1									
Muslim Fakeer (individuals)											
Yawkoopshaw (9b)[5]	1	1				1					
Gulahmoodin Saib (1)									1		
Mullah Saib (1)									68		

Table continues

Table 1 continued

Recipients	Number of			Tax-free Land				Dues			
	H	V	L	a	b	c	d	e	f	g	h
Sooltaun Fakeer (1)					1	1		2			
Sydda Moostapah Fakeer (1)					3	10					
Sydda Esmall Fakeer (1)					1	1					
Pandarum, temple servants											
Tope pandarum	2	2						19	21		
Veera pandarum	14	7							2		
Veerasica pand.	51	1									
Flower-garden p.	1	1									
Pandarum	7	6						1	3		
Sidda Pandarum	3	2				4	1	4	9		
Shiva pandarum								1	3		
Simcoo pandarum									1		
Amman pagoda servant	1	1									
Pagoda servant	4	2							2		
Brahman pagoda servant				2							
Wochun	7	4						4	4		
Panisiver	9	6		4				30	29	4	
Sautinian	3	2						1	1		
Valloovun								1	50		
Pandarum (individual), Conjiveram											
Paulia pandarum										60	
Trader											
Chitty	62	28									
Caumity	3	3									
Cornmeter				65				58	67		52
Chunam-seller	2	2							2		
Toddy-seller	3	3									
Oil-seller	15	11							2		
Weavers, other artisans											
Weaver	39	5									
Putnaver	1	1									
Cotton-refiner	6	6									
Stone-cutter	2	2									
Fisherman	6	1									
Woodcutter	41	13							3		
Tailor								1	1		
Pagodas											
Ammumpadar								1	1		
Annumar								1	1		
Conjiveram							2			67	2
Connegul									1		
Deivarara								1			
Durmarajah					1			3	6		

Table continues

Table 1 continued

Recipients	Number of			Tax-free Land				Dues			
	H	V	L	a	b	c	d	e	f	g	h
Eishveram and Permal (1)	1					3	6				
Eishveram and Permal		2					3	4	2		
Eishveram (125)							1				
Eishveram (1)										1	
Eishveram (33)						1	2			1	
Eishveram, Terikitchcoonum		2									
Eishveram			1	26	1			18	9		
Landholders Permal								1	1		
Munnarsaumy				1				1			
Nookalamun				1					2		
Ommun Padari				3				45	54		
Ommun				2				2	1		
Padari				3				3	12		
Permal and Eishveram							4	6			
Permal (131)			1		12						
Permal		2		18				5	2	2	
Permal (1)										1	
Pooliar				6				14	20		
Pagoda (1)										1	
Pagoda (33)			1								
Uncaulamun								2			
Veerabuddra				1				2			
Village								6	48		
Viravum								2			
Special-purpose funds											
Yary										27	
Pagoda Tope					1						
Tope				1		1		2	1		
Flower garden								1	1		
Pagoda flower garden		2					14	5			
Ammahdoor spring					1						
Channel (35)							1				
Channel (33)							1				
Channel				1				4	4		
Choultry (1)						1					
Choultry				1		2		5	9		
Cutcherry lamp oil								2			
Mudda Yary						1					
Channel (43)						1					
Others											
Singers									1		
Dancing-girl (131)					1						
Pagoda-sweeper								2	4		

Table continues

Table 1 continued

Recipients	Number of			Tax-free Land				Dues			
	H	V	L	a	b	c	d	e	f	g	h
Channel-cutter						1					
Others, unknown											
Bondelian	1	1									
Brazier	1	1									
Himbayan	1	1									
Tiviseremaker	1	1									
Toondaninad	1	1									
Muddoomadar										1	
Bownian								3	5		
Tickytawoman					1			2	1		
Unknown			3								
Total	782		57								
Frequency		48	10	33	4	19	13	45	59	15	7

Source: Jaghire, Barnard's Survey Accounts of Salavakam, vol.69.
Notes: (1) The following are expansions of the entries as marked: [1] Allaugoo Singaraja Brahman; [2] Raugoovumputter Brahman; [3] Seenivasa Brahman; [4] Tautacharia Brahman; [5] Yawkoopshaw Fakeer at Seetaupoorum (9b). (2) Numbers in parentheses show locations, according to the village numbers listed in Table II-2. (3) The columns show the numbers, in each category, as follows: (H) houses; (V) villages; (L) landholders; (a) old tax-free land belonging to the village establishment; (b) new tax-free land belonging to the village establishment; (c) old tax-free land belonging to strangers; (d) new tax-free land belonging to strangers; (e) dues paid before the treading of the corn; (f) dues paid before the measuring of the corn; (g) dues paid half by the state and half by the cultivators; and (h) dues paid by the state alone.

Table 2: *List of villages in Salavaucum Pargana recorded in the Barnard Report*

No. Village	No. Village	No. Village
Magan no 1	*Magan no.4*	93. Nelly
1. Salavaucum	45. Culliapettah	95. Uggraharum
3. Cruumberry	47. Vitchoor	*Magan no.7*
5. Poolipaucum	49. Molaghinnamainy	97. Torioor
7. Ahlapaucum	51. Chittenacouoor	99. Vellapootoor
9a. Siddundy	53. Tundry	101. Carrikilly
9b. Seetaupoorum	*55. Annahdoor	103. Woduntongel
11. Cullacaudy	*Magan no.5*	105. Chittahtoor
Magan no.2	57. Perramatoor	107. Vinaiaganelloor
13. Cuddungherry	59. Tooketory	107. Tundelum
15. Neercoonum	61. Codytundilum	111. Kinipaucum
17. Naicoonum	63. Tiroovundavaroo	113. Terootollum
19. Maumboodoor	65. Perroongurry	*Magan no.8*
21. Mayoor	67. Siroopunnioor	115. Undavaucum
23. Paulaishverum	69. Maumbaucum	117. Wodavaucum
25. Coonauvaucum	71. Chumbahdynelloor	119. Pooriddivaucum
27. Ahlanjerry	*Magan no.6*	121. Vellacoonum
29. Totenovel	73. Mungalum	123. Chittahlamungalum
Magan no.3	75. Palliuggerum	125. Poodaputtoo
31. Crumunjery	77. Chumbaucum	127. Pashimboor
33. Sautanunjerry	79. Woraoor	129. Sautummay
35. Pannioor	81. Kirnaugarachery	*131. Wyouoor
37. Tiroomooccoodul	83. Yailapaucum	*133. Naicoopy
39. Arrooncoonum	85. Vellatodoo	135. Coonavaucum
41. Purriavailey	87. Coonumgolatoor	137. Coonvelum
43. Pairanacauoor	89. Kietundelum	139. Yarivaucum
	91. Cauttoopoottoor	141. Totenovel

Source: Jaghire, Barnard's Survey Accounts of Salavakam, vol.69.
Notes: (1) Names of the Magan were as follows (with alternative spellings in parentheses): 1. Salavaucum; 2. Cuddungherry (Cadangara); 3. Sautanunjerry (Satenjary); 4. Calipettah (Culliapettah); 5. Palamatoor (Perramatoor); 6. Mungalum; 7. Vallapootor; 8. Undawaukum (Undavaucum). (2) The village numbers correspond to the page numbers of vol.69 (Salavakam) of the Barnard Report. (3) The asterisks indicates a shrotrium village, held on special terms.

Table 3: *Land categories in Salavaucum Pargana*

Percentages:	1	2	3	Cawnies
A. *Tax-free land (free-gift land)*				
Irrigated	60	0	6	1,302
Unirrigated	40	6	4	873
Sub-total	100	16	10	2,175
B. *State land (circar land)*				
Irrigated, cultivated	41	34	21	4,800
Irrigated, uncultivated	15	13	8	1,814
Unirrigated, cultivated	25	21	13	2,940
Unirrigated, uncultivated	19	16	10	2,253
Sub-total	100	84	52	11,808
Total, land fit for cultivation		100	62	13,983
Land for public use			38	8,513
Total area in Salavaucum Pargana			100	22,496

Source: Jaghire, Barnard's Survey Accounts of Salavakam, vol.69.
Notes: (1) Column 1 gives percentages of land in each of categories A and B; column 2 gives percentages of the total cultivable land (A+B); column 3 gives percentages of the total land in the pargana. (2) The final column shows the area in cawnies; 1 cawnie = 100 coolie = 240 sq.ft. = 1.32 acres.

Table 4: *Land use in Salavaucum Pargana*

Land Use in Salavaucum Pargana	Cawnies	Acres	%
Irrigated cultivated state land	4,800	6,336	21
Irrigated uncultivated state land	1,814	2,394	8
Unirrigated cultivated state land	2,940	3,881	13
Unirrigated uncultivated state land	2,253	2,974	10
State (circar) land, sub-total	11,807	15,586	52
Wood and bad ground	1,770	2,337	8
Yary	1,486	1,962	7
Wood	921	1,215	4
Hill	610	805	3
Muddoor	531	701	2
Yary and tongel	501	661	2
Town	476	629	2
Tanks etc.	421	555	2
Bad ground	404	533	2
River	385	508	2

Table continues

Table 4 continued

Land Use in Salavaucum Pargana	Cawnies	Acres	%
Topes etc.	269	355	1
Paracherry and burying ground	139	183	1
Topes	81	107	0
Bad grass	79	104	0
Stony ground	78	103	0
Channel	69	91	0
Tank	67	88	0
Paddy place	55	73	0
Great road	33	43	0
Pagoda etc.	25	32	0
A part of Arrimbillioor Yary	23	30	0
Pagoda	22	28	0
Flower garden etc.	14	18	0
Fort	14	18	0
Paddy place etc.	11	15	0
Burying ground	9	12	0
Road etc.	7	9	0
Tongel	6	8	0
Spring	5	7	0
Parachery	4	5	0
Well etc.	2	3	0
Land for public use, sub-total	8,513	11.237	38
Irrigated new tax-free land, stranger's	100	132	0
Irrigated new tax-free land, village	15	20	0
Irrigated old tax-free land, stranger's	95	125	0
Irrigated old tax-free land, village	1,093	1,442	5
Unirrigated new tax-free land, stranger's	2	3	0
Unirrigated new tax-free land, village	1	1	0
Unirrigated old tax-free land, stranger's	15	20	0
Unirrigated old tax-free land, village	855	1,128	4
Tax-free land, sub-total	2,175	2,871	10
Total area	22,496	29,694	100

Source: Jaghire, Barnard's Survey Accounts of Salavakam, vol.69.

Table 5: *Major recipients of the 'old free-gift land belonging to the village establishment' in Salavaucum Paragana* (in cawnie)

	Village sample (N=72)	Irrigated	Unirrigated	Total	%
Poligars/Talliars	71	343	267	610	31
Conicoplies	71	259	218	478	25
Artificers	70	92	101	193	10
Shroffs	69	49	58	107	5
Cornmeters	65	65	65	130	7
Landholders	34	84	60	144	7
Village chiefs	25	55	43	98	5
Sub-total		948	811	1,759	90
Other recipients		145	44	189	10
Grand total		1,093	855	1,947	100

Source: Jaghire, Barnard's Survey Accounts of Salavakam, vol.69
Note: 1 cawnie = 100 coolie = 240 sq.ft. = 1.32 acres

Table 6: *Major recipients of 'old free-gift lands the property of strangers' in Salavaucum Paragana* (in cawnie)

	Village sample (N=72)	Irrigated	Unirrig.	Total
Permal Pagoda, Wyouoor	12.0	12.75	7.5	20.25
Town Conicoply	12.0	18.725		18.725
Sydda Moostapah Fakeer, Salavaucum	3.0	18.5		18.5
Ramia Brahmin, Panauyoor, Dist.Chief	2.0	3.0		3.0
Yawkoopshaw Fakeer, Seetaupoorum	1.0	8.75	5.25	14.0
Sydda Esmall Fakeer, Salavaucum	1.0	5.0		5.0
Ramia of Panauyoor	1.0	4.0		4.0
Sub-total		70.725	12.75	83.375
Other recipients		24.125	2.5	26.725
Grand total		94.75	15.25	100.0

Source: Jaghire, Barnard's Survey Accounts of Salavakum, vol.69.
Note: 1 cawnie = 100 coolie = 240 sq. ft. = 1.32 acre.

Table 7: *Major recipients of 'new free-gift land, the property of strangers' in Salavaucum Paragana*

	No. of villages (Total = 72)	Irrigated	Unirrig.	Total
Sydda Moostapah Fakeer, Salavaucum	10.0	41.5		41.5
Conjiverum Padoga	2.0	5.0		5.0
Sooltaun Fakeer, Salavaucum	1.0	30.0		30.0
Sydda Esmall Fakeer, Salavaucum	1.0	5.0		5.0
Sub-total		81.5		81.5
Others		18.375	2.0	20.375
Grand total		99.875	2.0	101.875

Source: Jaghire, Barnard's Survey Accounts of Salavakum, vol.69.
Note: 1 cawnie = 100 coolie = 240 sq.ft. = 1.32 acre.

Table 8: *Major recipients of 'dues paid previous to reading the corn' in Salavaucum Pargana* (in cullum, marakkal, measures)

Recipients	Village sample (N=72)	Range of Dues Min.	Max.	Average
Artificers	71	0.03.0	1.04.0	0.08.0
Barbers	71	0.00.3	0.05.0	0.02.0
Conicoplies	70	0.02.0	0.10.0	0.06.0
Shroffs	69	0.01.0	0.08.0	0.04.0
Washermen	69	0.00.2	0.05.4	0.02.0
Snake doctors	63	0.00.2	0.05.0	0.01.0
Cornmeters	58	0.02.0	1.00.0	0.03.0
Poligars and Talliar	58	0.06.0	2.00.4	1.00.0
Panjangum Brahmans	55	0.00.2	0.04.4	0.00.6

Source: Jaghire, Barnard's Survey Accounts of Salavakam, vol.69.
Note: 1 cullum = 12 marakkal; 1 marakkal = 8 measures.

Table 9: *Major recipients of 'dues paid previous to measuring the corn'
in Salavaucum Paragana* (in cullum, marakkal, measures)

Recipients	Village sample (N=72)	Range of Dues Min.	Max.	Average
Landholders	71	0.03.0	5.04.02	0.06.0
Barbers	68	0.00.2	0.01.00	0.00.4
Panjangum Brahmans	68	0.00.2	0.05.00	0.00.4
Cornmeters	67	0.00.4	0.10.00	0.04.4
Washermen	67	0.00.2	0.01.00	0.00.4
Cultivators' servants	62	0.01.8	5.10.03	2.06.0
Snake doctors	61	0.00.2	0.01.00	0.00.4
Shroffs	60	0.00.2	0.01.00	0.00.6
Conicoplies	57	0.00.6	1.05.00	0.04.0
Ommun Padari Pagoda	54	0.00.1	1.02.00	0.01.0
Artificers	53	0.00.4	0.02.00	0.01.0
Valloovuns	50	0.00.2	0.01.00	0.00.4
Village Pagoda	48	0.03.0	0.08.40	0.04.4
Potmakers	32	0.00.4	0.01.00	0.00.4

Source: Jaghire, Barnard's Survey Accounts of Salavakam, vol.69.
Notes: (1) 1 cullum = 12 marakkal; 1 marakkal = 8 measures. (2) Exceptionally high rates were found in four villages (nos.19, 21, 23, 25) located to the west of Salavaucum town. The reason is not clear. The rate was also exceptionally high in Wyouoor (no.131) and Naicoopy (no.133), both of which were shrotrium villages held on special terms. The shrotrium-holder was Permal Pagoda in both of the villages, which may explain the high rate.

Table 10: *Major recipients of dues paid half by cultivators and half by the state in Salavaucum Pargana* (in cullum, marakkal, measures)

Recipients	No. of villages (Total = 72)	Range of Dues Min.	Max.	Average
Conigo of the distrist	*69*			0.11.7
	62	0.11.7		
	4	0.11.5	0.11.6	
	1	1.04.0		
	1	1.114		
	1	3.01.4		
Dovetraw	*68*	1.114	2.01.0	2.00.0
Mullah Saib of Salavaucum	*68*			0.11.7
	58	0.11.7		
	3	1.11.4	1.11.7	
	1	1.01.4		
	1	1.00.4		
	1	0.11.2		
	4	0.11.5	0.11.6	
Conjiverum Pagoda	*67*			0.11.7
	62	0.11.7		
	2	1.04.0		
	3	0.11.6		
Cultivators' servants	*67*			4.09.1
	61	4.09.0	4.09.2	
	1	6.07.8		
	2	5.05.7		
	3	6.03.0		
Daismooc of the district	*66*			
	60	1.11.3	1.11.7	1.11.5
	3	2.00.0		
	2	0.11.7		
	1	1.04.0		
	1	0.11.6		
Paulia Pandarum of Conjiverum	*60*			
	59	0.05.7	0.06.2	0.06.0
	1	0.00.7		
Yary fund	*27*			
	18	0.11.5	0.11.7	
	8	1.11.6	1.11.7	
	1	1.06.0		

Source: Jaghire, Barnard's Survey Accounts of Salavakam, vol.69.
Notes: Figures in italics indicate sub-totals; 1 cullum = 12 marakkals; 1 marakkal = 8 measures.

Table 11: *Major recipients of dues paid by the circar (state) alone in Salavaucum Pargana* (in cullum, marakkal, measures)

Recipients	No. of villages (Total = 72)	Range of Dues Min.	Max.	Average
Artificers	71	0.00.6	1.06.06	0.08.0
Cornicoplies	70	0.07.4	2.03.0	1.05.0
Shroffs	70	0.00.6	1.00.4	0.06.0
Poligars	68	0.02.0	6.02.2	1.02.0
Cornmeters	52	0.01.1	1.00.4	0.05.0
Tailliars	24	0.03.4	1.04.2	0.09.0
Poligar and Talliars	16	0.01.7	2.11.5	1.05.0
Barbers	7	0.01.0	0.01.4	0.01.4
Panisever	4	0.10.6	0.10.8	0.10.7
Conigo of the district	2	0.00.1	1.02.0	0.01.0
Conjiverum Pagoda	2	0.10.5		0.10.5
Daismooc of the district	2	1.09.2		1.09.2
Dovetraw	2	1.09.2		1.09.2
Landholders	1	0.05.7		0.05.7
Washermen	1	0.01.0		0.01.0

Source: Jaghire, Barnard's Survey Accounts of Salavakam, vol.69.
Note: 1 cullum = 12 marakkal; 1 marakkal = 8 measures.

Table 12: *Landholding castes in Salavaucum Pargana*

Caste Name	Owned villages	Villages of residence	Households
Brahman	*46*		
Gentoo Brahman	24	17	46
Vistnoo Brahman	18	20	143
Siva Brahman	4	15	54
Vellala	*17*		
Tooliva Vellala	10	14	89
Puccolum Vellala	6	18	109
Conidighetty Vellala	1	2	2
Pally	9	41	211
Conicoply	2	27	66
Cow-keeper	1	35	109
Raja	1	1	1
Pagoda	*7*		
Eishverum Pagoda	3		
Permal Pagoda	3		
Seetanunjerry Pagoda	1		
Unknown	3	3	

Source: Jaghire, Barnard's Survey Accounts of Salavakam, vol.69.
Note: Inset entries are sub-totals within their main heads (also totalled).

Table 13: *Accounte of 'unauthorised collection' by Gramattan in Acolagramum village of Tindivanam*

Collection	P.	F.	C.
Amount of unauthorised collections made by the Gramattan			
or Puttah Monigar	45.	12.	65

Disbursements by Puttah Monigar of his order by cash
Revenue administration

Paid 12 Mahatady peons on their coming to collect the money	16.	23.	58
Taking by the Gramattan as batta on his going to cutchery	2.	22.	40
Paid for olahs (palm-leaf) to write the curnum's account	0.	24.	0
Paid for lamp-oil on account of this village cutchery	0.	26.	0
Paid kist on cutcherry Iawabnevess as bribe	1.	5.	50
Paid kistnungar rossum Iawabnevess as bribe	0.	26.	57

Law and order

Paid to peons of Vatavalum Bundary, to apprehend thieves	1.	22.	40
Paid Permal Naig (poligar?) as charity	0.	6.	0

Irrigation

Paid as charity to repair a tank at Tindivanam	0.	24.	0
Paid a peon for watching a tank	1.	22.	40

Religious affairs, rituals

Paid to a church Bramin for his pagoda	2.	0.	0
Paid to a Malabar schoolmaster on Dasara feast	0.	14.	0
Paid to two Fakeers at Mahurrum feast	0.	25.	0
Paid to repair the church (Hindu Pagoda) at Sendamungalum	1.	0.	0
Paid for the expense of the church:			
Abshagum	0.	30.	0
on account of the feast of Shravanum	0.	16.	0
on account of the feast of Nurratrie	2.	0.	0
on account of the feast of Southy	0.	32.	0
oil for Abshagum	1.	0.	0
Abshagum made on account of the eclipses	0.	22.	40
Dupawaley feast	0.	22.	40
Kawithikee feast	0.	22.	40
Pongal feast	1.	22.	40
Typoosem feast	0.	11.	0
Rannaowmey feast	1.	22.	40

Artificers

Paid carpenter Aroonachellum	1.	0.	0
Paid Rungungar doctor for funeral ceremony	1.	22.	40

Personal and others

Paid a merchant as a debt due to the Gramattan to him	1.	18.	0
Paid a Panchangum Bramin to celebrate his marriage	2.	0.	0
Paid alms to a Bramin	0.	12.	0
Total	45.	2.	65

Table continues

Table 13 continued

Source: Letter from the principal collector, Southern Division of Arcot, 15 December 1805, Madras Board of Revenue Proceedings, 2 January 1806.
Notes: (1) 'The revenue of the village in 1804-5 was 1,033P.4F.73C., so that the proportion of the collection in the total was 4.16.21.75 per cent' (in original). (2) The italicised headings have been added. (3) Church = Hindu pagoda (temple). P.F.C. = pagoda, fanam, cash (gold currency); 1 pagoda = 45 fanam; 1 fanam = 80 cash. The prevailing exchange rates differed from area to area in the eighteenth century.

Table 14: *Juncan and fees in Tiruvendiupuram*

Juncan	Fees	Goods
3.25	1.55	Sandalwood, broadcloth, nutmeg, mace, cloves, lead, tin, tuthenaigur, washing stuff, mahmoties, iron, steel, betel nut, turmeric, long pepper, pepper, chilli, saffron, cooper, musk, ginger garlic, quinter seeds, chayroot, jaggary, sugar salt petre, physic salt, brim-stone, hing, mustard seeds, mint seeds, common seeds, garlinger
6.50	3.30	Cotton thread, cotton, ghee, oil
6.50	3.30	Calico brought into the bounds
7.70	3.30	Calico passing from south to north
-		Calico brought for the Company
7.70	3.30	Indigo, tobacco
16.50	3.30	Ganjee
2.49	1.55	Gingili seeds, lamp-oil seeds, horse gram, cotton seeds, sanigalo, wheat, almonds, callivances, green gram, black gram, red gram, indigo seeds, jayara seeds
1.73	1.27	Warago without husk, rice
1.06	0.69	Small quantity of paddy, combo, warago with husk, natcheny, shama, cholum, tena, caudacuuny
29.20	35.60	110 small oxen-loads of paddy, combo, warago with husk, natcheny, shama, cholum, tena, caudacuuny
2.24	0.56	Coconut, jaggary ball, plantain, lime, jack
2.00	1.00	Betel leaves
8.00	-	Piece-goods carried in baskets

Source: Reports and Accounts of the Old Farm of Tiruvendipuram, Selections from the Records of the South Arcot District, no.IV, 11-12.
Note: Figures are in fanam and cash; 1 fanum = 80 cash.

Table 15: *Recipients of fees collected at the customs house*
(in fanam and cash)

1 pagoda of Tiruvendipuram	20
1 pagoda of Tremanycuyly	20
1 pagoda of Trepapolore	20
3 inhabitants formerly Nattars	20
1 Head conccoply of the farm	10
1 Naudashary Brahman	10
1 Shingrachary Brahman	10
1 Shamchary Brahman	5
1 Timmenachary Brahman	2.5
1 Ragavachary Brahman	2.5
1 Vencatachary Brahman	2.5
1 Vizeangar Brahman	2.5
1 Oppaniengar Brahman	2.5
1 Veragapermal Iyengar Brahman	2.5
1 Watiar Annaviangar Brahman	2.5
1 Fa(c)keer Abdulla	1.25
1 Pandarum of a Choultry at Sharady	1.25
Total	1.55

Source: Reports and Accounts of the Old Farm of Tiruvendipuram, Selections from the Records of the South Arcot District, no.IV:11.
Note: 1 fanam = 80 cash.

Table 16: *Taxes abolished in Sattavid in 1805*
(in pagoda, fanam, cash)

Toranadurasanum	5. 30. 72
Tax paid by Chitties	28. 1. 24
Tax paid by bullock people	2. 0. 0
Tax paid by oil-mongers	2. 0. 0
Tax paid by oil press	1. 31. 20
Loom tax	53. 21. 27
Cutnum (a gift to the superiors) on the looms	8. 0. 0
Hemp tax	0. 14. 76
Tax paid by the people who sell glassware	1. 34. 40
Tax paid by cow-keepers	1. 43. 10
Tax paid by ironsmiths	1. 20. 70
Tax paid by artificers	1. 16. 43
Tax paid by goldsmiths	0. 5. 60

Table continues

Table 16 continued (in pagoda, fanam, cash)

Tax paid by Putnavers	2. 28. 0
Tax paid by palanquin boys	0. 30. 0
Tax paid by Tookeries (Talliar)	2. 34. 40
Tax paid by chuckler	0. 11. 40
Cutnum on the mauniam of Calatty Easoovarer	5. 0. 0
Cutnum on cawny mauniams	215. 21. 45
Muctah (fixed tax) paid by Chitties	15. 5. 60
Muctah paid for weavers	1. 0. 0
Muctah for garden and tope	18. 22. 23
Neercooly paid by the mauniamdars	2. 17. 30
Tookery Neketum	2. 0. 0
Tax upon grass from Toty	3. 0. 0
Sauderevaree (expenses for office)	218. 3. 57
Cundayem (tax) by the inhabitants	254. 23. 34
Anuntaverdom by the inhabitants	2. 0. 0
Calavery by the inhabitants	59. 35. 40
Cutnum uponcawny mauniams	3. 23. 0
Ponvary (a commission for money exchange?)	16. 5. 60
Payment at the time of the Pongal feast	4. 32. 40
Adoocole	153. 19. 4
Cundayem paid by Chitties and Comutties	28. 35. 50
Cundayem paid by Palaputtada (mixed area) people	0. 23. 0
Cundayem paid by Beetleuttada gardener	5. 2. 0
Cundayem paid by weavers	35. 32. 61
Cundayem paid by oil-mongers	3. 33. 60
Cundayem paid by bullock people	0. 23. 0
Cundayem paid by cow-keepers	0. 36. 75
Cundayem paid by toddy-drawers	0. 11. 40
Cundayem paid by artificers	0. 6. 38
Cundayem paid by watchmen and ba[-n-?]ker	0. 2. 71
Cundayem paid by Talliar	5. 6. 66
Cundayem by Toty	3. 29. 32
Total (*sic*)	578. 15. 31

Source: Letter from the Collector, Zilla Chingleput, Madras Board of Revenue Proceedings, 4 July 1805, pp.4, 827.
Note: The conversion rate of gold currency (see Table 13) in this region is not known.

Table 17: *Change in landownership and land use in Reddimangudi*

	1864 Acres	%	1898 Acres	%	1924 Acres	%	1982 Acres	%
Total Extent	3707.4	100.0	4044.0	100.0	4046.9	100.0	4046.9	100.0
Cultivated	1071.1	28.9	1808.0	44.7	2441.9	60.3	2657.1	65.7
Inam	78.3	2.1	39.0	1.0	36.9	0.9	36.9	0.9
Waste	1744.1	47.0	1374.0	34.0	335.2	8.3	156.9	3.9
Unassessed		0.0		0.0	11.9	0.3	11.9	0.3
Poramboke	814.0	22.0	823.0	20.4	1221.1	30.2	1221.1	30.2
Reddiyar	823.4	77.6	1057.5	60.3	971.9	40.6	539.3	26.9
Gounder	2.3	0.2	35.6	2.0	276.3	11.5	369.2	18.4
Muthraja	81.6	7.7	294.6	16.8	383.3	16.0	441.5	22.1
Udsaiyar	21.3	2.0	73.7	4.2	170.8	7.1	149.5	7.5
S.C. (sub-total)	8.1	0.8	148.6	8.5	406.6	17.0	380.6	19.0
Hindu pariah	1.6	0.2	48.3	2.8	137.2	5.7	118.9	5.9
Chr'st'n pariah	2.3	0.2	57.8	2.2	108.1	4.5	103.8	5.2
Pallan	2.8	0.3	22.5	1.3	84.9	3.5	86.8	4.3
Sakkili	1.4	0.1	20.0	1.1	76.5	3.2	71.1	3.6
Muslim	20.1	1.9	11.5	0.7	66.5	2.8	39.4	2.0
Brahman		0.0		0.0	3.3	0.1		0.0
Pandaram	4.1	0.4	0.7	0.0	6.2	0.3	4.7	0.2
Chettiyar		0.0		0.0	26.9	1.1	35.5	1.8
Ottan Chetti		0.0	10.8	0.6		0.0	10.9	0.5
Kal Ottan		0.0		0.0		0.0	0.0	0.0
Asari	13.7	1.3	38.3	1.6	17.6	0.7	17.6	0.9
Vannan		0.0	8.1	0.5	7.6	0.3	2.8	0.1
Pariyari		0.0		0.0	10.0	0.4	5.3	0.3
Vanniyar		0.0		0.0	5.2	0.2	5.7	0.3
Nayakkan	2.7	0.3	23.2	1.3	6.0	0.3		0.0
Pillai	0.8	0.1	3.9	0.2		0.0		0.0
Konar		0.0	13.2	0.8	10.1	0.4		0.0
Kusavan		0.0	24.6	1.4	6.0	0.3		0.0
Others		0.0		0.0	13.7	0.6		0.0
Unidentified	82.8	7.8	17.8	1.0	6.6	0.3		0.0
Sub-total	1060.9	100.0	1753.4	100.0	2394.6	100.0	2002.0	100.0
Temple or inam	79.0		83.3		81.7		83.3	
Total	1139.9	30.7	1836.7	45.4	2476.3	61.2	2085.3	
(Other villagers)							(571.81)	

Source: Reddimangudi Settlement Registers, 1864, 1898, 1924, and the Chitta currently used.

Notes: (1) SC = scheduled castes; 'cultivated' includes all occupied land, and 'waste' all unoccupied. (2) The figures for castes for 1982 refer only to land within the village administrative boundary owned by inhabitants of Reddimangudi. Other caste figures include landowners living elsewhere.

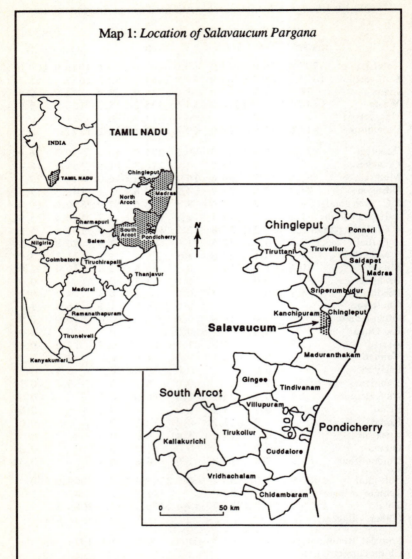

Map 1: *Location of Salavaucum Pargana*

Note: The district and taluk names as well as their boudaries are based on the 1871 Census maps.

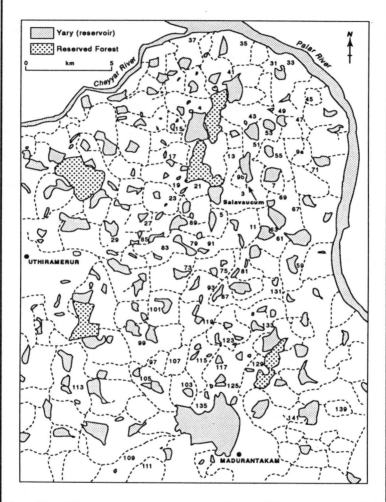

Map 2: *Location of the village in Salavaucum Pargana*

Note: The villages whose location cannot be identified have most likely been abandoned in the past two hundred years. The identification work has been done with the codified hamlet nemes taken directly from the one inch-one mile Taluk maps prepared in the 1960s.

References

Appadurai, A., 1977. 'Kings, sects and temples in South India, 1350-1700 AD', *Indian Economic and Social History Review* 14, 1.

Dewey, C., 1972. 'Images of the village community: a study in Anglo-Indian ideology', *Modern Asian Studies* 6, 3.

R.E. Frykenberg (ed.), 1977. *Land Tenure and Peasant in South Asia* (Delhi).

Gough, K., 1981. *Rural Society in Southeast India* (Cambridge).

Hall, K.R., 1980. *Trade and Statecraft in the Age of the Colas* (Delhi).

___, 1981. 'Peasant state and society in Chola times: a view from the Tiruvidaimarudur urban complex', *Indian Economic and Social History Review* 18, 3 and 4.

Heitzman, J., 1987. 'State formation in South India, 850-1280', *Indian Economic and Social History Review* 24, 1.

Karashima, N., 1984. *South Indian History and Society—Studies from Inscriptions A.D. 850-1800* (Delhi).

___, 1985. 'Nayaka rule in North and South Arcot Districts in South India during the sixteenth century', *Acta Asiatica* 48.

___, 1986. 'Vijayanagar rule and Nattavars in Vellar Valley in Tamilnadu during the 15th and 16th centuries', *Toyo Bunka Kenyuusho Kiyou* 101.

___, 1989. 'Nayaka rule in the Tamil country during the Vijayanagar period', *Journal of the Japanese Association for South Asian Studies*, 1.

___, 1992. *Towards a New Formation. South Indian Society under Vijayanagar Rule* (Delhi).

___ et al., 1988. *Vijayanagar Rule in Tamil Country as revealed through a Statistical Study of Revenue Terms in Inscriptions* (Tokyo).

Kotani, H., 1985. 'The Vatan system in the 16th-18th-century Deccan—towards a new concept of Indian feudalism', *Acta Asiatica* 48.

Mizushima, T., 1980. 'Village records on landholding in South India and the ways for processing them', *Studies on Agrarian Relations in South Asia* 5 (Tokyo).

___, 1983. 'Changes, chances, and choices. The perspective of Indian villagers', *Socio-cultural Change in Villages in Tiruchirapalli District, Tamil Nadu, India*, part 2-1 (Tokyo).

___, 1986. 'Nattar and the socio-economic change in South India in the 18th-19th centuries', *Study of Languages and Cultures of Asia and Africa*, monograph series no.19 (Tokyo).

___, 1987. *Minami-Indo Zaichi Syakai no Kernkyuu* (A Study of Local Society in South India) (Tokyo).

___, 1990a. 'Shohyou: Kotani Hiroyuki Indo no Chuusei Syakai' (review article: H. Kotani's *Medieval Society of India*), *Rekishi-Gaku Kenkyuu*, no.605.

___, 1990b. 'A Study of Local Society in South India', *Regional Views*, no.3 (Komazawa).

___, 1991. *18-20 Keiki Minami Indo Zaichi Syakai no Kenkyuu* (A Study of Local Society in South India in the 18th-20th Centuries) (Tokyo).

___ and T. Nara, 1983. 'Social change in a dry village in South India. An interim report', *Studies in Socio-cultural Change in Rural Villages in Tiruchirapalli District, Tamil Nadu, India*, no.4 (Tokyo).

Murton, B.J., 1973. 'Key people in the countryside: decision-makers in interior Tamilnadu in the late eighteenth century', *Indian Economic and Social History Review* 10, 2.

Saberwar, Satish, 1971. 'Regions and their social structures', *Contributions to Indian Sociology* (new series) V.

Stein, B., 1977. 'Circulation and the historical geography of Tamil country', *Journal of Asian Studies* 37, 1.

___, 1977. '"Privileged landholding": the concept stretched to cover the case', in Frykenberg, 1977.

___, 1980. *Peasant, State and Society in medieval South India* (Delhi).

___, 1982. 'South India: some general considerations of the region and its early history', in T. Raychaudhuri and I. Habib (eds.), *The Cambridge Economic History of India*, vol.I (Cambridge).

Stokes, E., 1977. 'Privileged land tenure in village India in the early nineteenth century', in Frykenberg , 1977.

Subbarayalu, Y., 1973. *Political Geography of the Chola Country* (Madras).

Thurston, E., 1909. *Castes and Tribes of Southern India*, vol.I (Madras).

Chapter 3

THE PEASANTRY OF NORTHERN BENGAL IN THE LATE EIGHTEENTH CENTURY[1]

Shinkichi Taniguchi

This paper deals with the peasantry of northern Bengal in the late eighteenth century with special reference to the economic and political role of the wealthy raiyats (often called under a general designation of *jotedars*). They formed the core not only of agricultural production but also of village-level politics, and in this sense we may call them the small local leaders. Much has been said about them by such scholars as N.K. Sinha, B.B. Chaudhuri, Rajat and Ratnalekha Ray, and Rajat Datta.[2] Ratnalekha Ray's *Change in Bengal Agrarian Society* may be

[1] This paper is based on Shinkichi Taniguchi, 'Structure of agrarian society in northern Bengal (1765 to 1800)', unpublished Ph.D dissertation, Calcutta, 1977. I am grateful to Professor Binay Bhusan Chaudhuri for his valuable comments and suggestions on the earlier version of this paper. If the present paper is readable at all, it is largely due to Dr. Peter Robb's editing. I am grateful to Dr. Robb for his trouble. The following abbreviations are used in the notes:

BRP	Proceedings of the Board of Revenue (1786-1806).
CRP	Proceedings of the Committee of Revenue (1781-86).
GG Rev	Proceedings of the Governor General in Council-Revenue, Department (1778-80, 1790).
HR	BRP, 22 March 1790, No.15 (Main Report) (Harrington's Report).
IESHR	*Indian Economic and Social History Review*.
PCD	Proceedings of the Provincial Council of Revenue at Dinajpur (1772-74).
WBSA	West Bengal State Archives, Calcutta. Proceedings volumes were consulted at this archive.

[2] The major works in the controversy, excluding those mainly concerned with the jotedars of the nineteenth and twentieth centuries, are N.K. Sinha, *The Economic History of Bengal*, vol.2, (Calcutta, 1962); B.B. Chaudhury, 'Some problems of the peasantry before the permanent settlement', *Bengal Past and Present* 142 (1957), pp.136-51, 'Some aspects of peasant-economy of Bengal after the permanent settlement' in ibid., pp.137-49, and 'Rural power structure and agricultural productivity in eastern India, 1757-1947', in M. Desai, S.H. Rudolph and A. Rudra (eds.), *Agrarian Power and Agricultural Productivity in South Asia* (Delhi, 1984), pp.100-70; Rajit and Ratna Ray, 'The dynamics of continuity in rural Bengal under the British imperium: a study of quasi-stable equilibrium in underdeveloped societies in

considered the most vigorous attempt at situating them in the total structure of Bengal agrarian society during early British rule. Ray doubt-lessly offered a very attractive and far-reaching frame of reference, even if one does not fully accept her arguments. She identified seven major groups which comprised Bengal agrarian society. They were (1) sover-eign power (the state), (2) zamindars and independent talukdars, (3) local officers of both the government and the zamindars, (4) local gentry (holders of privileged tenures such as *brahmattar*, *aimadars*, some of the *haoladars*, *patnidars*), (5) village heads (*mandals*, *pramaniks*, *bosneahs*) and other rich peasants (jotedars, *gantidars* and some of the *haoladars*), (6) agricultural stock (ordinary raiyats), and (7) landless labourers (*krishans*, *mojurs*). Socially speaking, group (1) consisted of the British colonialists and their Indian collaborators, groups (2), (3) and (4) con-sisted mostly of high-caste Hindus and high-ranking Muslims, groups (5) and (6) consisted of agricultural castes such as Sadgop, Kaivartta, Namasudra, Rajbangshi and low-ranking Muslim peasants, and group (7) consisted of untouchables and tribal peoples. Thus Ray saw close correspondence between economic and political stratification and the social or ritual hierarchy.

Ray's other important contribution to the understanding of Bengal's agrarian structure was the clear-cut distinction she made between the revenue-receiving and the landholding structure. She contended that the so-called 'landed property' conferred upon the zamindar by the

---

a changing world', *IESHR* (June 1973), pp.103-28, and 'Zamindars and jotedars: a study of rural politics in Bengal', *Modern Asian Studies* 9, 1 (1975), pp.81-102; Ratnalekha Ray, *Change in Bengal Agrarian Society* (Delhi, 1979); Rajat Kanta Ray, 'The retreat of the jotedars?', *IESHR* 22, 2 (1988), pp.237-47; and Rajat Datta, 'Merchants and peasants: a study of the structure of local trade in grain in late eighteenth-century Bengal', *IESHR* 23, 4 (1986), pp.379-402, 'Agricultural production, social participation and domination in late eighteenth-century Bengal: towards an alternative explanation', *Journal of Peasant Studies* 17, 1 (1989), pp.68-113, and 'Rural society and the structure of landed property: zamindars, taluqdars and *la-khiraj*-holders in eighteenth-century Bengal', in Datta, Peter Robb and Kaoru Sugihara, *Agrarian Structure and Economic Development: Landed property in Bengal and theories of capitalism in Japan* (Occasional Papers in Third-World Economic History, no.4, London, 1992), pp.34-60. During the preparation of Taniguchi, 'Agrarian society', under the supervision of Professor B.B. Chaudhuri, I was unaware of the Rays' work.

introduction of the permanent settlement (1793) was in fact no more than the rent-receiving right which presupposed the revenue-receiving rights of the state, and that the state could not give the zamindars land-holding rights or actual possession of lands which the state itself had hardly ever possessed. Thus the frequent and large-scale sales of the zamindars' estates (zamindaris) immediately after the permanent settle-ment under the vigorous 'sunset law' did not affect the landholding structure within the villages. It was the preserve of the village oligarchy composed of the local gentry, village heads and rich peasants. In combination effectively they prevented the new zamindars who purchased lots at public auctions from coming into direct contact with the actual cultivators of lands through various kinds of manoeuvre. Within the village, they cultivated their considerable holdings by letting out their lands to the sharecroppers and also by the use of landless labourers, who mostly belonged to the untouchable castes and the tribal peoples. The rich peasants often advanced money and grain to the sharecroppers and agricultural labourers and, thus, exercised great influence over them.

Ray also provided us with important clues to the regional character-istics of Bengal agrarian society. A full set of the above seven groups was typically found in the districts of Burdwan, Bishunupur and Dhaka where a high concentration of high-caste Hindus was observed. Such high-caste Hindus formed the core of the local gentry. In the districts of Dinajpur and Rangpur where high-caste Hindus were largely lacking, local gentry were not to be found. In these districts, the village heads and/or the rich peasants constituted the village oligarchy. In Midnapur, the powerful agricultural caste (Kaivartta) provided the rent-receiving class, the village oligarchy, and the landholding body. In this way Ray implicitly showed that her conceptual frame was relevant to almost all the localities in Bengal with certain modifications. She argued for the omnipresence of rich peasants in Bengal. She further asserted that the position and function of this important group remained unchanged during the whole period of colonial rule. The appropriateness or otherwise of these points will be discussed in the concluding part of this paper.

Recently, a radically different interpretation of Bengal agrarian society has been put forward by Rajat Datta.[3] In spite of Datta's criticism, the present author thinks that Ray's conceptual frame is still useful for our research. Datta's argument on the jotedars in the eighteenth-century Bengal may be summarised as follows. (1) The so-called jotedars or rich peasants existed only in the frontier districts, Dinajpur and Rangpur, to the north, and Jessore to the south, where vast arable waste lands were easily available. But outside these three districts, Datta argues, it is difficult to show the domination of rich peasants with large landholdings exceeding, say, 55 acres each. Therefore it is not justifiable to characterise eighteenth-century Bengal as jotedar-dominated. (2) Agriculture in eighteenth-century Bengal should be characterised as small-peasant dominated, and the peasantry may suitably be classified into two groups: the poor or inferior peasant, and the middling peasant whose landholding hardly exceeded 10 acres. (3) The limiting factor of production in those days was not land, but the agricultural stock (subsistence fund). (4) Jotedari holdings as found in the above three districts were created by the zamindars for convenience in the management of their estates. They neither posed a challenge to the zamindars' control in the countryside nor compelled the zamindars to acknowledge the jotedars with their large holdings as a powerful and rising class of rich peasants. (5) The frequent peasant uprisings and disturbances of eighteenth-century Bengal, which were wrongly taken by other researchers as signs of the rise of rich peasants, in fact expressed the agricultural decline of certain districts because of natural disasters. Datta asserts that such disturbances were concentrated in Birbhum, Rajshahi, Purnea and Dinajpur. He further argues that such disturbances took place only in 1787/8. (6) Polarisation of the peasantry was also blocked by the existence of customary codes which allowed the old settled peasant (*khudkashta* raiyat) to claim agricultural lands in preference to a new occupant even after an absence of many years from the village. (7) The real masters of the countryside were the zamindari officials and the grain merchants. The former obtained influence by issuing cultivating leases (*patta*) to the raiyats,

---

[3] See note 2 above.

and the latter because they provided the peasants with badly-needed loans, the most scarce factor of production in the countryside.

Let us examine Datta's argument as summarised above. It will be necessary to give a definition of the rich peasant or jotedar as conceived by the present author. A rich peasant may be defined as any farmer who held fairly large landholdings, the cultivation of which could not be carried out by family labour alone, and who had to employ subordinate labour force in the form of annual labourers, sharecroppers and other types of under-tenants. He received shares of crops from his sharecroppers and under-tenants, which far exceeded his family needs and rent costs. The surplus production thus put at his disposal might be lent out to his needy neighbours or sold to the grain merchants or ploughed back into agriculture. The actual size of such holdings might be 30 acres or 55 acres or even 500 acres. On this rough definition, rich peasant or jotedar holdings, as the present paper establishes beyond doubt, existed in Birbhum and Purnea, and we have reason to say that the same kind of rich peasants existed in highly-cultivated Burdwan, one of the richest districts of Bengal. We can multiply such instances. For example, a rich raiyat of Burdwan, Godhadur Mandal, cultivated 5 bighas of sugar-cane lands, and his total holding was suggested to be 60 bighas. He employed a few annual labourers. He gave evidence that rich peasants like him often grew sugarcane in this district.[4] Thus it is difficult to consider the rich peasant with large holdings as exceptional. It is true that jotedari tenure was created or sanctioned by the zamindar, but it does not automatically prove that there was no contradiction of interests between the two. Contrary to Datta's assertion, the present paper shows that the jotedars extended their holdings at the cost of both the zamindar and the poorer peasants. As will also be shown, Datta's argument that peasant uprising were more or less confined to the year 1787/8 is hardly sustainable. We rather doubt its generality of the customary code protecting khudkashta raiyats, judging from a patta in the district of Dinajpur which clearly forbad the claims of an absconding raiyat.[5] It would

[4] Anon., *Bengal Sugar: An account of the method and expense of cultivating the sugar-cane in Bengal* (Calcutta, 1794), pp.75-90.
[5] BRP, 31 October 1791, no.37.

be possible to argue that it was the existence of a customary code, which obliged the Collector of the district to insert such a provision in the new 'regulation'. But, even if we admit that khudkashta raiyats were thus privileged in many districts, this hardly precludes the acquisition of extensive holdings by the rich peasants. As to the great influence of the zamindari officials and the grain merchants in the countryside, we feel it necessary to make some reservation. First, many of the low-ranking zamindari officials possessed raiyati lands as well as lands held as *chakran* (service tenure). It was the custom of the country that the officials would return the chakran lands when they resigned from the zamindar's service. Therefore, it becomes rather difficult to distinguish these people clearly from the peasants, especially after their retirement. Second, the authority of issuing pattas was often exercised also by the rent-farmers of the villages. And we have instances where the superior raiyats became the rent-farmers of their own village. Third, it is hard to believe that outside grain merchants could penetrate deep into every corner of the country and were able to provide the whole agricultural population with necessary loans. The advance-payments system set up by the merchants prevailed only in such localities where commercial crops were grown extensively. It was not possible for the grain merchants to cater for all the loan requirements of the great mass of needy peasants everywhere. The well-to-do of the neighbourhood would supply loans to the poor where the grain merchants were not available.

The present paper is not concerned so much with the current controversy over the nature of the jotedar in the late eighteenth century, as with a detailed and analytical description of the socio-economic conditions of the peasantry. We need to know them more concretely before we can construct a well-founded 'theory'. This paper's first section will try to reconstruct the composition of the peasantry in northern Bengal mostly on the basis of two famous reports, that is, Harington's report on the zamindari of Swaruppur (1790) and Buchanan's report on Dinajpur (1807/8). In the second section, an attempt is made to locate the wealthy raiyats in the agrarian society in connection with their economic as well as political roles. And in the last section, the household economy of each gradation of raiyats is examined. Here we also make a

special study of the management of the landholdings of the wealthy
raiyats in Rangpur with reference to the sharecroppers (*adhiyars*) working under them. Overall this paper lays stress on the historical development of a very important social group during the period of political
confusion and chaos of the early British rule.

I. *Composition of the peasantry: analysis of some village statistics*

During our period, land was abundant, while the distribution of agricultural resources or means of subsistence was quite uneven.[6] The size of
the holdings held by the raiyats was mainly determined by the size of
agricultural stocks possessed by them. This allows a classification of
the raiyats on the basis of self-sufficiency (or the minimum size of
holding whose produce would enable a raiyat and his family to subsist).
Accordingly, the rural population could be classified as follows: (1)
Those who had no agricultural stock. They could not undertake cultivation of any size on their own account and therefore hired themselves out
as wage labourers. Needless to say, they had no rights to the lands, and
cannot be considered as peasants (raiyats). (2) Those who had small
agricultural stock, but too little to provide them with subsistence even
at a minimal level. They had to supplement the deficiency either by cultivating the lands of the wealthy neighbours who provided them with
the necessary production loans or by working as hired labourers. We
may call them 'the raiyats below the level of self-sufficiency'. (3)
Those who had just enough agricultural stock to undertake on their own
the cultivation of the minimum area of lands needed for their subsistence. They may be called 'the small self-sufficient raiyats'. (4) Those
who had large agricultural stock and rented extensive lands from the
zamindar. They cultivated the lands by subletting to their under-raiyats
for rents both in kind and in money and/or by hiring agricultural
labourers. We may call them 'the wealthy farmers' or 'the rich peasants'
or jotedars.

  The peasantry could be thus classified in terms either of produce or
of money.[7] Let us take a small self-sufficient raiyat with an average

---

[6] See Taniguchi, 'Agrarian Society', ch.5.
[7] The calculation that follows is based mainly on data taken from northern

family size of five members: father, mother and three children (or two children and an old dependent relation). We may suppose that one male adult consumed three quarters of a seer of rice per diem and the other members half a seer each.[8] The annual consumption of grain of this family would be 25 maunds of rice or about 33 maunds of paddy.[9] Other necessary items of their diet (oil, pulses, vegetables, and ash which was used as a substitute for the salt by the poorer people) were supplied from their small gardens and nearby forests. A small quantity of cotton and tobacco was also grown for home consumption.[10] For the replacement cost of agricultural implements and the future purchase of cattle and so on, at least Rs.2 or 8 maunds of paddy had to be set aside.[11] For seed, the usual requirement was 20 seers per bigha.[12] Besides these, the raiyat had to pay, in money, basic rent (*asal jama*), impositions (*abwabs*) and village expenses (*mathots* and *gram karcha*). A fair estimate of these would be Rs.1-8 or 6 maunds of paddy per local bigha.[13] The small raiyat's income was derived entirely from the culti- vation of his holding. According to J.H. Harington, the average produce of paddy per local bigha was 10 maunds 29 seers on clay (*khya*r land) and 7 maunds 12 seers on loam (*pali* land).[14] Considering that loam could be double-cropped, 10 maunds per local bigha for both sorts of paddy land would be a reasonable estimate of output. To sum up, from the produce of one bigha of paddy land, the raiyat had to set aside 6.5 maunds for rent, impositions, village expenses and seed, and could retain 3.5 maunds for his own use. To obtain, from this balance, the

Bengal. Land fertility, agricultural prices and rent rates differed considerably from locality to locality; it may be valid only for northern Bengal.

[8] Buchanan estimated the daily consumption of rice by a male adult at 0.5 seer, while Paterson estimated it at 1 seer. We may take 0.75 seer as a fair estimate. Francis Buchanan, *A Geographical, Statistical and Historical De- scription of the District or Zila of Dinajpur in the Province or Subah of Bengal* (Calcutta 1833), pp.115-32; BRP, 15 June 1789, no.86.

[9] One maund of paddy equals 30 seers of cleaned rice; Buchanan, *Dinajpur*, p.180. One maund = 40 seers = approximately 37kg.

[10] Buchanan, *Dinajpur*, p.232; BRP, 15 June 1789, no.86.

[11] Calculated on the basis of Buchanan's data. Buchanan, *Dinajpur*, pp.215-6 and 237.

[12] Buchanan, *Dinajpu*r, p.237.

[13] One local bigha was about 1.8 Calcutta bigha or nearly 0.6 acre.

[14] HR.

necessary 41 maunds of paddy (33 maunds for food and 8 maunds as the
replacement cost of agricultural implements and so on), he would apparently have to cultivate 12 local bighas of land. However, this seems an
overestimate, since a peasant of Dinajpur having one plough could
hardly have cultivated more than 8 local bighas in a year.[15] The calculation presupposed that all his holdings were duly assessed at rent-rates as
fixed in the table of rates (*nirkhbandi*). But most raiyats possessed more
land than the zamindari records showed, as was clearly established by
the survey of J.H. Harington in 1790. This could reduce the actual burden of rent and village expenses to less than one rupee per local bigha.[16]
On that supposition, we may estimate the minimum size of the self-
sufficient holding at about 7 local bighas (12 Calcutta bighas).

We can reach a similar result by calculating in money terms: 41
maunds of paddy would have cost about Rs.10 at the market price of
those days. Rent and other impositions would have come to 14 annas
and seed to 2 annas, a total of Rs.1 per local bigha. The gross produce
of one bigha of paddy land was Rs.2-8 which implies that the raiyat
retained Rs.1-8 per local bigha as net income. Once again, to obtain the
necessary Rs.10, he would have had to cultivate about 7 local bighas of
paddy land.[17] Possession of a holding of this size therefore marked an
approximate dividing line between the poor and the self-sufficient raiyats.

The upper line of demarcation, between self-sufficient and rich raiyats, cannot be so clearly drawn, as it varied according to family size. As
one raiyat could cultivate little more than 8 local bighas in this locality, this may be considered as the point after which he would have to

---

[15] Buchanan, *Dinajpur*, p.234; HR.

[16] J.H. Harington found that, on average, the raiyats of pargana Swaruppur (Sooroopoor), also in Ghoraghat, possessed lands twice as large as
those recorded in the village accounts; HR. See also Table 5 below.

[17] Buchanan estimated Rs.22-11 as the usual living expenses of a family
of the lowest order. However, this seems to have been an overestimation.
He ignored in his calculation that the raiyats of this gradation produced almost all the necessaries of life by themselves, and that they consumed
coarse rice which was much cheaper than that for sale or for export. The Collector of Jessore observed in 1788 that an income of Rs.8-4-2 would be
'fully adequate for his [the raiyat's] necessaries'. Buchanan, *Dinajpur*, and
BRP, 11 July 1788, no.14.

employ labour. But, as the economic condition of a family became better, it tended to become larger. Many families of small self-sufficient raiyats had two adult workers—for example, a father and a grown-up son or an unmarried male relation. Such a family could occupy 16 to 17 local bighas without depending on hired labourer. We may tentatively set the demarcation line between the second and the third categories at 17 local bighas or 30 Calcutta bighas. Buchanan's observation that small farmers 'have 1 or 2 ploughs and seldom employ servants' would support this estimate.

Next we shall analyse four sets of village statistics to identify the characteristics of peasant stratification in north Bengal in the late eighteenth century (see Table 1). In the middle of 1770, J. Grose, the Superintendent of Rangpur, sent his assistant, G. Robertson, to Govind Gunge in the zamindari of the nine-anna division of Edrackpore, to enquire into the real assets of the country. Robertson acquired the papers of the village accountant (*patwari*) in Ryampore taluk, and prepared a 'Statement of collections made from Ryampore village in Talook of the Aumil'.[18] This statement shows the names of the villagers, the quantity of land held by each, their *sadar jamas* and impositions, and the actual sum realised by the amil in the preceding year. Table 1 is based on Robertson's statement after making necessary adjustments,[19] and is arranged according to the criteria set out above. Out of 93 holdings belonging to the group of households presumed to be 'below the level

[18] Letter copy book of the Resident at the Durbar at Murshidabad, 23 July 1770, no.8, Enclosures. An amil was an officer deputed by the government to take charge of the management of a defaulted zamindar's estate.

[19] Despite the great value of these data, we cannot use them without reservation. First, Robertson did not mention the size of the local bigha in use in that locality; secondly, Robertson had hardly any means to check the authenticity of the patwari papers, and J.H. Harington clearly established that forgery of village accounts by the patwaris in collusion with the raiyats was a common practice in the countryside. Fortunately, we have a detailed account of the local bighas in use in different parganas of sircar Ghoraghat, of which Edrackpore constituted a major part. According to this, in pargana Edrackpore (9-anna division of Ghoraghat) the current unit of measurement (*gaz*) was 34.25 and '52 Buz constitute the Begah Cord in the Pullee Mahals, and 48 Guz in the Khear Mahals'. From this we may conclude that one local bigha used in Robertson's statement equalled about 1.5 Calcutta bigha; HR.

of self-sufficiency', 80 were smaller than three Calcutta bighas. Raiyats with such holdings could not have been independent peasants, and must have worked as under-raiyats or hired labourers under the wealthy raiyats.

The following three abstract statements are based on data collected by J.H. Harington during his survey in the zamindari of Swaruppur in 1790. He selected three villages and had them measured under his direct supervision. We may therefore regard his data as much more accurate than Robertson's.[20]

The first, Radhanagar, was a fairly big village with 50 registered raiyats and five *huzuri* jotedars.[21] The total village area was 5,041 local bighas. Of these, 846 bighas were waste lands and 4,195 bighas were cultivated lands and house-sites. Of the cultivated lands 1,121 bighas were held under special tenures (*bazi zamin*), and the remaining 3,073 bighas were ordinary rent-paying (raiyati) lands. Of the bazi zamin, 282 bighas were rent-paying service lands (*jama* chakran lands) and the remainder were various kinds of rent-free lands such as *lakhiraj*, zamindars' household lands (zamindari *khamar* lands) and rent-free service lands (*bejama chakran*). As the five huzuri jotedars held 2,741 bighas of raiyati lands, the rent-paying lands under the jurisdiction of the village amin[22] were reduced to 605 bighas. Harington did not investigate the internal conditions of the huzuri jotedars' holdings, and the zamindari records did not furnish information on them. Table 2 concerned only the lands under the jurisdiction of the village amin.

As one cultivator with a plough could cultivate not more than 8 local bighas, there must have been at least 400 to 450 under-raiyats (prajas) and agricultural labourers (krishans) working under the huzuri jotedars and the holders of the privileged tenures (bazi zamin). The

---

[20] The following description of the three villages depends on the Harington report and its enclosures (nos.1-28) and appendices (A-I), BRP, 22 March 1790, no.15-16. These villages nowadays belong to the Upazila Badarganj, District Rangpur, Bangladesh. For present conditions of this locality, see the detailed village report, in Shinkichi Taniguchi, *Society and Economy of a Rice-producing Village in Northern Bangladesh* (Tokyo, 1987).

[21] For the huzuri jotedars, see Section 2 of this paper.

[22] For more about these village officers, see Taniguchi, 'Agrarian Society', ch.1.

names of these poorest cultivators never appeared in the zamindari records. Part of such lands might have been cultivated by non-resident raiyats (*pahikasht* raiyats) from the neighbouring villages, but their number could not have been very large as Harington nowhere referred to them in his report. Therefore, by far the greatest portion of the population of this village must have been the prajas and the krishans.

Maheiskol (see Table 3) was a small village with only 14 villagers whose names were registered in the zamindari records. There were no huzuri jotedars in the village. The total village area was 342 local bighas of which 65 bighas were waste land and 277 bighas the cultivated lands and the house-site. Out of 277 bighas of the cultivated lands, 48 bighas were rent-free and 224 bighas were rent-paying.[23] The cultivated land per villager was 16 local bighas. The average size of holdings under the wealthy farmers comes to about 36 local bighas, considerably bigger than a cultivator could cultivate. This suggests the existence of more than 10 prajas and krishans in the village.

Finally, Rogonatpoor (see Table 4) was apparently a very small village with only 9 villagers whose names appeared on the zamindari accounts. Its total area was 452 local bighas, of which 31 bighas were waste land and 421 bighas cultivated lands and homesteads. Of the latter 151 bighas were under privileged tenures, and the remaining 270 bighas were ordinary rent-paying lands. The cultivated land per registered villager was 32 local bighas which far exceeded the size a cultivator could cultivate. Another interesting feature of the composition of the peasantry in this village was the high concentration of lands in the hands of three wealthy raiyats who held 82 per cent of the cultivated lands or about 78 local bighas in average. These figures strongly suggest that there was a considerable number of prajas and the krishans working under the wealthy raiyats, though we have no direct evidence to prove this.

Table 5 summarises figures given in earlier tables. The composition of the peasantry of northern Bengal may also be considered through a critical study of Francis Buchanan's descriptions. Here we rely mostly

[23] The difference of five bighas arose because of miscalculation in the original records.

on data he collected in Dinajpur in 1807 and 1808.[24] Though invaluable, his statistics are rather misleading and should be used cautiously. Buchanan divided the cultivators (or the people engaged in agricultural production) into five groups: krishans (agricultural labourers), adhiyars (share-croppers), small-farmers,[25] middling-farmers and large farmers. The krishans had no agricultural stock to enable them to take up and cultivate lands on their own account. They usually hired themselves out to middling and wealthy farmers by the day, month, half-year or year. There was plenty of work for them. Each month they received 8 annas in money, and food and clothes which cost 12 annas to 1 rupee. Their monthly income, therefore, did not exceed Rs.1-8. Unlike the adhiyars who held small areas of land and grew vegetables and other necessities (even a small quantity of cotton), the krishans were wholly dependant on their employers and on purchases. In comparison with an adhiyar, a krishan required more money to maintain his family, and had a distinctly lower standard of living. Buchanan thought there were 80,000 krishans in Dinajpur district.

The adhiyars were those who 'cultivate land for a share of produce'. As 'they have not stock sufficient to enable them to cultivate without the assistance', they could not rent land to the full capacity of their family labour. They had 'in general 2 or 3 bighas, for which they pay rent and employ their leisure time in cultivating land for their neighbours for one-half of the produce'. The adhiyars generally used their own plough-unit (ploughs and bullocks) in cultivating the lands on an *adhi* (share-cropping) arrangement.[26] The seeds were often furnished by the wealthy farmers who sublet lands to them. Though the adhiyars possessed agricultural implements and bullocks, their stock of grain was so small that 'for six months in the year they would starve, did not the wealthy farmers advance them grain to eat'. Many of the adhiyars were indebted to their employers for more than the 'whole value of their stock'. Such adhiyars were called 'under-raiyats' by Colebrooke in his

---

[24] F. Buchanan, *Dinajpur*, pp.234-6.
[25] Buchanan used the word 'farmer' in the sense of a peasant.
[26] *Adhi* means half, hence the sharing of produce by half and half.

Purnea Report in 1790.[27] There was another type of adhiyars who rented lands from the agent of the zamindar like ordinary raiyats under the regular local (mufassil) settlement, but paid their rent in kind.

These adhiyars were not the under-raiyats of wealthy farmers, but cultivated lands on their own account. The *khamar* lands[28] of the zamindar were often cultivated in this way. The adhi engagement was preferred on such lands when the soil was inferior or the harvest precarious, since the loss from unfavourable seasons was borne by both the adhiyars and the zamindars.[29] Buchanan wrote that 'the number of the Adhiyars is very considerable...it is probable that there are above 150,000 families'.

The small-farmers, who constituted about half of the farmers, had one or two ploughs and seldom employed servants. If they had insufficient land to employ their ploughs over the whole year, they would cultivate extra land as share-croppers. But they paid higher rates of rents than the wealthy raiyats. Many of them were obliged to borrow from their richer neighbours, and not infrequently would be reduced to the rank of under-raiyats.[30] The middling farmers, who had '3 or 4 or 5 ploughs, form perhaps $7/16$ths of the whole: these are not exempt from holding the ploughs, but hire servants to make up for the deficiency in the number of labourers'. We may consider them as the lower portion of our third class (the wealthy raiyats). The wealthy farmers were described thus:

About one farmer in 16 may rent from 30 to 100 acres (or about 50 to 170 local bighas). These seldom labour with their own hands, but keep as many ploughs as they have dependent relations, or hire 2 or 3 additional men. The remainder of their lands, they give to people who cultivate for share. These men have, in general, large capitals and advance money or grain both to those who cultivate for a share and to their own necessitous neighbours to

[27] BRP, 18 June 1790, no.8.
[28] Khamar land was land retained by the zamindar for his provisions. For more on this, see Taniguchi, 'Agrarian Society', Appendix (c) to ch.1.
[29] See ibid.
[30] For more detailed treatment of this problem see ibid., Section 1-(2)-d of ch.6.

enable them to live while the cultivation is going forward....[31]

In his statistical data on the peasantry, Buchanan gave the figures summarised in Table 6. In addition, Buchanan mentioned 150,000 adhiyars and 80,000 krishans. He estimated the total strength of labour in the Dinajpur district at 442,000. He thus used two different classifications in his Dinajpur report; the five groups explained earlier and the above eight groups. As he did not indicate the criteria on which he classified the latter, the eight groups are better rearranged into his five groups on which he provided adequate explanation.

Table 7 shows the result of such a regrouping of Buchanan's statistics. It also shows a similar regrouping of his eight gradations into our three economic groups. In Table 7, the recalculation omits the krishans, as our criteria do not apply to the landless labourers. Harington's village accounts (Tables 2, 3, 4 and 5) also did not include the landless.

Table 7 refutes Buchanan's own assertion that the middling farmers constituted seven-sixteenths of the whole farmers. And if we include the adhiyars among the farmers, his other assertion that small farmers constituted half of the total farmers also becomes quite doubtful. However, his statistics reveal a pattern of the composition of the peasantry similar to that of Harington; both Buchanan (Table 6) and Harington (Table 5) agree, in that the first category (those below self-sufficiency) constituted nearly half of the peasant population.

From the foregoing study of statistical data, we can confidently say that agrarian society of north Bengal in the late eighteenth and early nineteenth centuries was highly stratified. These data show that more than half the peasantry, or, if we include the krishans, nearly two-thirds of the rural population, were lacking in sufficient agricultural resources to cultivate the minimum size of holding needed for self-sufficiency. Such people had to depend on loans in money and in kind. Thus the majority of the rural population were under-raiyats (prajas) and agricultural labourers (krishans), in abject subjugation to the wealthy raiyats (the suppliers of agricultural stocks). Secondly, we should note the smallness of the class of petty self-sufficient raiyats. Harington's vil-

[31] Buchanan, *Dinajpur*, p.235.

lage accounts show that in the zamindari of Swaruppur they constituted only one-sixth of the 'registered' raiyats. Buchanan's statistical data show that they formed one-third of such raiyats. Furthermore, we have reason to say that the numbers in this class was steadily decreasing during our period.[32] Thirdly, our analysis of the four village statements corroborates the well-known assertion that the rates of rent paid by the upper class of raiyats were considerably lower than those of the common and the lower raiyats. As Table 5 shows, the average rent per Calcutta bigha for the first class was 7-19 annas, for the second 6-16 annas, and for the third 5-4 annas.

II. *The small local leaders*

The local small leaders consisted of the wealthy raiyats, the principal or superior raiyats (mandals, paramanicks, pradhans, and bosneahs), the small local under-farmers (kutkinadars and hudadars), and the holders of privileged tenures. Among them by far the most numerous and important were the wealthy and the principal raiyats. They were to be found widely all over Bengal, including in the Purnea district of Bihar, the eastern half of which once belonged to the Bengal subah. We shall first study the size of their holdings and wealth, and the degree of their influence.

In Dinajpur, such people held 90 to 300 bighas of lands and possessed a considerable capital 'of from Rs.5,000 to Rs.20,000, but perhaps 5,000 may be about the average'.[33] In Rangpur, they often acted as the village heads (paramanicks, pradhans or bosneahs).[34] One Habshee Mandal of Culleah Muderam in Coondy pargana had been the farmer of his village for ten years and paid Rs.325 in rent for the land and a market-place (*hat*) held by him. He had 'the reputation of a man of great influence over the raiyats throughout the district and one whom the zamindars particularly respect on account of that influence'.[35] Shaik

[32] For evidence of this, see Taniguchi, 'Agrarian Society', Section 1-(2)-d of ch.6.

[33] Buchanan, *Dinajpur*, pp.235-7.

[34] On their role in the zamindari administration, see Taniguchi, 'Agrarian Society', Section 1-(5)-c of ch.1.

[35] Proceedings of Committee of Revenue, 10 June 1784, no.17, WBSA.

Caubil of Coondy pargana had been paramanick of his *calleah*[36] for ten years and was responsible for the payment of Rs.1,600 to Rs.1,900 as the rent of his 'zillah'.[37] Shaik Bauker was a hereditary paramanick and paid rent of Rs.105.[38] In the same pargana, there were eight to ten principal raiyats who kept other raiyats under their control.[39]

In the pargana of Swaruppur in Rangpur, we find principal raiyats of a similar type. Mohammed Shuffeck Paramanick called himself the pradhan of Rampoorah village. He was also a huzuri jotedar. He held 317 bighas and paid Rs.221 as his tahud jama (an engaged sum of rent). Khider Paramanick of village Rogonatpore held 60 bighas in the village and 450 bighas separately as huzuri jote. His rent for these 510 bighas was Rs.537. Kirmool Paramanick of village Kalakapoor held 90 bighas as huzuri jote and paid Rs.88 for it. In Radhanagar village, there were five huzuri jotedars who held 3,348 bighas and paid, separately from the village, Rs.1,549 as their rent. The sizes of their respective jotes were 1,014 bighas, 810 bighas, 623 bighas, 231 bighas and 61 bighas.[40]

In the district of Rajshahi also we find that the wealthy raiyats held large holdings and exercised great influence in the countryside. The Collector of Rajshahi, Peter Speke, observed that 'Head Mundles are become the real masters of the land and that the first object with the zamindar should be to affect a gradual reduction of their power'.[41] He remarked in another letter:

Such of the wealthy ryots of Rajshahi as were holding lands in Khosh Bass [residing at pleasure] were men of some substance and took large tracts of lands and who both from the respectability and from their being obliged to hire labourers are allowed easier terms and to pay in one sum.... This valuable class of tenants, it would be one certain means of bringing much of the waste land into cultivation and turning money to that mode of engage-

[36] The meaning of *calleah* is not very clear. It may be 'Arable land in general, from its usually being of a dark colour: black soil or mould, cultivable land of a superior quality', H.H. Wilson, *Glossary of Judicial and Revenue Terms* (reprint, 1968), p.252.
[37] Proceedings of Committee of Revenue, 10 June 1784, no.19, WBSA.
[38] Ibid., no.21.
[39] CRP, 5 June 1784, nos.42-50.
[40] HR.
[41] BRP, 24 June 1788, no.20.

ment.[42]

We may take some examples also of the principal raiyats and the mandals in Birbhum district, though it is outside our study area. Modun Mandal was a mandal of his village, Cotoolgossah, and held 54 bighas of land at a jama of Rs.52-9. His property was attached by the gomastha of the village for a balance due from the village amounting to Rs.250; this property consisted of 150 maunds of paddy stored in his own house and 150 maunds more deposited in the house of Binood Chakerbutty, along with 2 jars of rice (about 12 maunds), 15 maunds of jaggery (gur), 2 maunds of cotton, 2 brass plates and a brass pot, 12 maunds of paddy for seeds, 7 cawns of straw, and fish from his tank. All these were worth, at the market prices of those days,[43] about Rs.127 exclusive of fish and the brass utensils.[44] His several bullocks and ploughs being added to the list, the total amounted to nearly Rs.150. This was probably less than principal raiyats would have possessed on average, considering that Binood's landholding was rather small, and that he would no doubt have concealed some of his valuable articles such as silver, ornaments, before they were attached.

Among others in Birbhum, was Cossinaut Surma who was once the accountant (patwari) of his village, Aditpore. He held 49 bighas of land for which he paid Rs.63-6. Lackeen Samant held 121 bighas of land at a rent of Rs.108-12. He was one of the ringleaders of the disturbance in 1783.[45] Ruttun Mandal of Bongang village held 118-11 bighas of land at a rent of Rs.133-6. He was one of the principal raiyats of Hudah (a smaller unit of the estate consisting of a few villages, which was often farmed out to rent-farmers in a similar way to kutkina of Dinajpur). Kabil Mohammed was a mandal of his village, Sonseel. He formerly held 436 bighas of land 'which being more than he could cultivate, he let it to some other ryots who paid their rents to him'. He paid Rs.213-

[42] Ibid., 13 April 1787, nos.88-89.
[43] For prices of rice, paddy and gur, see 'Price Current' of Birbhum and Bishenpur from May to October, West Bengal District Records, Birbhum (1786-97 and 1855), pp.19, 43. For cotton, Rs.5 per maund would be a fair estimate. Buchanan, *Dinajpur*, p.200.
[44] CRP, 11 August 1783, no.6.
[45] CRP, 11 August 1786, no.6, ibid., 15 September 1783, nos.23-31.

9 for the land.[46] As the total jama of the village was Rs.481, he occupied nearly half of it.

In the district of Purnea also we come across such wealthy raiyats. They often farmed villages and accumulated lands through various artifices. The following examples illustrate the size of their holdings and their relations with the poorer families. In the pargana of Dhurrumpore, there was a big village with a jama of Rs.6,800, that was ruled by 15 or 20 wealthy raiyats with about 400 bonded labourers and under-raiyats.[47] In the pargana of Usjah, one Seromun Raiot almost doubled his holdings in less than ten years (from 432 to 720 bighas between 1780/1 and 1788/9).[48]

The economic position of the wealthy raiyats largely explains why they could act as local leaders. It was also important that they played a role in the zamindari administration. Zamindars tried to control the villages through their principal raiyats. In districts such as Dinajpur, Purnea and Birbhum, such village leaders often acted as rent farmers; they were known as *kutkinadars*. In other estates, like Swaruppur, Coondy and Bitterband, the zamindars employed their own people (village amin or village gomastha) to assess and collect rent, but they too transacted business in collaboration with the mandals, paramaniks and pradhans, who represented the villagers. The office of village head (mandal) tended to become hereditary and to be held by the most influential family which was supposed to be the oldest and descended from the village founder.[49] As far as revenue matters were concerned, their authority in the village was nearly absolute. In answering a question from the Collector, the Sub Recordkeeper (*naib kanungo*) of Rangpur, Hurram, replied:

Should ten ryots of a hundred come and prefer a complaint without previously acquainting either of the Kurrumcharries and Principal ryots, and get any cause settled and sign the settlement themselves, the kurrumcharries and

---

[46] Ibid.

[47] BRP, 1 February 1790, no.36.

[48] See Section 3 of this paper. The source has 1188 to 1196 Mulki Year, an era then current in Purnea;1188 M.Y. = 1187 B.S. = 1780/1 A.D.

[49] PCD, 1 December 1778; CRP, 3 April 1786, nos.3-49.

the Principal ryots can and have the authority to set aside their signature.[50]

Such authority might often derive from the fact that the principal raiyats were responsible for the collection of rent. However, they could not allocate the rent demand between the villagers as that responsibility lay either with the village amins or with the rent farmers. Profits could be derived from the process, which would account for the eagerness of the principal raiyats to become local rent farmers.[51] In Purnea, this authority was held by the village farmer-cum-principal-raiyat.[52]

The wealthy raiyats also played important role in the reclamation of the waste lands,[53] and more generally as suppliers of agricultural loans. Buchanan's statement concerning the conditions in Dinajpur is worth quoting:

It may indeed be said that their stock carried out at least half of the whole cultivation of the country, most of the adhiyars and small farmers are more indebted to them than the whole value of their stock...the wealthy farmers advance them grain to eat. It is they who even furnish the seed...the landlords do not like this class of men, but it is evident that they are absolutely necessary unless the landlords themselves advance the money to their necessitous tenantry, but it is only practicable in very small estates.

He further observed that,

a rich man, in place of a capital which can be realised, acquires a number of necessitous dependents, to whose wants he must administer, in order to procure a share of their labour in place of interest and these dependents are reduced to perhaps one of the worst kinds of slavery, that of insolvent debtors.[54]

Heatly and Colebrooke found a similar situation in Purnea:

The ryots are in general very poor and the aid of government absolutely requisite towards enabling them to enter upon cultivation. In cases this is withheld, they are under the necessity of borrowing at a heavy rate or engaging themselves to others in more affluent circumstances.

[50] CRP, 10 June 1784, no.26.
[51] BRP, 8 July 1788, no.17.
[52] Ibid., 1 February 1790, no.36.
[53] Taniguchi, 'Agrarian Society', Section 1-(2) of ch.5.
[54] Buchanan, *Dinajpur*, p.235.

and 'Their [the principal raiyats'] readiness to assist the poorer ryots with loans is the cause which had hitherto retained them in subjection to the principal ryots.'[55] Such a situation was seen extensively in many parts of Bengal. The reports of the Collectors and Commissioners from the districts of Rajshahi, Rangpur, Nadia, Jessore and Tipperah tell the same story explicitly or implicitly.[56]

Loans to needy raiyats were also provided by wealthy residents in the capacity of under-farmers (kutkinadars). One Colley Chand Doss, an under-farmer of Curkadom village in the Panjara pargana in Dinajpur zamindari, 'advanced Taccavy and greatly improved the village'.[57] In Surjapore pargana in Purnea, Mahommed Warris held in farm Paunkuy village where he resided. He 'paid the malguzary without balance, and advanced money to the raiyats, and supplied them with seed for cultivation'.[58] Thus, wealthy raiyats supplied loans to needy raiyats as their rich neighbours, or to their under-raiyats as their masters, or to villagers in general as village rent farmers. Such necessary loans offered a means of great influences.

Wealthy raiyats, as the Collector of Dinajpur observed, were the main suppliers of the grain to the merchants who sent grain from the villages to the towns. They received considerable quantities from their under-raiyats and kept it in their store-houses (*golahs*). They also purchased grain from their neighbours and sold it at the most advantageous time. They hoarded grain for years in their golahs which might well serve as a stock in the case of scarcity.[59]

The small local leaders played a significant part in agrarian disturbances, to which there are many references in records pertaining to the late eighteenth-century Bengal. To identify general characteristics of such disturbances, we may take as a representative case resistance by the

[55] BRP, 24 April 1787, No.48, ibid., 18 June 1790, no.8.
[56] Ibid., 29 June 1792, no.15 (Rajshahi); 22 March 1790, no.15 (Rangpur); 10 April 1787, no.4, and 18 February 1793, no.31 (Nadia); 11 July 1788, no.4, and 19 June 1793, no.7 (Jessore); and 7 May 1789, no.35 (Tipperah).
[57] PCD, 31 March 1778.
[58] Ibid., 14 March 1780.
[59] Taniguchi, 'Agrarian Society', Section 3 of ch.6.

raiyats of Surjapore pargana in Purnea, each January between 1776 and 1779. At this time, 'when they are to pay their rents', the raiyats would leave their homes, and 'they assemble in a body and repair to the Sudder where they prefer false complaints of oppression..., hoping by this means to avoid paying their rents'.[60] In 1779, the raiyats went as far as Calcutta to complain of the oppression of the zamindar, and obtained an order in their favour. They 'paraded through the pargana with the above letter'.[61] The interesting point here is that the leaders of these disturbances were the under-farmers of the pargana, and that most of them were the principal raiyats of the villages. A similar disturbance occurred in the district of Birbhum. The Collector reported:

such disturbances become almost annual custom for the ryots headed and excited by those Mundoles (in whom they cannot discover the latent cause of every imposition) to assemble in arms, in the month of Augran and Poose (November, December and January) and put a stop to the collection till they have brought the farmers to terms. And the revenue...can never be realised without the presence, at this time, of a military force to check such commotions.[62]

These two instances show that the principal raiyats or mandals often organised disturbances at the time of the heaviest collection (December and January) by assembling the inferior raiyats and thus bargaining for better terms. The inferior raiyats, the Collector of Birbhum suggested, scarcely gained by their participation.

The raiyats of Kajirhaut (Carjeehat) pargana in Rangpur 'have long been famous for their turbulence of their dispositions'.[63] In 1761, they rose in arms against the exaction of Nawab Mir Kasim.[64] In the great uprising of the raiyats of Rangpur against the public farmer, Devy Sing, in 1783, Kajirhaut was one of the centres of the rebellion.[65]

[60] PCD, 16 January 1776.
[61] Ibid., 1 March 1779.
[62] GG Rev, 10 February 1790, no.49.
[63] BRP, 24 October 1786, no.15.
[64] Fifth Report from the Select Committee of the House of Commons on the Affairs of the East India Company, 28 July 1812, vol.1, p.16, Para.62.
[65] N. Kaviraj, *A Peasant Uprising in Bengal, 1783* (Calcutta, 1972), pp.20, 21, 34.

Again from 1786 to 1788, they joined in disturbances three times.[66] In April 1788, 'the ryots of the district of Soobaugunge (in the zamindari of Kajirhaut) had assembled in a tumultuous manner and setting all authority at defiance, had attacked the Cutchery of that place and declar-e[d] that they would pay no more revenue'. On examination, the Collector found that the cause of the disturbance was 'the intrigues of the Busneahs [head ryots] who were not satisfied with the deduction' from the revenue demand granted to them on account of the inundation of 1787. He argued that 'the [ordinary] ryots had no cause of complaints whatever, but that they were betrayed into their present excess by the intrigues of the Busneahs'.[67] In this case, one Dewannoo was the ringleader and seven or eight Busneahs with a large body of their raiyats had participated in the action at the instigation of 'the Civil Council of Dewannoo'. The Collector of Rangpur wrote in another letter that go-masthas of the zamindar often entered into 'intrigues with the Head ryots, who have the lower class entirely at their devotion'.[68]

In the following three instances, we can once again observe principal raiyats leading lesser raiyats in agrarian disturbances in order to obtain particular objectives. In 1788, the raiyats of Bangong pargana in Raj-shahi demanded a reduction of Rs.4,000 in their rents, as had been granted in the preceding year on account of the disastrous season of 1787. At the instigation of 'the munduls, [they] have dispossessed the Farmer now full ten days and yet cannot be brought to obedience'. The Collector did not apply for the assistance of a military force, knowing from the past experience that 'the Munduls will keep themselves out of danger', while they would 'urge on the mob [so as to] compel' the sepoys to fire. And:

if the arms are used, and some people killed, the pargana is deserted, the crops are carried away. The loss of the lives makes that of revenue only of no consideration. Enquiry ceases, the leaders escape, and their end is obtained... The object is to lower the Jumma and as long as a Tahud Millan-nee Hastabud [assessment on mutual agreement of both parties] continues,

[66] BRP, 9 May 1788, no.27.
[67] Ibid.
[68] Ibid., 24 October 1786, no.15.

when once lowered, it is extremely difficult to raise it.[69]

Such disturbances had frequently occurred in many parts of Rajshahi since 1781 when Lieutenant Kinlock's party of sepoys fired at armed raiyats and many lives were lost. At that time the raiyats had risen against the exactions of the public farmer, Nundlol Roy. This incident seems to have shown them their potential power. It led to the ousting of the prestigious public farmer who had enjoyed the full support of the Company government. After this, the raiyats of Rajshahi employed gunmen (*barkandazes*) who were paid either from voluntary contributions by the raiyats or from forcible collections by the mandals.[70]

In 1788, J. Sherburne, the Collector of Birbhum, tried to introduce a new rent-roll (*jamabandi*). It was designed to lessen the rent burden on the inferior raiyats, which, of course, involved an increase of the rent of 'mundules and superior orders' because of the equalisation of rates of rents. The latter formed a powerful combination and set the whole of the district in arms.[71] The mandals summoned the raiyats of the district and selected representatives who were authorised by a bond (*mooktayar nama*), signed by 200 mandals, to plead the raiyats' point of view with the government. A complicating feature was that some of the lately dismissed officers of the zamindari supported the rebels. The Board of Revenue was, at last, obliged to remove Sherburne, the Collector, and appoint C. Keating as his successor to pacify the country. The new jamabandi established by Sherburne remained in force only for a short while and was totally ineffective, largely because of the intrigues of the mandals and the 'superior orders'.[72]

The zamindari of Baharband in Rangpur was granted to Lockenaut Nandi (alias Cantoo Baboo) at a very low assessment in 1774. The 'ancient ryots' who claimed to be the descendants of the original cultivators of the locality started an agitation against the zamindar in 1786. They entertained 400 to 500 armed men and expelled the zamindari offi-

[69] Ibid., 24 December 1788, no.27.
[70] Ibid.
[71] Ibid., 17 February 1789, nos.43-8.
[72] Ibid., 25 July 1791, no.12; ibid., 25 September 1795, no.32; West Bengal District Records, Birbhum, 22 August 1795, pp.108-9.

cers from the pargana. The occasion was an attempted general measurement of lands by the zamindar. The raiyats asserted that their rent on lands which they held as *khudkasht* raiyats had been fixed for many years and was not liable to any alteration through measurement. They went as far as to claim a kind of proprietary (*talukdari*) right over their lands. The zamindar argued that he had the right of measuring land at any time he pleased, and he rejected the raiyats' claim, saying that under the custom of the country the raiyats could never acquire a talukdari right. He also added that the poorer raiyats welcomed the measurement and that it was only such wealthy raiyats as had been enjoying a privileged rent who opposed it. Despite such chances of benefit the poor could not foil the wealthy raiyats' attempt to prevent it and, indeed, participated in the violent action against the zamindari officers.[73]

The following instance of the disturbance in Shupore pargana in the zamindari of Rocunpore in the Murshidabad district seems to have been one of the most striking among the agrarian disturbances. Among others, Soupan Sha, an inhabitant of Mouza Parsan, pargana Shupore, had collected 200 to 300 gunmen and footmen (*paiks*), and, for several years past, had been 'constantly throwing obstructions in the way of collections'. In 1784, he assembled a large force and dispossessed the zamindari officer (*naib*) of his office and turned him out of the pargana. Soupan Sing fought a battle with a party sent for him under the direction of Peraun Mudgemdar, a zamindari officer. Three or four men were killed and the pargana was desolated as a result. For five months, he defied the authority of the zamindar by force. Due to his obstructions a revenue balance of Rs.22,200 remained due from the pargana from 1779 to 1783. He himself held Parasan, Chuckgong and other mouzas in farm, for which he paid Rs.2,000. In the pargana, he assumed the authority of holding a 'punnya' (the annual ceremony for the settlement of the new year's rent and the start of cultivation). He reinstated a former zamindari naib in the office, and appropriated Rs.4,100 collected from the raiyats. The Committee of Revenue decided to assist the zamindar in retaking possession of the pargana; the Collector of

[73] CRP, 3 April 1786, Nos.43-9; BRP, 16 June 1786, nos.30-1.

Shilberris sent a party of sepoys, and put an end to this pretended local authority.[74] It had been an agrarian disturbance headed by raiyats in combination who retained several hundreds of armed men. They successfully defied the authority of a zamindar who was by no means small or weak zamindar. They completely controlled the locality for five months or more. The incident shows that some of the superior raiyats were willing and able to supersede the authority of the zamindar, albeit for a short period.

We may conclude for the records that the agrarian disturbances of the late eighteenth century were, without exception, led by small local leaders, especially wealthy raiyats, mainly for the purpose of protecting or increasing their own economic interests at the cost of both the zamindar and the inferior raiyats. By exercising their influence the principal raiyats persuaded their inferiors to take up arms. They thus improved the conditions on which they held their lands, and increased or consolidated their privileged holdings in many parts of Bengal. Attempts by the zamindar or the government to undermine their position, they furiously resisted. The frequent agrarian disturbances in the late eighteenth century were in fact part of a wider historical process: the increase of the wealthy raiyats' privileged holdings and a greater differentiation within the section of small self-sufficient raiyats.[75]

The wealthy raiyats constituted about 20 per cent of the peasant population and the small self-sufficient raiyats and raiyats below the level of self-sufficiency constituted about 80 per cent. What were the economic relations between the two?

Contemporaries were nearly unanimous in accusing the wealthy raiyats of exploiting the poorer raiyats. The following extract from a letter of the Collector of Nadia in 1787 might be taken as an example:

The high class of ryots now find it their interest to be employed by the renters as the instrument of their oppression, who living at ease, and enjoying particular indulgences upon the great body of the indigent and industrious ryots whose general poverty and the abject subjection they are kept in by their oppressors are insuperable obstacles to their extricating them-

[74] CRP, 16 August 1784, nos.7-8; ibid., 9 December 1784.
[75] Taniguchi, 'Agrarian Society', Section 1-(2)-d of ch.6.

selves from their present unhappy condition.[76]

Manipulation of rent-rates by the wealthy could place extortionate demands on the poorer raiyats, though this was not the only cause of discriminatory rates in favour of the 'superior orders'. As the Collector of Murshidabad observed, the practice of giving rent-free lands to Brahmans was regarded by the people as 'far from criminal',[77] and it was also not unpopular that high-caste Hindu raiyats should be exempted from the payment of impositions and expenses.[78] The exemption was necessary, the high-caste raiyats and the 'superior orders' asserted, because of the caste ban on their holding the plough: to maintain their respectability, they had to hire labourers to cultivate their lands.[79]

On the other hand such manipulation was exercised to an extent unjustified by cultural considerations. The Collector of Birbhum observed in 1787, that, on the inspection of the detailed accounts of the villages:

it will be found that the Mundoles with their connections, and the wealthy ryots have collusively sunk one half of their Jumma and that they are too frequently exempted from the payment of the local charges of collection which are consequently ammulated [sic] on those who are least able to support them...this extensive abuse...has been gradually increasing for many years probably from the Company's accession to the Dewanny.[80]

Similarly the Collector of Murshidabad wrote in 1793 that, on comparing the jama of each raiyat, 'so great a difference appeared in the rates of assessment owing to the higher class of ryots being entirely exempted from all abwabs'.[81] The Commissioner at Rajshahi reported in 1792 that 'from the known influence of the Munduls and other principal ryots the general rate of the rents paid by them is reputed to be much lower than that paid by the inferior ryots'.[82] Buchanan also found the same situation in 1807 in Dinajpur. The zamindar won over the wealthy raiyats by special favours in order to impose abwabs on the poorer tenantry.

[76] BRP, 10 April 1787, no.4.
[77] Ibid., 12 May 1795, no.51.
[78] Taniguchi, 'Agrarian Society', Section 1-(3) of ch.1.
[79] BRP, 25 July 1791, no.12; Ibid., 17 December 1793, no.28.
[80] Ibid., 28 August 1787.
[81] Ibid., 17 December 1793, no.28.
[82] Ibid., 17 December 1792, no.17.

The wealthy raiyats persuaded the poorer ones to accept the zamindar's terms, and the latter seldom protested because of their unavoidable dependence on the former for loans.[83] In view of the fact that impositions and expenses formed about a half to five-sixths of the total rent obligation,[84] we can see how striking the difference could be between the total rent obligation of the higher and poorer raiyats.

The wealthy raiyats sometimes had recourse to more objectionable practices. The Collector of Rajshahi observed in 1790:

Many ryots in this district through collusion with the Mofussil Currumcharries [local officers] have caused a considerable part of their Jumma lands to set down in the accounts as Pallateeka or forsaken and fresh leases at the lowest rate of assessment to be granted for those lands to their own dependents...several individuals of substance have by collusive means obtained remissions of revenue which the lower class of ryots are obliged to make up for by contribution.[85]

Colebrooke wrote of similar practices in Purnea. In the pargana of Sirreepore, the principal raiyats succeeded in reducing their rent through dishonest means and, to make up for the deficiency thus created, levied an imposition of 4.5 annas in the rupee on the basic rent (assal jama) of the raiyats in the pargana. Since the basic rent of the poorer raiyats was already disproportionately high, the additional burden was intolerable for some of the raiyats. They deserted the land and the assets of the pargana were thus reduced.[86]

The zamindars often extorted increased payments by threatening the raiyats with measurement of their lands. Even in such cases wealthy raiyats could take advantage. Colebrooke reported that in Sultanpor pargana they made a compromise (bonaup) with the farmer whenever he tried to measure the pargana by accepting a special additional rate for the whole village. The majority of the raiyats had to agree, mainly because of the threats and persuasion of the principal raiyats. Given that inferior

---

[83] Buchanan, *Dinajpur*, p.236.
[84] BRP 22 March 1790, nos.15-16, Appendix D; Proceedings of the Superintendent of the Khalsa Records, 26 July 1776, no.134, WBSA.
[85] BRP, 28 May 1790, no.8.
[86] Ibid., 18 June 1790, no.18.

raiyats paid more and probably concealed less additional land, they presumably felt the increase more severely, and again some were compelled to desert the village.[87] In Birbhum district we have a similar instance:

In 1187 B.S.(1780 A.D.), a measurement was made by Mohammed Wooly, the zamindar's dewan, as the basis of an equal and general assessment, but this the Mundoles and superior ryots evaded by a contribution and new taxes of Ootubundee Nerick Beshee[88] which were made to fall on to the lower class of ryots were substituted.[89]

The rent-farming system, in which wealthy raiyats figured prominently at village level, became another source of misery for the poorer raiyats. Sherburne, the Collector of Birbhum, observed in 1788 that 'The mundoles have proposed to be responsible for the Jama of the respective villages, provided they were permitted to regulate the assessment.'[90] Thus they would have nullified the effects of the new jamabandi (revenue settlement) which he was about to introduce. Heatly, the Collector of Purnea, was more candid in saying that 'a wealthy ryot taking a village in farm, by oppression, forced the poorer cultivators to desert their grounds which were suffered to be waste for one or two years, and then by a bribe obtained on low rated pottahs'.[91] In the same district the alternative of *lokhbhara* (revenue settlement with the raiyats collectively) also adversely affected the poorer raiyats. Such settlements were made when the zamindar failed to find suitable persons as rent-farmers for the parganas or mahals. The responsibility was supposedly collective, but wealthy raiyats would pretend to make an equitable distribution of the assessment: they would measure the lands and overrate the land resources of the poor. The wealthy would gain, and the poor would suffer, under the lokhbhara.[92]

The consequences of such an economic relationship are underlined in a brief remark by Sherburne. In the district of Birbhum 'the lands

[87] Ibid., 7 July 1790, no.30.
[88] *Utbandi nirikh beshi* probably means an increase of rent on such lands where cultivation actually took place.
[89] GG Rev, 10 February 1790, no.49.
[90] BRP, 8 February 1788, no.17.
[91] Ibid., 1 February 1790, no.36.
[92] Ibid., 18 June 1790, no.18.

collusively held by the Mundoles, belonging to the ryots they have excited to desert...', he noted, 'such lands are held under the mofussil term, Benamy (in fictitious name), or Pyecast (non-resident), with the usual indulgence to such ryots.'[93] As Colebrooke explained, more elaborately, in regard to the parganas of Dhurrumpore, Usjah, Sultanpore and Cuttiar, the wealthy caused the poorer raiyats to desert their rented lands, which were eagerly taken on by their 'fortunate neighbours' at the low rates usual for abandoned holdings. This being done, the expelled raiyats were recalled to cultivate their former lands, not any longer as the *pattahdars* (raiyats who had *pattahs* or leases), that is 'the registered raiyats', but as under-raiyats. In this way stronger raiyats elevated themselves to the status of an intermediary tenantry. Colebrooke concluded: 'the accumulation of the land in the hands of a few wealthy ryots made a rapid progress through their such artifices'.[94]

The statement (in Table 8) of the holdings of several raiyats in Purnea shows how rapidly their holdings expanded over 20 years.[95] The total quantity of lands held by these 8 raiyats increased 2.3 times within 20 years, while their total rent (jama) slightly decreased. Thus, in 1771 they paid on an average 6 annas 2 gandas per bigha but in 1790 they paid only 2 annas 9 gandas.

As lands were accumulated by a few wealthy raiyats, there were fewer small self-sufficient raiyats (our second category) and larger numbers of under-raiyats and agricultural labourers. The decline of small self-sufficient raiyats was inevitable in the context of an acute shortage of labour. Mere accumulation of lands did not mean much, without the labour necessary for their cultivation. The wealthy raiyats needed to turn poorer raiyats, the only source of such labour, into dependant agricultural workers. They organised these workers to cultivate the privileged holdings which they accumulated. Colebrooke wrote:

they (the wealthy raiyats) have been annually adding to their quantity of land, taking in Waste lands, as they have been able to encrease the number of their under Raiots, or by various arts compelling the poorer Raiots to

[93] GG Rev, 10 February 1790, no.49.
[94] BRP, 7 July 1790, no.30.
[95] Ibid., 7 July 1790, no.32.

resign their own Pottahs & cultivate for their benefit.

He further observed,

A combination of a few raiots whose union and wealth give them great power & influence in the pargana has completely established this system within no very distant period. The consequence of which is that the cultivator instead of paying the moderate & equitable Rates which government would exact either pays to an intermediate tenant, a half the produce as his Addyadar, or a high Nuckdee rent (money rent) as his Colait Raiots....[96]

III. *The management of the raiyati holdings*

The three levels of the peasantry of northern Bengal in the late eighteenth century each had different ways of managing their holdings in relation to the fundamentals of agricultural production (land, labour and agricultural stock).[97] For convenience, we study first the management of holdings in our second category, those of the small self-sufficient raiyats, who constituted about one-sixth of the peasant population.[98] Most of the holdings of this class came under regular local settlement and formed part of the village. They had to pay full rent and other impositions. In some localities, the mode of assessment was *fasli* (according to the crop cultivated); in others it was *tahudi* or *thika* or *hari* or *chukani* (according to the contracted quantity of land). In the former case, the raiyats had to pay at a different rate if they changed what was cultivated.[99] In the latter, the raiyats could cultivate any crops they chose. Not infrequently, both types of assessment co-existed in the same village and the raiyats possessed both at one time.[100] Cultivation was, as a rule, by family labour. The owners of these holdings had the minimum in resources needed to cultivate without taking loans from rich neighbours or merchants.

[96] Ibid., 18 June 1790, no.8.
[97] See Taniguchi, 'Agrarian Society', ch.5.
[98] See Section 1 of this paper.
[99] BRP, 6 November 1792, no.10A.
[100] In Swaruppur we find the coexistence of ordinary raiyati lands under regular local settlement and chukani lands with fixed assessment; ibid., 22 March 1790, No.15. In Nadia also a raiyat possessed both the ordinary raiyati lands and *hari* lands with fixed assessment; ibid., 5 August 1794, nos.10-16.

What was this minimum? Two different calculations should be made; firstly, when a raiyat started cultivation as a new settler, and secondly, when he was established as a settled raiyat and possessed a house and agricultural implements. Buchanan observed that a new settler should have agricultural stock as set out in Table 9.[101] His total seems to be an over-estimate. Buchanan wrote elsewhere that a humble house would cost 9 annas if the raiyat built it himself.[102] Similarly, the expenses on the ornaments, furniture and clothing could be reduced to Rs.3-11. The expense of food for six months would be Rs.8-5 according to his own calculation of the diet for the family of a common labourer.[103] The above sum is thus reduced to Rs.29-6 or about Rs.30.

The settled raiyat required only such stocks as needed to be supplemented or replaced annually. The replacement cost of durable implements would be about one rupee per annum. Most of the implements of a perishable nature would have been produced on his holding, and only those which required skill in manufacturing would have been purchased. For the replacement of plough-oxen, he might save a rupee a year. Seed was available from the previous year's paddy. For the repair and replacement of the house, furniture, ornament and clothes, Rs.3 would be enough. Generally, a raiyat owned both clay (*khiar*) and loam land (*pali*). The latter was capable of producing the early crops like spring paddy (*boro*), summer paddy (*aus* or *bhadai*), and cotton (*kapas*). The sale of these crops would have paid rent which fell due during the early part of the year, and also furnished the family with food until the winter paddy, which was the staple crop, was harvested in November, December and January. The stock needed by a small self-sufficient raiyat at the beginning of the year, would thus be less than Rs.6 (half of the replacement cost, half of clothes and seed). A settled raiyat, therefore, could continue his cultivation from year to year with a small fund of Rs.6 (including seed) or Rs.2-10 (excluding the seed).

Now let us examine how holdings of this kind were utilised. The following examples are taken from Harington's report on the zamindari

[101] Buchanan, *Dinajpur*, p.237.
[102] Ibid., p.130.
[103] Ibid., p.131.

of Swaruppur. Munsoor, a raiyat of Hurrupoor, held 5-10 local bighas of land and cultivated summer paddy and mustard. Kaloo of the same village held 6 local bighas of land and cultivated summer paddy and mustard.[104] Dyaram, a raiyat of Roganatpoor, was a fairly big cultivator possessing 18 local bighas. His tahud jama was Rs.32, but, in fact, he paid Rs.50 in 1788 (1195 B.S). The main products of his land were paddy, mustard and *kalai*.[105] He also grew betelnut (*pan*), sugar-cane, mulberry, cotton and tobacco but to a very small extent. He stood on the border of the second and the third class of raiyats.

Harington prepared a statement (Table 10) showing, for a typical holding, the standard varieties and quantities of the several articles of cultivation, on the basis of the information collected during his survey in the Swaruppur zamindari. The table shows that the gross cropped area of this typical holding was 21 bighas, considering the double cropped land. Of the 21 bighas, 14 bighas were devoted to the production of paddy (winter paddy 8 bighas and summer paddy 6 bighas), 2 bighas to mustard, 2 bighas to *masur*, 2 bighas to *thakuri kalai*, and 1 bigha to *arahar* (the last three being kinds of pulse). The summer paddy was of inferior quality and mainly used for home consumption. After a deduction for home consumption, there was a surplus which was sold locally. Most of the winter paddy must have been sold, as it was of fine quality and fetched a good price. Table 11 shows the average produce per bigha and the price of several articles in the zamindari of Swaruppur.

The holding is not shown as producing sugarcane, tobacco, cotton or jute, but it is probable that it did so, for home consumption, given its size (16 local bighas). Similarly, almost all kinds of vegetables would have been produced as garden crops near to the dwelling-places. Buchanan found that 'a man who has a plough cultivates a certain extent of land, in which he has a proportion of many of the different grains above mentioned' (that is to say, paddy, barley, pulses, oil-seed, cotton, and so on, in rotation).[106]

Raiyats at this level produced almost all the necessities of life on

---

[104] BRP, 22 March 1790, no.15.

[105] *Kalai* is a general name for common pulses.

[106] Buchanan, *Dinajpur*, p.185.

their own holdings. They did not have much use for cash, except to purchase iron implements, salt, and cattle, and to pay rent which was the largest item in their budget that was disbursed in money. Let us suppose that they paid Rs.1 per local bigha inclusive of all charges. The holder of a typical holding (as above) would pay Rs.16. The sale of winter paddy from eight bighas would have brought in Rs.28 which was more than enough to pay the rent and all other household expenses. Six bighas of summer paddy would produce on average about 44 maunds of paddy or 32 maunds of rice—enough for the family's home-consumption for one year.[107] In addition to the income from the sale of winter rice, this raiyat could expect a certain amount of money from the sale of surplus portions of the oil-seed, pulses, vegetables, tobacco, cotton, and so on produced in his holding. In short, a raiyat holding 16 local bighas could get along with ease, and accumulate a certain amount of capital if he was cautious in his expenditure. He could thus take up extra lands using additional labour.

There is no direct evidence on the way the holdings of small self-sufficient raiyats were managed. Let us suppose a holding of 8 local bighas cultivated as in Table 10. In this case, the sale of winter rice would have brought in Rs.14, out of which Rs.8 would be paid as rent and about Rs.3 set aside for forthcoming expenses of cultivation. The summer rice (16 maunds) would have been insufficient to meet the needs of home consumption. Therefore, the raiyat would have had either to consume part of his winter rice or to purchase coarse rice from the local market; and he must generally have preferred the latter. The necessary 9 maunds of summer rice (13 maunds of paddy) would have cost Rs.1-12, while the sale of the same quantity of winter rice would realise Rs.4-5. But all the other necessities could be supplied from a holding of 8 bighas. Such a raiyat could also save money by using ashes instead of salt. Thus he was not in such an easy condition as a cultivator with 16 local bighas, but he could pay his rent and get along without difficulty. He could even expect a small surplus, in addition to the reserves he

---

[107] This family presumably contained two male adults, because the size of land held by them required two ploughs for cultivation. Therefore, 32 maunds of rice were just enough for its home-consumption.

needed for the ensuing year.

In our first category of raiyats were those who fell below the level of self-sufficiency, and who constituted more than half of the peasant population. Generally, they possessed the minimum agricultural stock; they had a plough and a few bighas of rented lands where they built their houses and grew summer paddy, mustard, pulses, vegetables, jute and so on. They had to depend for subsistence on loans from their neighbours, a circumstance which prevented them from renting more land, in spite of the abundance of arable wastes. In order to support their families, they had to work as hired labourers or cultivate lands belonging to other people on share-cropping (adhi) contract. When cultivating as share-croppers (adhiyars), they almost always borrowed seed and repaid it with interest at a rate of 100 per cent.[108] They were then called the prajas, which suggested a kind of subjection.

The following description of a raiyat in a pargana in eastern Bengal (Tipperah) , taken from Paterson's report on Baldacaul, gives us a vivid idea of the way the people of this type managed :

During the time he can spare from his own little field, he hired himself out as a labourer, whilst his wife employs herself at home in spinning of cotton. A Riatt of this description can not afford to pay above 2 cannies[109] of land at furthest, & to do that he must be allowed Tuccavi (loan). The produce of his land without the above occasional resources could not maintain him, especially if he has a family.[110]

We have good reason to believe that this description was generally applicable in northern Bengal. Buchanan also observed that the income of wives as cleaners of rice and spinners of cotton was indispensable for the subsistence of their families.[111]

Let us examine the income and disbursement of the raiyats of this class. We cannot depend on Buchanan's data, as he entirely ignored the produce from their small rented lands. Moreover his assumption that the

[108] Francis Buchanan, *An Account of the District of Purnea in 1809-10* (Patna 1928), p.445.

[109] One kani in Baldacaul was 130 sq.yds smaller than a Calcutta bigha which was 1,600 sq.yds or 0.33 acre.

[110] BRP, 7 May 1789, no.35.

[111] Buchanan, *Dinajpur*, p.244.

adhiyars would get half of the gross produce of their lands was unrealistic. For our estimate, we may choose a raiyat possessing 2 local bighas of rented land and 6 local bighas of adhi land. On the rented land, he would cultivate summer paddy as the first crop, and various kinds of pulses, oil-seed, vegetables, jute and so on as the second crop. On the adhi land he would cultivate paddy exclusively. Two bighas of shalee-lal (winter paddy only), two bighas of shalee-do (winter paddy and summer paddy) and two bighas of bhadai-lal (summer paddy only) would be a usual pattern for adhi land. From the rented land, the raiyat would get all the necessary pulses, green vegetables, jute and oil-seed for home consumption, and about 14.5 maunds of summer paddy. From the adhi land, he would receive one third of the gross produce; about 14 maunds of winter paddy and 10 maunds of summer paddy. He would sell the winter paddy and obtain Rs.4-11. The 24.5 maunds of summer paddy which he would get from both rented and adhi lands would have been inadequate for home consumption. Therefore, he would have had to purchase 11.5 maunds from the local market at a cost of about Rs.2-7. He would have paid Rs.2-8 as rent for his rented land, and 1 rupee for seed which was advanced to him.[112] He would have had to set aside Rs.2 for supplementing and replacing the worn-out implements and for the future replacement of his plough-oxen. For clothes, repairs to his house, ornaments, and so forth, Rs.3 would be sufficient.[113] To sum up, his total expenditure would be about Rs.10, while his income amounted only to Rs.4-11. The deficiency had to be made good by the earnings of his wife who, if she laboured very hard throughout the year, would earn Rs.7-12 from cleaning rice and Rs.2-8 from spinning. This family, thus, could get along with the income of husband and wife , but their's was a very precarious existence. The raiyat's wife would become pregnant every few years and then the husband's income alone would be far from adequate. He would be obliged to borrow from his rich neighbours. The repayment of such loans with what was known to be

---

[112] He received seed for the adhi land from its lender, and repaid the seed with interest at 100 per cent.

[113] Buchanan, *Dinajpur*, pp.130-21.

exorbitant interest,[114] would involve him in recurring debts.

Below the adhiyars, there were agricultural labourers (krishans) whose economic condition was still worse. According to Buchanan's statistics, they formed 18 per cent of the rural population. They had not enough stock to keep a pair of plough-oxen, and were unable to rent any land on their own account. They depended wholly on the income derived from wage labour. Buchanan calculated that their income totalled Rs.16-4 (Rs.6 from the man's wages, and, from the woman's, Rs 7-12 for beating rice, and Rs.2-8 from spinning). Their total expenses he estimated at Rs.22-11, of which 10 annas for clothing and Rs.5-8-9 in food were provided to the man as wages in kind. This left a balance of Rs.16-6-3 to be found from money income.[115] A krishan could thus earn just enough for subsistence, though the earnings of his wife again were indispensable. As almost all his food and other necessities had to be purchased, even a slight rise in the prices of grains affected him greatly. It would have been almost impossible for him to get along without loans from the rich neighbours who often also employed him.

Our final category was formed by the wealthy farmers who constituted about 20 per cent of the peasant population. They held 60 to 76 per cent of the raiyati lands, most of which were cultivated for them by krishan and adhiyars.[116] They often possessed different types of tenure at the same time. As we have seen, the huzuri jotedars in the zamindari of Swaruppur owned lands under the regular local settlement and also separately from the villages, assessed at privileged rates of rent. Table 12 provides a statement of the holding of a wealthy raiyat in Burdwan, Kishen Mandal. He held lands under four kinds of tenures. The khud-kasht land was presumably under the regular local settlement and paid the full impositions. Other lands were under different privileged tenures and paid much less. The apparently high pitch of rent was deceptive. It was a common practice for raiyats to hold more lands than were registered in the zamindari records. They were thus able to pay apparently high rates of rent. Kishen held 9-5 bighas under the regular local settle-

---

[114] See Taniguchi, 'Agrarian Society', Section 3-(3) of ch.6.
[115] Ibid.
[116] See Section 1 of this paper.

ment and 38 bighas under privileged assessments of various kinds, probably as a result of his efforts to evade payments at full rates.

The wealthy raiyats usually possessed the best, most productive land in the locality. For example, *shalee* land of the first quality produced, according to the survey of J.H. Harington, 15 maunds 2 seers per local bigha, the second quality 10 maunds 6 seers, and the third quality 7 maunds.[117] Thus one bigha of the first quality shalee land was more valuable than two bighas of shalee land of the third quality. The best shalee land produced 15 maunds of aman rice which could be sold for Rs.5, while the third quality summer land produced 4 maunds 36 seers of summer rice sold at only Re 1-1 only. Moreover, good shalee-do land could support highly profitable crops such as sugar-cane and mulberry.

The wealthy raiyats held more lands than they could cultivate using family labour. Therefore, they employed krishans and sublet surplus lands to needy raiyats on adhi contract. The loans advanced by the wealthy raiyats enabled them to procure labour at a much lower rate than would have been possible otherwise.[118] Possession of the best land at low rates of rent and access to labour at a cheaper price were the economic foundations of their large privileged holdings.

Now let us examine the nature of land-utilisation by the wealthy raiyats. Buchanan quoted three statements selected from among many submitted to him by raiyats. Table 13 summarises these data. As said above, some adjustments are needed. Buchanan deducted one half of the gross produce as the expenses of cultivation, whereas, according to the present author's calculation, it hardly exceeded one third. Secondly, the recorded rates of rents were provided by the raiyats themselves and Buchanan himself doubted their authenticity. We have just remarked how wealthy raiyats tended to pay at lower rates. Finally, though this shortcoming was not Buchanan's fault, agricultural prices at the time of his survey were considerably higher than those of our period, perhaps by more than 20 per cent. Thus Buchanan's raiyats obtained much higher profits than those of our period.

[117] BRP, 22 March 1790, no.15.
[118] This situation is still observable in Rangpur. See Taniguchi, *Society and Economy*, pp.87-91.

Table 13 suggests that various farming strategies were adopted by wealthy raiyats during our period. Firstly, the table establishes beyond doubt the importance of the cultivation of paddy. It accounted for 65 to 90 per cent of the cropped area of each holding, and constituted 67 to 95 per cent of the gross value of output. Its cultivation was heavier where the proportion of the clay was larger. Secondly, the quality of land had other distinctive effects. In holding (A), clay and loam occupied nearly the same proportion. As double-cropping was extensive on loam, the intensity of cultivation on this land (cropping intensity) was more than 160 per cent. A high level of utilisation had a good effect on the management of the holding. The value of the gross produce per labourer in this holding was the largest among the three shown, though the value of the gross produce per cropped bigha was the lowest. Clay land formed three-quarters of holding (B), and, therefore, the intensity of cultivation was very low. The produce was almost entirely paddy. However, the monocultural cultivation did not mean that management of this type of holding was disadvantageous. Thanks to a shorter period of cultivation, the cost of cultivation was much lower than for the other two holdings, and the profit per man-day was almost twice as high. In holding (C), the cultivation of only one bigha of sugarcane greatly increased profitability. Both the profit per cropped area and that per plough were the highest among the three, irrespective of a rather low intensity of cultivation (120 per cent). The value of output from one bigha of sugarcane land constituted 22 per cent of that for the entire holding of 30 bighas.

Thirdly, we can calculate the factor distribution of the gross produce of these holdings. In holding (A), one hired labourer produced Rs.57-8 per annum, while he received only Rs.18 in the same period. In holding (B), one hired labourer produced Rs.43-8 in 6 months, and received Rs.9 for his labour. In holding (C), one hired labourer produced Rs.54-8 and received Rs.18 in a year. Considering that Rs.18 was the highest estimate for an agricultural wage and that Buchanan was inclined to assess it at Rs.15,[119] the labourer's share seems to have been 17 to 28

---

[119] F. Buchanan, *Dinajpur*, p.244.

per cent of the gross produce.

Fourthly, Buchanan reported that these three holdings were cultivated by two family members and between one and three hired labourers. Though they were sufficient for the daily transactions, at the peaks of labour demand outside labourers had to be called in; this occurred at times of transplantation and harvest. If such additional labour was not easily available, the raiyats had no alternative but to sublet part of their holdings on adhi contract. Buchanan does not report this happening in the three holdings, and we may tentatively suppose that additional labourers were available when necessary. But it is very likely that wealthy farmers whose holdings were bigger than those indicated would usually have sublet the greater part to under-raiyats. We shall see such cases in the following sub-section. Fifthly, it is apparent that mustard and pulses were indispensable crops. They were grown mostly for home consumption, and otherwise would have been purchased in the market. From mustard, oil was obtained for cooking, ointments, and lighting.

As we have seen, wealthy raiyats often had to depend on under-raiyats for the cultivation of their extensive holdings. The following instances show the relationship between the holders of some privileged raiyati-holdings (*gatchdars*, huzuri jotedars and so on) and their under-raiyats (prajas and adhiadars) in Rangpur and Purnea. Such relationships were not confined to these few instances. In Dinajpur, Birbhum[120] and Jessore, we have evidence that similar kinds of relations existed. They occurred throughout Bengal wherever poor raiyats (those below the level of self-sufficiency) formed the majority of the population. As H.P. Foster, the Collector of Tipperah, remarked, of Bengal, '...near two-thirds of the land is cultivated by under-ryots'.[121]

J.H. Harington's report on Swaruppur gives us a detailed account of the relationship between the head raiyats (huzuri jotedars) and their under-raiyats (prajas):

It was a general usage that the Pirja or inferior ryot cultivated lands rented from the zamindar by Head Ryots on condition of giving a proportion of the

[120] On Dinajpur, see BRP, 11 April 1788, No.58C; on Birbhum see GG Rev, 10 February 1790, no.49.

[121] Ibid., 28 February 1794, no.32.

produce in kind. The head raiyats advanced as much grain as may be necessary to the Pirja for seed and receives from him double the quantity when the
crop is cut, the former takes the Bissoree [a deduction from the gross
produce to meet rent and other charges], or a certain number of Done per
Bessee, varying from 2 to 5 according to agreement on the whole remaining
crop [Bissorree itself excepted]. After which the surplus is equally divided
between both. Thus supposing the produce of a Bega of Ground to be 10
Bessee, 10 Done given account Seed & 3 Done taken for the Bissorree, the
Division would be:

	Bessee Done	
Total produce		10- 0
Receivable by Head Ryot		
Double quantity of Seed	1- 0	
Bissorree at 3 Done per Bessee on 8 Bessee (1 Bessee deducted account Bissorree)	1- 4	
Half of Surplus	3-18	6- 2
Receivable by Pirja		3-18.[122]

The above calculation shows that though the under-raiyats were sometimes called 'addea' (half), in fact they received only one third of the
gross produce. Harington observed that 'a considerable proportion of
grain lands in Sooroopoor is cultivated on these terms'. The prajas 'in
general had small spots of land from the zamindar, on which they reside
and cultivate Pulse, the Castor Oil, Shrub, and other articles at a rent
payable in money'.[123] The poorer raiyats thus held small plots of land
under the regular local settlement and cultivated a greater quantity as
under-raiyats on the share-cropping basis.

What was the profit of the head raiyats under this system? From
Table 4, we may take 8 annas per local bigha (or 4 annas 8 gandas per
Calcutta bigha) as the average rate of rent paid by the head raiyats in
Swaruppur. The gross income of the head raiyat in grain would be 6
bessee 2 done, from which should be deducted the zamindar's rent at 8
annas per bigha (1 bessee 10 done, at the rate of 3 bessees per rupee)
and the expense of the seed advanced to the cultivators (1 bessee). The
head raiyat thus obtained a clear profit of 3 bessee 12 done, plus interest
on the seed advanced, at 10 done (100 per cent)—in all, 4 bessee 2 done

[122] BRP, 22 March 1790, no.15. One bessee = 20 done = 1 maund.
[123] Ibid.

in his capacity both as intermediary landholder and as supplier of the seed. The paddy of 4 bessee 2 done was equal to Rs.1-5 at the rate of 3 bessee per rupee. The adhiyar received paddy equivalent to Rs.1-2. But he had to deduct from this replacement costs for his implements and bullocks, which would have amounted to 4 annas per bigha. Thus his net income was only 14 annas from the cultivation of one bigha of adhi land. If the zamindar demanded more rent, the number of the bissorree would be raised proportionately so that the burden of the enhanced rent should fall equally on the head raiyat and the under-raiyat. An independent raiyat could expect from the cultivation of the same quantity of land a net income of 7 bessee 9 done or Rs.2-8 after deducting the rent and costs of cultivation.[124]

Colebrooke's report on Purnea district showed a similar relationship between head raiyats and under-raiyats. In thinly-populated areas, especially in the foothills, a tenure called *gatchdari* prevailed. The gatchdar rented a tract of land from the zamindar at a fixed sum. Under the gatchdar, there were many under-raiyats called 'Addia Raiots, who pay in kind to the Gutchdar'. The adhiadars cultivated their lands on a verbal agreement with the gatchdar. The payment of rent was made 'not by an estimation of the crop valued at the market price, but by actual division of the grain'. Some customary deductions were made for the disbursement of common expenses before the division of the crop. The following deductions as found in the Surjapore pargana will serve as an example:

(1) The quantity of seed sown repaid twofold to the person who furnished it.
(2) Two Cottahs [a cottah = a twentieth of a bessee] per Bessee on the quantity remained (after the above deduction) to Etamamkar or person employed to watch the crop.
(3) Thirteen Cottahs to the blacksmith who repaired the plough.
(4) 0.25 Cottah per bessee to Kyal who measured off the crop.
(5) Niaze & Pajeh, expenses of sacrifices. Two Cottahs per plough at each harvest.
(6) One Cottah per plough for Dhaumee, fee for the Conjurer for protecting

---

[124] The total output was worth 10 bessee. Rent would be deducted at 10 annas per local bigha (given the higher rates for ordinary raiyats). There were 3 bessee per rupee.

the crop.

After the above deductions were made, the 'remainder of the crop is equally divided between the Gutchdar and Addi Raiot'.[125]

If we suppose that one bigha of land produced 10 bessee of paddy, the share of the gatchdar and adhidars would be 3 bessee 19 cottahs each. The rent to the zamindar was paid out of the share of the gatchdar, while the expenses of the community, the repayment of the seed advanced, and the repair of agricultural implements fell equally on both sides. Therefore, the adhidars here enjoyed better conditions than the prajas under wealthy raiyats in the zamindari of Swaruppur. But, nevertheless, an adhiadar in Purnea was 'obliged to borrow in the early months of the year for the maintenance of his family and receiving the loans in kind at usurious rates'. It was one of the objects of the special enquiry by Colebrooke to depress the high rates of interest on the loans given to the adhiadars by the gatchdar. The custom, in some pargana, was that the debtors had to pay 'half of the original loans if repaid from the Buddye [summer paddy] & a quarter if from the Augonee [winter paddy]'; in other parganas 'the loan [was] repaid two-fold'. Colebrooke observed that 'the Addidar's share of the produce is seldom adequate to liquidate his Debt which thus annually accumulating, he becomes enslaved to his Creditor, the Gutchdar'.[126]

In Purnea, there was another type of under-raiyat called *kolait*. They were found in the plains and more densely populated areas. To quote from Colebrooke's report:

I found that the Rents received by the zamindar and under-renters are not drawn immediately from the actual cultivators of the soil, but from a few principal Raiots who cultivate a small portion with their own ploughs & the remainder by Colait & by Addee Raiots. By the first are understood in this pargana (Usjah), the Raiots to whom the land is under-leased by the Pottah-holders on Neckdee Terms (or payment in cash). The Addyadars pay half the produce to the Raiots, who employ them.[127]

The following instance gives a clear idea of such management of

125 BRP, 18 June 1790, no.13.
126 Ibid.
127 Ibid., no.8.

privileged raiyati holdings in Purnea:

Seromon Raiot had a Pottah in 1188 Molky for 432 Begas at 8 Begas per Rupee.... He had a Pottah in 1196 for 720 Begas 6 Cottahs at the same rate, which with the Abwabs & the Julkur & Phulkur...amounts to the Jumma of Rs.15-4. From his acknowledgement, it appears that he keeps only two ploughs with which he cannot cultivate more than 80 Beegas, the remaining 640 are cultivated by Colait & Addyar....[128]

Colebrooke observed that the kolait raiyats and adhia raiyats were indebted to the wealthy raiyats, 'which has hitherto retained them in subjection to the principal ryots...the Colait slavery'.[129]

IV. *Concluding remarks*

We shall now briefly summarise the main findings and arguments of this paper. According to a very simple economic standard of self-sufficiency, its basic structure of the rural population of northern Bengal was found to have been as shown in Table 14. Both Harington and Buchanan agree that more than half of the population belonged to the landless and to the classes below self-sufficiency, and that about 20 per cent belonged to the classes which enjoyed a surplus. Harington's data showed that more than two-thirds of the village lands were concentrated in the hands of these wealthy raiyats. We were surprised to find too that the small self-sufficient raiyats represented less than one-third of the rural population. In short, rural society in those days was highly stratified.

These findings establish beyond doubt the great importance of the wealthy raiyat in the rural economy, politics and society. How did this social group come to the prominence? It seems that during the days of political confusion and chaos following the Nawab Mir Kasim's desperate attempt to restore an authoritative Mughal provincial government in Bengal, the zamindars' local administration greatly deteriorated. They were no longer able to maintain an elaborate and well-defined system of rent collection based on regular local assessment with annual investiga-

[128] Ibid. In Purnea, a plough was drawn by four to six oxen, and cultivated 32-36 bighas of land; Buchanan, *Purnea*, pp.432-4.
[129] BRP, 18 June 1790, no.8.

tions into the actual crops grown (*fasli* settlement). Instead they were obliged to adopt a *tahudi* settlement with the representatives of the villages who gave undertakings to pay a lump sum fixed on the respective village. Under this weakened estate management by the zamindars, the old established rent-rates (*nirikh*) became obsolete, and impositions of an arbitrary nature accumulated to an alarming level.[130]

Taking advantage of the zamindars' loose and summary system of rent collection, the wealthy raiyats greatly extended their land-holdings and had them cultivated by prajas and krishans many of whom they had contrived to reduce from the position of small self-sufficient peasants. The wealthy raiyats, having thus consolidated their economic and political position, appeared as contenders for local power in many parts of Bengal. The Company's government, taking the side of the zamindars, adopted elaborate measures to counteract these movements from below with a view to stabilising the agrarian system of Bengal. These efforts culminated in the permanent zamindari settlement of 1793 under Cornwallis. Though the small local leaders lost their political vigour under concerted pressure from the Company's government and the zamindars, they seem to have retained their position as wealthy raiyats in the villages.

If we return now to the conceptual frame of Ratnalekha Ray it will be seen that the evidence given fully support her contention that the jotedar class was present over almost all of Bengal. But her view that the basis of the jotedar's domination remained unchanged during the whole period of British rule needs serious rethinking. As B.B. Chaudhuri rightly pointed out, there was a very remarkable rise of 'new' jotedars in the late nineteenth century as a consequence of the further involvement of the peasant economy in commercial agriculture and the

---

[130] Because of the shortage of space, full discussion of this important process of deterioration in the system of zamindari rent-collection could not be made in this paper. For an extended treatment, see Taniguchi, 'Agrarian Society', ch.6, section 1-(2)-d, and do., 'Structure of estate-control by the northern zamindars under early British rule', a paper read at a seminar on the local history of Dinajpur organised by the Bangladesh Itihas Samity, 1 and 2 December 1994 at Dinajpur.

market economy at a time of increasing pressure of population.[131] These 'new' jotedars acquired their jotes through land-markets of various types, while the 'old' jotedars extended their holdings mainly through political pressure within the village and against the zamindars. The reclamation of waste lands was undeniably important in the formation of large landholdings such as the gatchdari tenures in the hilly parts of Purnea, but most of the 'old' jotedars in the settled parts of the plains acquired or extended their jotes through non-market mechanisms. After the establishment of a well-defined police system by the colonial government, the disappearance of the frontiers of unoccupied or under-used land, and the appearance of a market in raiyati holdings, it seems that the channel for the emergence of the 'new' jotedars became highly market-oriented.

Ray's attractive theory of a close relation between the agrarian structure and the social or ritual hierarchy also needs to be treated cautiously in view of the fact that the areas with a particular concentration of high Hindu castes formed only a small proportion of the whole of Bengal.[132] Thus a 'social' theory has only very limited validity. The present author feels strongly that study of internal differentiation or disintegration of the agricultural castes and the 'semi-Hinduised aboriginals' (to adopt the term used in the 1872 Census Report) should be more seriously pursued, considering their numerical strength[133] and their importance as actual cultivators of the soil.

[131] Chaudhuri, 'Rural power', pp.120-8.

[132] Statistical figures of the preliminary Bengal Census (1872) show that out of 27 districts covered, high concentrations of high-caste Hindus (say, above 7 per cent of the population) can be observed only in five districts; the all-Bengal average was 3 per cent.

[133] According to the 1872 Census, these two classes occupied 24 per cent of the indigenous population of Bengal or 47 per cent of the non-Muslim population.

Table 1: *Composition of the peasantry in Ryampore village*

Holdings (by size)	Number	%	Total area	%	Rent (Rs.)	Rent (Rs.) per bigha
0 to 12	93	69	279	20	157	0-9-00
12 to 30	30	22	587	41	281	0-7-14
over 30	12	9	548	39	257	0-7-10
Total	135	100	1,414	100	695	0-7-17

Source: See note 18. *Note*: Areas are given in Calcutta bigha.

Table 2: *Composition of the peasantry in Radhanagar village*

Holdings (by size)	Number	%	Total area	%	Rent (Rs.)	Rent (Rs.) per bigha
0 to 7	35	70	107	18	78	0-11-13
7 to 17	7	14	83	14	46	0-08-17
over 17	8	16	415	69	176	0-06-16
Total	50	100	605	101	300	0-07-19

Source: BRP, 22 March 1790, No.17, Enclosures Nos. 2 and 21.
Note: Areas are given in local bighas.

Table 3: *Composition of the peasantry in Maheiskol village*

Holdings (by size)	Number	%	Total area	%	Rent (Rs.)	Rent (Rs.) per bigha
0 to 7	7	50	17	8	13	0-12-05
7 to 17	2	14	28	13	10	0-05-14
over 17	5	36	179	80	81	0-07-05
Total	14	100	224	102	104	0-07-08

Source: As for Table 2. *Note*: Areas are given in local bighas.

Table 4: *Composition of the peasantry in Rogonatpoor village*

Holdings (by size)	Number	%	Total area	%	Rent (Rs.)	Rent (Rs.) per bigha
0 to 7	3	33	14	5	13	0-14-17
7 to 17	3	33	38	13	26	0-11-00
over 17	3	33	235	82	145	0-09-17
Total	9	99	287	100	184	0-10-05

Source: As for Table 2. *Note*: Areas are given in local bighas.

Table 5: *Peasant stratification in north Bengal*
A. *Ryampore, Radhagar, Maheiskol and Rogonatpoor;*
B. *Radhagar, Maheiskol and Rogonatpoor*

Holdings (by size)	Number	%	Total area	%	Rent (Rs.)	Rent (Rs.) per bigha
A.						
0 to 12	138	66	525	15	261	0-7-19
12 to 30	42	21	855	25	363	0-6-16
over 30	28	13	2,024	59	659	0-5-04
Total	208	100	3,404	99	1,283	0-6- 01
B.						
0 to 12	45	62	246	12	104	0-6-15
12 to 30	12	16	266	13	82	0-4-19
over 30	16	22	1,478	74	402	0-4-08
Total	73	100	1,989	99	588	0-4-15

Source: As for Table 2. *Note*: Areas are given in Calcutta bighas.

Table 6: *Buchanan's estimates of peasant stratification*

	Number	Average area	Total area
Principal farmers	6,600	165	1,089,000
Great farmers	8,800	75	660,000
Comfortable farmers	11,000	60	660,000
Easy farmers	19,800	45	891,000
Poor farmers	55,000	30	1,650,000
Needy farmer	111,000	15	1,650,000
Total farmers	212,000		6,600,000

Note: Areas are given in Calcutta bighas.

Table 7: *Rearrangements of Buchanan's figures*

	Number	%
A. According Buchanan's five categories		
Krishnans	80,000	18
Adhiyars	150,000	34
Small farmers	138,500	31
Middling farmers	62,700	14
Wealthy farmers	11,000	3
Total	442,000	100
B. According to the three gradations		
Raiyats below self-sufficiency	150,000	41
Small self-sufficient raiyats	138,500	38
Wealthy raiyats	73,700	20
Total	362,000	100

Note: In list A, the small farmers include 'needy farmers' and the lower half of 'poor farmers'; middling farmers possessed 3 to 5 ploughs, and therefore include the upper half of the 'poor farmers' plus 'easy' and 'comfortable farmers' and the lower half of the 'great farmers'; wealthy farmers comprised the upper half of the 'great farmers' plus the 'principal farmers'. In list B the wealthy farmers included the 'middling' and 'wealthy farmers' of Table 6.

Table 8: *Comparison of the jama paid by sundry raiyats in 1771 and 1790 (1179 MY and 1198 MY)*

Names	1771		1790[1]		1790 Jama @
	Land	Jama	Land	Jama	1771 rates
Munbodh of Kenoo, raiyat, Urora	205	95	360	58	119
Deer Sjing Rai, son of Anop[2]	55	31	366	65	126
Dhola, raiyat, Mohunpore	244	94	273	33	93
Shankeer Ullah, ditto	49	14	248	52	85
Seduck, raiyat, Dehta	222	20	291	28	76
Sydal, ditto	113	65	484	47	119
Manullah, raiyat, Buttorbaree	38	26	55	17	24
Durkakee, son of Cala Mundle, ditto	20	16	138	41	56

Notes: [1] In the original written as 1179, but evidently a mis-copying; [2] a raiyat of Pirwa Cooree.

Table 9: *Agricultural capital needed by a newly settled raiyat*

Agricultural implements	Rs.2-12
Two plough-oxen	Rs.6
Seed for 15 Calcutta bighas	Rs.3-6
House	Rs.4
Furniture	Rs.3-4
Clothing and ornaments	Rs.4-0
Food for six months until his first crops	Rs.15
Rent for six months at 10 annas a year	Rs.4-11
Total	Rs.47-1

Table 10: *Statement showing a typical holding in Swaruppur*

Kinds of Crops	Area (bighas)	Value (Rs. per bigha)	Total (Rs.)
Shalee-do (winter paddy and summer paddy)	2	4	8
Shalee-lal (winter paddy only)	6	3.8	21
Bhadai dhan (summer paddy)	1	1.8	1.8
Serree-do (mustard and summer paddy)	1	2	2
Serree-lal (mustard only)	1	1	1
Takuri-do (takuri pulse and summer paddy)	1	2.1	2.1
Takuri-lal (takuri pulse only)	1	1.1	1.1
Wehur (*arahar* pulse only)	1	1.4	1.4
Masur-do (*masur* and summer paddy)	1	1.12	1.12
Masur-lal (*masur* pulse only)	1	0.12	0.12
Total	16		40.6
Deduct seed			2.6
Remaining produce (Rs. 2.6 per bigha)			38.0

Source: BRP, 22 March 1790, nos.14 and 15, Main Report.
Note: *Masur* is a kind of common pulses. Amounts are given in rupees and annas.

Table 11: *Average produce and prices of articles of cultivation in Swaruppur in 1790*

Crops	Average output (maunds per local bigha)	Price (maunds per rupee)	Average value (Rs. per bigha)
Shalee (winter paddy)	10.29	3.0	3.9
Bhadai (summer paddy)	7.12	4.30	1.9
Mustard	1.35	1.30	1.1
Takuri kalai	1.39	1.30	1.2
Wehur (*arah*ar)	1.20	1.05	1.5
Masur	1.39	2.25	0.12

Source and Notes: As for Table 10.

Table 12: *An abstract of accounts of Kishen Mandal, raiyat of Culana village, Ranahutty pargana, Burdwan for 1180 B.S.*

Kinds of tenure (or jama)	Area (bighas)	Rent (Rs) A.	B	Additions (Rs.) A	B	Total (Rs.) A	B
Khudkasht	9-5	9-03	1-00	21-02	2-04	30-05	3-04
Tika	17-0	12-12	0-12	25-06	1-08	38-02	2-04
Patkur or pahikasht	15-0	14-15	1-00	13-04	0-14	28-03	1-14
Karij cossipoor	6-0	3-01	0-08	5-14	1-00	8-15	1-08
Total	47-5	40-00	0-14	65-10	1-06	105-12	2-04

Source: WBSA, Proceedings of the Superintendent of the Khalsa Records, 26 July 1776, No.134, Enclosures.
Note: 'Karij cossipur' is land cultivated in another village. Amounts in rupees show the original rent, the additional impositions (abwabs), and the total of the two, first (A) in toto and then (B) per bigha.

Table 13: *Land-utilisation on the holdings of some wealthy raiyats in Dinajpur (1807-1808)*

Class	Soil	Paddy					Mustard				
(bighas)		Area	%	Value	%	Rs/bg	Area	%	Value	%	Rs/bg
A.	c	24	30	120	42	5-0	0	0	0	0	0
(55)	l	28	35	112	39	4-0	20	25	40	14	2
Total		52	65	232	81	4-7	20	25	40	14	2
B.	c	30	75	140	81	4-11	0	0	0	0	0
(40)	l	6	15	24	14	4-0	1.10	4	3	2	2
Total		36	90	164	95	4-9	1.10	4	3	2	2
C.	c	8	22	24-8	15	3-1	0	0	0	0	0
(30)	l	19	53	85	52	4-8	6	17	12	7	2
Total		27	75	109-8	67	4-1	6	17	12	7	2

Class		Khesari					Tobacco				
(bighas)		Area	%	Value	%	Rs/bg	Area	%	Value	%	Rs/bg
A.	c	0	0	0	0	0	0	0	0	0	0
(55)	l	7	9	10-8	4	1-9	1	1	5	2	5-0
Total		7	9	10-8	4	1-9	1	1	5	2	5-0
B.	c	0	0	0	0	0	0	0	0	0	0
(40)	l	3-5	9	5	3	1-7	0	0	0	0	0
Total		3-5	9	5	3	1-7	0	0	0	0	0
C.	c	0	0	0	0	0	0	0	0	0	0
(30)	l	1	3	1	1	1	1	3	5	3	5-0
Total		1	3	1	1	1	1	3	5	3	5-0

Class		Sugarcane					Total crops		
(bighas)		Area	%	Value	%	Rs/bg	Area	Value	Rs/bg
A.	c	0	0	0	0	0	24	120	5-0
(55)	l	0	0	0	0	0	56	167-8	3-0
Total		0	0	0	0	0	80	287-8	3-10
B.	c	0	0	0	0	0	30	140	4.11
(40)	l	0	0	0	0	0	10	32	3-3
Total		0	0	0	0	0	40	172	4-7
C.	c	0	0	0	0	0	8	24-8	3-0
(3)	l	1	3	36	22	36	28	139	5-0
Total		1	3	36	22	36	36	163-8	4-9

Summary (values shown in rupees and annas)

	Labour			Costs			Net income			Profit	
	F	H	V/M	Cult.	Rent	Bg.	Crop	Plough	Total	Crop	Plough
A	2	3	57-8	153-12	78-12	55	0-11	11	119-12	1-8	23-14
B	2	2	43	89	60	29	0-12	9-15	84	2-2	21
C	2	1	54-8	81-12	52	29-12	0-13	9-11	76-4	2-2	25-7

Table 13 continued.

Source: Buchanan, *Dinajpur*, pp.241-3.

Notes: The table shows three raiyati holdings, (A) one of 55 bighas, (B) one of 40, and (C) one of 30. Figures are local bighas (bg), or rupees and annas. The kinds of land are distinguished (*c* = clay, *l* = loam). The summary shows first the labour used (F = family; H = hired) and the value of output per worker (V/M), second the costs of cultivation and rent, with the total per bigha (Bg.), third the net income per crop and plough and in toto, and finally the profit (corrected net income) per crop and per plough. Figures are modified from Buchanan's, supplemented where his were unclear and corrected where he miscalculated. Buchanan deducted half of the gross produce as expenses. The expenses above were recalculated by the present author.

Table 14: *Rural stratification in northern Bengal*

Category	Terminology	Percentage of total:	A	B	C
Landless	Krishan		-	-	18
Below s-s	Adhiyar, kolait, praja; under- or inferior raiyat		62	41	34
Small s-s	Raiyat, inferior raiyat		16	38	31
Above s-s	Pradhan, paramanick, bosneah, mandal, jotedar, gatchdar; wealthy, superior, principal or head raiyat		22	20	17

Note: A are Harington's estimates for the proportion of the total in each category, B are Buchanan's estimates (1), and C are Buchanan's (2); s-s = self-sufficiency or self-sufficient.

Chapter 4.1

ELEMENTS OF UPWARD MOBILITY FOR AGRICULTURAL
LABOURERS IN TAMIL DISTRICTS, 1865-1925[1]

Haruka Yanagisawa

Introduction

In the conventional view of the changes in Indian rural society under
British rule, the development of commercial agriculture and the monetisation of the economy led to the impoverishment of cultivators and the
growth of stratification in rural society; landownership became concentrated in the hands of moneylenders and large farmers.[2] However,
Dharma Kumar's work using land revenue statistics for the Madras
Presidency concluded that a clear tendency toward the concentration of
land did not exist in South Indian districts.[3]

Our analysis of the settlement registers of the villages in Lalgudi
taluk, Trichinopoly (Tiruchirapalli) district, for the two benchmark
years of 1895 and 1925, revealed that two different trends seem to have
been at work in this taluk, especially in the wet and intermediate
zones.[4] The first trend was a gradual deterioration in the traditional
pattern of dominance in landownership by higher castes. In the villages
in these zones, the high-caste Mirasidars tried to prevent their lands
from being transferred to the lower-caste villagers, especially to
depressed-caste members.[5] However, our research has shown that

[1] I am grateful to the participants in the SOAS workshop for their valuable
comments on my paper. I owe special thanks to Dharma Kumar, M.S.S.
Pandian, Peter Robb, K. Sugihara, Clive J. Dewey, David Ludden, David
Washbrook, B.B. Chaudhuri, Benedicte Hjejle and T. Mizushima.

[2] Surendra J. Patel, *Agricultural Labourers in Modern India and Pakistan*
(Bombay, 1952).

[3] Dharma Kumar, 'Landownership and inequality in Madras Presidency:
1853-54 to 1946-47', *Indian Economic and Social History Review* 12, 3
(1975).

[4] Haruka Yanagisawa, 'Mixed Trends in Landholding in Lalgudi Taluk:
1895-1925', *Indian Economic and Social History Review* 26, 4 (1989).

[5] Various measures taken by Mirasidars to prevent the landless from obtaining landownership are described in G.O., No.590, Revenue, 13 April

landownership in the Brahman community decreased between 1895 and 1925, and the decrease was sharper in larger groups of landowners. On the other hand, there was an emergence of new landholders among those lower caste-Hindus and depressed-caste members who had been working as agricultural labourers or tenants in the fields held by members of higher castes, reflecting the gradual development of a sense of independence and self-reliance in these communities. These changes indicate that high-caste landowners were unsuccessful in retaining such influences as they had asserted traditionally in the control of landownership in the village, though they still owned the largest share of landholdings. The rapid increase in the number of small landholders—that is, the 'subdivision of landholdings'—between 1895 and 1925 was mainly due to the emergence of these small holders among the lower-caste people.

The second trend was the concentration of wet lands in the hands of the larger non-Brahman landholders and the transfer of lands to some of the rich non-Brahman non-agriculturists. These non-agriculturists were generally outsiders living in towns, and were better placed to exploit new economic opportunities which developed under British colonial rule. If the change in the land distribution was to be analysed only according to the size of the holding, without taking into account the land distribution among different communities, this growth of stratification among non-Brahman communities might not be easily noticed as it is shadowed by the sharp decrease in large landholdings in the Brahman communities.

As the second change is an aspect stressed by the conventional view, its process and implications have been well discussed by many scholars. In contrast, the process and implications of the gradual deterioration of dominance in landownership by higher castes and the emergence of lower-caste small landholders have yet to be clarified. This chapter aims to examine the first trend, particularly its historical implications in comparison with agricultural change in Japan.

Tables 1-A, 2-A and 3-A show the distribution of the land by caste (title of the landholder) in 14 wet villages in Lalgudi taluk for 1865,

1875, p.1377.

1895 and 1925 respectively.[6] Though a comparison of Tables 1-A and 2-A indicates many interesting facts which need explanation,[7] we shall focus our attention on the change in the landownership by lower castes: The extent of land owned by depressed castes increased from only 4 acres in 1865 to 39 acres in 1895; Areas owned by Muttiriyans, who are presently categorised as members of backward classes by the government, also increased, from 131 acres to 356 in this period. The increase in landownership by these lower castes is also noticed in intermediate and dry zone villages between 1865 and 1895: Areas owned by depressed-caste members increased from 30 acres to 96 acres in intermediate zone villages and from nil to 86 in dry villages; those owned by Muttiriyans, from 107 acres to 181 acres in the intermediate zone and from 289 to 1,105 in dry zone.[8] This means that the increase in areas owned by lower castes was a long-standing trend, not only between 1895 and 1925, as indicated by our previous work, but also between 1865 and 1895, though the decrease in the Brahman landownership became apparent after 1895.

To clarify the characteristics of land acquisition by lower castes, the cases of land transfer to depressed-caste members between 1895 and 1925 in a village in Lalgudi taluk have been tabulated. In this village depressed-caste members owned a total of 18 pattas in 1925, of which 16 cases are shown in Table 4 to illustrate the land transfer to the depressed castes. Interestingly, these 16 examples do not include any land which was registered in 1895 as poramboke (lands for public purposes), village common lands and plots without a registered pattadar. Almost all of these plots had been owned by members of other castes.

[6] For the sources and the way of processing them, see Yanagisawa, 'Mixed Trends'

[7] The 'Nadan' changed their title to 'Nadavan' between 1865 and 1895 and therefore, these two titles denote the same community. Though Tables 1A-3A point to a change in the landownership by this community, it may be inappropriate to regard this change as indicating a general trend in the wet zone as a whole, since the majority of their lands were concentrated only within two villages.

[8] Haruka Yanagisawa, *Minamiindo Shakaikeizaishi Kenkyu* (Socio-Economic Change in South Indian Rural Society) (Tokyo, 1991), pp.67, 183, 262 and 278.

While the acquisition of lands by depressed-caste members in some other villages may have been partly due to the government's policy of assigning land which had been 'uncultivated waste' in 1895 registers, many of the land transfers to the depressed castes are likely to have been through purchase. More important, the land owned by a depressed-caste member in 1925 in many cases consisted of plots which had been owned by different persons, sometimes of different communities, in 1895. This also indicates that land acquisition by depressed castes was mainly not through non-market transactions, such as gifts from their landlords, but through buying lands bit by bit on the land market. In other words, a considerable number of the depressed-caste families came to own small bits of lands through normal market transactions.

The purpose of this chapter is to discuss factors that promoted this trend in landholding, that is the acquisition of land by these lower-caste members and the gradual decrease in higher-caste landownership. It is argued here that this same trend was observable in other Tamil districts.

Intensive cultivation and small scale farming

The change in agricultural practice in Tamil districts that began in the latter half of the nineteenth century was one of the factors that promoted the change in landholding. My hypothesis is that agriculture in the period showed a gradual tendency toward intensity in which more attention and care was paid by the cultivators to obtain a better yield. This change in agricultural practice may have been underpinned by an increase in the demand for agricultural products, particularly that for paddy, both from the foreign market and the domestic market in the period between the late nineteenth century and the 1920s. A rising trend in the prices of these products was also noticed in this period.[9]

First, the number of wells increased remarkably in the later half of the nineteenth century. Between 1852 and 1890, areas irrigated by private wells increased by 138 percent in the Madras Presidency.[10] The

[9] Christopher John Baker, *An Indian Rural Economy 1880-1955: The Tamilnadu Countryside* (Oxford, 1984), pp.186 and 241; Dharma Kumar, 'Agrarian relations: South India,' in Dharma Kumar (ed.), *The Cambridge Economic History of India*, vol.2 (Cambridge, 1983), p.231.

[10] S. Srinivasa Raghavaiyangar, *Memorandum on the Progress of the*

three decades after 1891-92, as shown by Table 6, witnessed an increase of 50 percent in the area irrigated by wells in Tamil districts.

In the Salem district, the number of wells increased from 51,000 to 54,000 between 1871 and 1876.[11] The settlement report of the district in 1903 stated that a remarkable feature in the agricultural history of the last 30 years was the increase in the number of wells.[12] The Gazetteer indicated an enormous increase in number of wells in Madura (Madurai) district.[13] An increase in the number of wells was also reported in the Coimbatore and Tinnevelly (Tirunelveli) districts.[14]

According to the settlement report of Trichinopoly district, areas irrigated by wells almost doubled, from 51,000 acres in 1905-06 to 100,500 acres in 1919-20.[15] Table 5 is compiled from village census data given in the settlement registers for 60 villages in Lalgudi taluk, Trichinopoly district. The villages are grouped into three categories, that is, wet, intermediate and dry zones. The average number of wells per village increased from 18 to 33 between 1894 and 1925.

Second, the increase in number of wells and the improvement of canal irrigation resulted in an increase in the gross area of irrigated land. In addition to the 138 percent increase in lands irrigated by private wells between 1852 and 1890 mentioned above, the increase in the area of cultivation in this period was 25 per cent in unirrigated lands and 41 percent in lands irrigated by sources of irrigation maintained by the government.[16] As shown in Table 6, the total irrigated area further increased by about 36 percent in the three decades after 1891.

The increase in the irrigated area was accompanied by a rapid expansion of paddy cultivation. The area cropped with paddy in Tamil

---

Madras Presidency during the Last Forty Years of British Administration (hereafter *Progress*) (Madras, 1893), p.48.

[11] *Manual of Salem District* (Madras, c.1882), p.153.

[12] P.B.R., No.205, 15 June 1903, p.17.

[13] *Gazetteer of Madura District* (Madras, 1906), p.131.

[14] P.B.R., No.1760, 26 June 1878, p.5735; P.B.R., No.565, 4 November 1910, p.5; *Manual of Tinnevelly District* (Madras, 1879), p.25. See also, David Ludden, *Peasant History in South India* (Princeton, 1985), pp.147-48.

[15] P.B.R., No.86, 26 November 1923, p.27 (para. 22).

[16] *Progress*, p.48.

districts increased by 68 percent between 1880-82 and 1932-34.[17] An expansion of paddy cultivation at the expense of ragi and other cereals was noticed in the villages in Lalgudi taluk between 1894 and 1925, as shown in Table 5, particularly in the intermediate zone villages, where the land under paddy cultivation increased by 109 acres overweighing the decrease of 21 acres in the cultivation of sugarcane.[18]

Third, and even more important, the double-cropped area extended considerably in the Tamil district (see Table 7).[19] The double-cropped area in Tamil districts roughly doubled between the 1890s and the 1940s. The data in Table 5 confirms the increase in double-cropped area in Lalgudi taluk, Trichinopoly district, even though 1925 was a year when Tanjore (Thanjavur) and Trichinopoly were among a few districts which suffered most seriously from the insufficient supply of water.[20]

Fourth, there were important improvements in agricultural practice. A rapid if uneven spread of the practice of transplanting paddy around the turn of the century was noted by Baker, though the source of this information was not mentioned.[21] Sources from Tanjore indicate a spread of the use of manure in the district. According to the Manual of the district published in 1883, 'As a rule, in the case of irrigated nansei [wet land] or rice cultivation, it is only the fields on which two crops

[17] The average area cultivated with paddy per year in 1880-82 in Tamil districts was 2,753,000 acres, whereas it jumped to 4,612,000 in 1932-34; *Administration Report of the Madras Presidency during the years 1880-81, 1881-82 and 1882-83; Season and Crop Reports for Fasli 1342 (1932-33), 1343 (1933-34) and 1344 (1934-35)*.

[18] Trichinopoly was one of four districts, where the reduction of sugarcane cultivation was most marked in 1925, because of an insufficient supply of water in the irrigation. Therefore, though the figures for 1925 in Table 5 suggest a deduction of the sugarcane cultivation in the Lalgudi villages, it may be inappropriate to accept the figures for 1925 as indicating the general trend in sugarcane cultivation. A clear trend of decline is not discernible in the area under sugarcane cultivation in Trichinopoly district after 1891-92, when it increased from about 2,000 acres in 1880-81 to about 6,000 acres (*Season and Crop Reports*).

[19] Baker, *An Indian Rural Economy*, pp.176, 221-2 and 574 (n.135).

[20] *Season and Crop Report for Fasli 1335 (1925-1926)*, pp.3-11. *Madras District Gazetteers, Statistical Appendix for Trichinopoly District* (Madras, 1905), p.9, Table VIII; *Madras District Gazetteers, Statistical Appendix for Trichinopoly District* (Madras, 1933), p.15, Table IX.

[21] Baker, *An Indian Rural Economy*, p.176.

are raised that are manured every year, and even these not invariably, for it is not always that the required manure is available. Single crop lands are generally manured once in five years; in some cases not at all....'[22] The situation changed during the next 20 years. As the Gazetteer of the district wrote in 1906, 'the ryots have found that it is not impossible to secure manure for their fields, and enquiries now reveal that its use has greatly spread. It is now apparently not an exaggeration to say that there are few wet fields which are not manured every year'.[23]

The spread of the practice of using manure was reported from other districts. According to the settlement report of Chingleput district in 1909 a large amount of manure was brought from Madras, and its continued application had greatly improved soils, many of the red soil taking on the appearance of black loams.[24] As for Trichinopoly district, Puckle stated in 1860 that manure was very seldom used under river irrigation, as the deposit left by the Cauvery water was considered to be sufficiently fertilising.[25] Thirty-five years later, the 1895 settlement report described the change in this situation. After citing Puckle's 1860 statement, the report wrote: 'Enquiries now made show that this is still true to a large extent so far as the first crop is concerned, but that for the second crop the practice of using leaf manure is more or less general, and that this practice is more extensively followed now than was formerly the case'.[26] The Gazetteer of the district also reports a large increase in the use of manure during the last quarter of the last century.[27] A remarkable increase in the application of manure between 1916 and 1957 was reported in Gangaikondan village in Tinnevelly district.[28] In Vadamalaipuram village in Ramnad district, chemical fertilisers, which had not been used in 1916, were reported as being

[22] *Manual of Tanjore District* (Madras, 1883), p.348.
[23] *Gazetteer of Tanjore District* (Madras, 1906), pp.101-2.
[24] G.O., No.2240, Revenue, 14 August 1909, pp.104-5 (para.80).
[25] *Papers Relating to the Survey and Settlement of the Trichinopoly District*, Selection from the Records of the Madras Government, No.L (Madras, 1876), p.9.
[26] P.B.R., No.2, 4 January 1895, p.18.
[27] *Gazetteer of Trichinopoly District* (Madras, 1907), p.141.
[28] V. B. Athreya, *Gangaikondan 1916-1984: Change and Stability* (Madras Institute of Development Studies, Madras, 1985), mimeo, p.74.

applied in 1936.[29]

The application of manure into the dry lands also increased. In Tanjore district, dry lands were treated a little less generously than wet lands, but still received plenty of attention.[30] As the Director of Agriculture noted in 1911, 'in the Southern Districts much more capital is now being put into dry lands. The soils are being improved by carting silt on to them while much greater efforts are being made to adequately manure such lands'.[31]

The increase in the use of manure was facilitated by the local trade by cart among the neighbouring districts. Green-manuring was especially popular in Trichinopoly district. The favourite leaves used in the district were those of wild indigo, which were carted in great quantities from Coimbatore and Salem districts.[32] Wild indigo was the favourite green manure in Tinnevelly district also and was grown on the red soils of the dry taluks and often brought 20-30 miles to the river valley.[33] The settlement report of Coimbatore district stated in 1910 that a rise in the price of paddy enabled those raiyats who cultivated lands irrigated by the Kalingarayan channel to increase their expenditure on manure; hence there was an expansion in the radius within which employment was provided in plucking and carting wild indigo.[34] In Tanjore district the movement of cattle among the neighbouring districts enabled the raiyats to secure manure. A shortage of cattle, sheep and goat manure owing to the absence of pasture land in Tanjore district was noted around 1880,[35] but the Gazetteer stated in 1906 that herdsmen from the Marava country brought cattle to the district to fertilise the fields in the cultivation season and that the lack of local

[29] V. B. Athreya, *Vadamalaipuram: A Resurvey* (Madras Institute of Development Studies, Madras, 1984), mimeo, p.65. It was reported that 'evidence is not wanting to show that much greater efforts than formerly are being employed to manure such lands [wet lands]...much greater pains are taken than formerly to exploit every source of manure supply'; *Season and Crop Report for Fasli 1321 (1911-1912)*, p.4.

[30] *Gazetteer of Tanjore District*, p.102.

[31] *Season and Crop Report for Fasli 1321 (1911-1912)*, p.4.

[32] *Gazetteer of Trichinopoly District*, p.141.

[33] P.B.R., No.94, 1 April 1907, p.24 (para.34).

[34] G.O., No.102, Revenue, 10 January 1910, p.9.

[35] *Manual of Tanjore District*, p.348.

grazing land was made up for by the raiyats' driving their cattle across the Coleroon into the forests of South Arcot.[36] Later, in the 1930s, an increase in the cultivation of wild indigo for the use of green manure purposes was reported from Tanjore and Madura districts.[37]

The impact of these changes was not uniform across the sub-regions. The increase in the number of wells was more remarkable in the dry area than in the wet area. In a highly irrigated zone where the land was already highly double-cropped by 1895, such as the wet zone in Lalgudi taluk, while the expansion of the double-cropped area was not so dynamic as other areas, the change in the application of the manure is likely to have been more remarkable. In spite of this diversification across the sub-regions, each change was more or less noticed in each of the sub-regions and the evidence reveals a basic similarity in the nature of change within Tamilnadu in this period.

The increase in the number of wells, the expansion of double-cropped area and the increase in the application of manure indicate that South Indian agriculture exhibited a trend toward intensive cultivation, which required more labour input per acre and a greater degree of care and attention.

Information from Tanjore district indicates an increase in labour input per acre. The Manual of the district published in 1883 stated that the area of land which could be cultivated by means of one plough with a pair of bullocks and one farm labourer was assumed to be six-sevenths of a Tanjore veli (1 veli = about 6.5 acres), equal to 5.67 acres.[38] Thirty-eight years later, in 1921, it was stated that 'At least two men are required for the efficient cultivation of one veli of land'.[39] As remarked by S. S. Raghavaiyangar in 1893, 'lands also are believed to be much more carefully cultivated now than in the old days'.[40] 'My enquiries tend to show that, under the stress of necessity and the additional incentives to individual exertion promoted by the break up of the joint family system, greater care is now bestowed on cultivation of

[36] *Gazetteer of Tanjore District*, p.102.
[37] K. Ramiah, *Rice in Madras: a popular handbook* (Madras, 1937), p.84.
[38] *Manual of Tanjore District*, p.382.
[39] P.B.R., No.28, 12 Febuary 1921, p.54.
[40] *Progress*, p.202.

lands in Tanjore district than in times past; and this is to some extent the case in other districts also'.[41]

This trend was also observed in a village survey in South Arcot district in 1916. Here, as the possibility of extending cultivation was exhausted, the raiyats were resorting to more intensive cultivation, and the construction of wells was being vigorously attempted.[42] The Gazetteer of Tinnevelly district also noted 'the growing tendency towards intensive methods' in 1918,[43] and the settlement report of Coimbatore district stated that the progress of the district since the last settlement had chiefly been in the direction of intensive cultivation.[44]

The importance of the care and attention paid by the farmer and his family members to cultivation is well illustrated by the following case from Tinnevelly district, where the amount of unused land available for cultivation was negligible in all taluks and the growing tendency toward intensive methods was reported to be the most satisfactory feature of its economic outlook:[45]

All these are the places where it is not enough that the landowner shall wait for the season and then give orders to a 'tenant'. The supply for the wet lands is precarious, and, if it does not come or comes out of season, resources and patience are needed; a gingelly crop must be sown in the nick of time in place of the paddy that has failed, or the last drop of water left in the tank must be husbanded and led to the field at the critical moment; the land is by the register entitled only to a dry crop, but the united labour of a family will lower it till water is bound to reach it, and fearless of 'penal rates'. the peasant owner will turn a dismal waste into a smiling garden. In many places the tank fills only during *pisánam* and to the non-cultivating owner the value of such land is but one crop in the year. The cultivating owner can make double, or even three times, that income; wells are sunk, and, by the steady labour of himself, his family and joint owners, precious crops—chillies, onions, brinjals and cholam, exceeding in value even the

----

[41] Ibid., p.56. In the Shiyali taluk (Tanjore district), more efficient cattle power was employed for ploughing, and the cattle were better fed and less vulnerable to epizootics (ibid., p.ccxcix).

[42] G. Slater (ed.), *Some South Indian Villages* (Oxford, 1918), p.225.

[43] *Gazetteer of Tinnevelly District* (Madras, 1917), p.192. ⋯

[44] G.O., No.102, Revenue, 10 January 1910, p.12.

[45] *Gazetteer of Tinnevelly District*, p.192.

paddy of *pisánam* —fill up the interval between the paddy seasons.[46]

The importance of farmers' personal participation and attention to cultivation is further evidenced by the following statement from the settlement report of Tinnevelly district:

Thus in one village in which at last settlement the Brahmans owned some 2,000 ares and now possess less than 200, a Brahman villager pointed out a Naicker who was going to his field, to plough his land taking his food for the day with him, saying that that man was worth about Rs. 10,000. He added 'How can the Brahman compete with the Naick? Suppose a Naick's land yields four pothis of cotton, the Brahman's close by of equal extent will give less than two.'[47]

On the other hand, the advantage enjoyed by a larger farm over the small peasant farm tended to lessen in this period. As a general rule, a large farm depending upon non-family labourers required a very strict supervision of the labourers by the farm-manager to make them work as industriously as family labour. Even with such strict control, it may have been difficult to secure from non-family labour the same quality and intensity of work as from family members, particularly when agriculture was becoming more intensive, and much more care and attention from labourers was needed.

In Eruvelliput village in South Arcot district, it was observed in 1916:

the customary method of transplanting is to put several plants together into one hole in the levelled mud of the rice field. Experiment has proved that if each seedling is separately planted and all are evenly spaced, there is a considerable saving of seed and a heavier crop. The new method is spreading, but as yet very slowly...the cooly whom they employ will not change their methods unless compelled to do so by strict supervision. They want to do their work in the semi-automatic manner attained by unchanging habit, and not to tax their brains and think about what they are doing.[48]

The single planting was reported to require more personal supervision. Those who cultivated the land themselves found it enough if they

[46] Ibid.
[47] P.B.R., No.565, 4 November 1910, p.15.
[48] Slater (ed.), *Some South Indian Villages*, p.5.

applied only 12 cartloads of manure, while others who cultivated with the help of labourers had to apply about 24 cartloads, for the land was not as well ploughed as in the other case.[49]

A source from Tanjore district indicates the difficulty a large farm encountered in securing enough labour at the peak season. 'After the grain has ripened, it is allowed to dry for two to three weeks before it is reaped; but in the case of large estates, the reaping is delayed a few days longer from the amount of labor required not being at once forthcoming'.[50] In Tinnevelly district, 'the method of cultivation followed by the small holder in this country, especially if he holds dry lands or wet lands under a precarious source, is as a rule better than that adopted by the owner of extensive safe wet lands'.[51]

[49] P.J. Thomas and K.C. Ramakrishnan (eds.), *Some South Indian Villages: A Resurvey* (Madras, 1940), pp.163 and 167.

[50] *Manual of Tanjore District,* p.379.

[51] P.B.R., No.200, 5 June 1909, pp.7-8. A. Satyanarayana revealed many important findings in his study on Andhra peasants (*Andhra Peasants under British Rule: agrarian relations and the rural economy 1900-1940*, Delhi, 1990), in which he stresses the technical superiority of the large farmers. He remarks as follows: 'The settlement officer of Guntur came across instances where "large ryots" carried on (paddy) cultivation more scientifically. They "imported" manure "in large quantities" from the neighbouring districts' (p.114). However the report of the settlement officer of Guntur, on which Satyanarayana bases his assertion, seems to indicate a different fact. 'In only two places did I find ryots who had actually purchased manure last year for their fields. One of them was a very large ryot at Karenchedu where cultivation is carried on more scientifically than in most places in the delta. He told me that he imported *pig* manure in large quantities from the Nellore district and proved to me that he sustained a loss on every acre of paddy he grew. (He explained this away afterwards). But even here in my inspection I came across a few acres of molakolukulu paddy which bore at least twice as good a crop as the neighbouring fields. It was explained to me that it belonged to some Gollas (shepherds) who are able to manure the crops well. Again and again I found a similar patch and was told that the owner had only two or three acres and so could manure all his lands. The custom undoubtedly is for a ryot to take up more lands than he can farm properly. I heard of a well authenticated case of a man who was compelled to sell half his land and obtained a much larger produce from the remaining half than he did for the whole farm before' ('Report of the Settlement Officer of Guntur', P.B.R., No.463, 17 December 1903, p.24.) The settlement officer clearly asserted, contrary to Satyanarayana's interpretation, that the large farm was less properly managed and that the yield of the smaller farm was higher than that of the larger farm. Satyanarayana also argues that in Coastal Andhra dis-

It is appropriate to conclude that, with the development of intensive cultivation, farm-managers' direct participation in and devotion to cultivation became an important factor in having a better yield; the large farms employing non-family labourers encountered difficulty in making them work as industriously as family members work on their small farms, thus lessening the advantage large farms possessed in the past.

Emigration of depressed classes and their emancipation

The emigration of the depressed castes and other low-caste members from South India to overseas countries such as Ceylon and southeast Asian countries seems to have played an important role in promoting the change in landholding in South India, in two ways: directly by providing them with a chance to save money with which to purchase the land in their villages, and indirectly by providing them with alternative employment opportunities and by promoting the development of a sense of independence among them.[52]

A professor in Madras stated at the Royal Commission of Labour in India:

To take the emigrants to Burma, it is largely true to say that many workers live there for more than two years, although every one of them hopes to come back to his native village sooner or later and to set himself up as an independent landowner, or fairly well-to-do tenant or a small business man.... Similarly, but for a few thousands of workers, who have made

---

tricts of Guntur and Nellore 'big ryots' increasingly used chemical manure (p.115). Though the evidence surely indicates the existence of a large farmer who used chemical fertiliser in Nellore district, the total amount of sales of chemical manure mentioned in the document shows that the number of such cases was very few, probably only one or two, in the district; *Royal Commission on Agriculture in India*, vol.3, *Evidence taken in Madras* (London, 1927), p.762. It may be an exaggeration to conclude from this evidence that big ryots in general increasingly used chemical manure.

[52] For the emigration from South India, see Dharma Kumar, *Land and Caste in South India* (Cambridge, 1965), ch.8; R. Jayaraman, 'Indian emigration to Ceylon: some aspects of the historical and social background of the emigrants', *Indian Economic and Social History Review* 4, 4 (1967); Barbara Evans, 'From agricultural bondage to plantation contract: a continuity of experience in Southern India, 1860-1947', *South Asia* 13, 2 (1990).

Ceylon their permanent country, the others always look forward to go back with their savings to their villages in Southern India.[53]

A report from Tanjore district attests to this trend. Here both tenant and labourer classes emigrated, though the latter category was dominant in number. The report notes that all returned with considerable savings, by means of which the porakudi became a landholder and the labourer set up as a tenant farmer.[54] In his report on a village in South Arcot district, Slater reported that 'their economic condition in Ceylon is vastly improved, and they have opportunities of saving money, and, if they choose, of returning to their native districts and buying land'.[55] In Madura district also, members of such agricultural labour classes as 'Kallars', 'Pallars' and 'Valayars' went to Ceylon and Malaya and with their savings bought a pair of cattle, and bid for lands as tenants.[56]

These pieces of evidence suggest that some emigrants managed to save money and became small landholders or tenants by purchasing lands or cattle in their native districts, though the extent of lands purchased by the lower-caste members in this way may not have been large since 'many of them' 'return[ed] poorer' and an increasing number of Tamil emigrants did not return to their native villages but settled in the estate with their family.[57]

A fact probably more important than this direct economic impact is that emigration had a great influence on the minds of the lower classes of rural society. They developed a sense of independence from the high-

[53] Royal Commission on Labour in India, *Evidence*, vol.VII, part I, *Madras Presidency and Coorg, Written Evidence* (London, 1931), p.322. See also, Kumar, *Land and Caste*, p.140. Benedicte Hjelle has already indicated the importance of the alternative employment available on the plantations as the real factor in the movement towards emancipation of low-caste agricultural labourers; Benedicte Hjejle, 'Slavery and agricultural bondage in South India in the nineteenth century', *Scandinavian Economic History Review* 15, 1 & 2 (1967).
[54] *Progress*, pp.145-46.
[55] Slater (ed.), *Some South Indian Villages*, p.9.
[56] K.G. Sivaswamy, 'Agricultural leases in ryotwari areas', in *The Madras Ryotwari Tenant* (hereafter, *Tenant*) (South Indian Association of Agricultural Workers, Madras, 1948), part 2, p.50.
[57] Slater (ed.), *Some South Indian Villages*, p.82; Jayaraman, 'Indian Emigration to Ceylon'.

caste large landholders.

According to the Census Report,

One social effect of emigration has been indicated above, viz., a growth in independence and self-respect on the part of the depressed classes who go abroad. This is all to the good. A man who, little removed from praedial serfdom [such as pannaiyal system] in Tanjore, finds himself treated on his own merits like every one else when he crosses the sea, paid in cash for his labours and left to his own resources, must in the majority of cases benefit from the change, and it is probably the existence of the emigration current that has contributed most to the growth of consciousness among the depressed classes in India....[58]

Slater's report mentioned above noted that one way for the padial to escape from the condition of servitude and poverty was emigration. This emigration was discouraged by their creditors (employers). It was impressed upon the padial that it was a point of honour for him never to leave his master; and he was given to understand that he could not legally do so.[59] Such a discouragement could not completely prevent them from migration, however. The advances paid them were reported to be not infrequently lost, as the labourers emigrated to Ceylon or the Straits without repaying them.[60] As noted by Slater,

emigration of padials or poor tenants or peasant proprietors to foreign cities, such as Colombo, Rangoon, Singapore and Penang, is common. As these people do not go there with any capital, they seldom make profits, and, as many of them go there through emigration depots they return poorer and emaciated. But their account of their wanderings and experiences in such foreign places (which have a halo of romance about them) usually induces people to try their chance and thus causes a general unrest in the village and its neighbourhood.[61]

As early as 1879, the Manual of the Tinnevelly district noted that 'the spirit of independence has reached the labourer, who carries his labour

---

[58] Census of India, 1931, vol.14, Madras, part 1, p.93. Weakening of the labourer's traditional dependence on the landlord was asserted by Kumar, 'Agrarian relations: South India', p.232.
[59] Slater (ed.), *Some South Indian Villages*, p.9.
[60] *Manual of Tanjore District*, p.311.
[61] Slater (ed.), *Some South Indian Villages*, p.8.

into the best market and does not scruple to leave his master if he thinks he can better himself'.[62]

As the depressed-caste people had been the nucleus of the agricultural labour force in wet areas hired by landowning higher-caste communities, the landowners came to feel difficulty in securing the necessary labour force to cultivate their farms. In Trichinopoly,

these [wage] rates are said to be much higher than those paid in the past, but they need to be higher yet to check the emigration which is now going on to other countries, and the supply of pannaiyáls is not at present equal to the demand.... Daily wages have also risen everywhere, but the supply of day labour is as unequal to the demand as that of pannaiyáls.... [T]he complaints of landowners of the scarcity of labour are loud.[63]

The scarcity of labour was also reported in 1914 in the Tinnevelly district, where the number of homestead labourers was comparatively small and was annually decreasing. Emigration to Ceylon was exceptionally easy, and the Tinnevelly Pallan knew that both there and on the tea estates of the high ranges he could command a good wage. 'I do not mean that labour is fully mobile, but that if a landlord does not treat his labourers fairly according to their ideas, the latter are not slow to seek employment elsewhere'.[64]

The situation was no different in Tanjore district, in most parts of which landholders complained bitterly of the defection of their serfs and the scarcity of labour in general. The result was that the remaining labourers could demand much better terms than their fathers. Everywhere the hours of labour were reported to have decreased.[65] The

[62] *Manual of Tinnevelly District*, p.30. See Kumar, *Land and Caste*, p.142.

[63] *Gazetteer of Trichinopoly District*, p.152. See P.B. Mayer, 'The penetration of capitalism in a South Indian District: the first 60 years of colonial rule in Tiruchirapalli', *South Asia*, new series, 3, 2 (1980).

[64] 'Note to G.O. Nos.3594-95, Revenue, 9th December 1914', p.76. See also, a report of the Director of Agriculture in *Season and Crop Report for Fasli 1316 (1906-1907)*, p.4; *Season and Crop Report for Fasli 1317 (1907-1908)*, p.4.

[65] *Gazetteer of Tanjore District*, p.111. The Director of Agriculture reported in 1906 that 'in the Tanjore delta, many complaints are heard from landowners of the growing scarcity and independence of the day labourers in consequence of emigration' (*Season and Crop Report for Fasli 1316 (1906-*

settlement report of 1921 confirmed this as follows: 'Returning to the delta pannaiyal, my inquiries show that increase in grain wages is distinctly exceptional. It is more generally alleged that pay has risen indirectly by shortening of hours of work or by an increase in toddy money'.[66] In Palakkurichi village in Tanjore district,

the landlords complain that it has become very difficult nowadays to extract work from pannayals, as they fight for more leisure. At the time of my visit I was told that one Pannayal had run away and the master was trying to get him back by force. There is no unemployment among agricultural labourers or educated persons.[67]

Both the increase of emigration and the scarcity of labour for repairs to roads and tanks and for coffee estates were reported by the Collector of Madura district in 1888.[68]

A rapid expansion in groundnut cultivation in South India accelerated this trend in some districts.[69] As the Director of Agriculture

---

1907), p.4). The shortening of the hours of work put in by agricultural labourers has been already discussed by M. Atchi Reddy, 'Work and leisure: daily working hours of agricultural labourers, Nellore District', *Indian Economic and Social History Review* 28, 1 (1991). While the shortage of labour caused a rise in the wage paid for the daily labourers, the grain wage paid to the permanent labourers was reported to be unaltered. However, the ryots had to offer various small additions in the shape of an occasional free meals, etc., in order to retain the services of the permanent labourers; *Season and Crop Report for Fasli 1317 (1907-1908)*, p.4; ibid. *Fasli 1321 (1911-1912)*, p.4.

[66] P.B.R., No.28, 12 Feb. 1921, p.28, para.31. See also *Season and Crop Report for Fasli 1316 (1906-1907)*, p.4.

[67] Thomas and Ramakrishnan (eds.), *Some South Indian Villages: A resurvey*, p.141. See also *Season and Crop Report for Fasli 1316 (1906-1907)*, p.4.

[68] G.O., No.366, Confidential, 27 May 1888, para.24.

[69] The groundnut cultivation in Tamil districts increased from 66,000 acres per year in 1880-82 to 1,229,000 acres in 1932-34. However, this expansion of groundnut cultivation did not accompany a reduction in the cultivation of labour-intensive crops. As was the case of paddy, the cultivation of many labour-intensive crops rather increased between 1880-82 and 1932-34; sugarcane, from 17,000 acres to 45,000 acres; plantains, from 26,000 acres to 68,000 acres; tobacco, from 31,000 to 55,000 acres; spices including chillies, from 100,000 acres to 195,000 aces (*Season and Crop Reports*). Considering the great increase of 4,654,00 acres in the total cultivated area in this period, it is likely that the groundnut cultivation contributed to the expansion of the acreage under cultivation and thus resulted

stated in 1906, 'at Palur [in South Arcot district] so many labourers now cultivate small patches of ground-nut on their own account that it is found very difficult to obtain labour for the picotas [an indigenous water lift] during the season of the irrigated ground-nut'.[70]

Thus, as a result of the increase of emigration, landlords began to feel not only the scarcity of labour but also the difficulty in making their pannaiyals work as hard as they used to. As reported by Sivaswamy, with the growth of opportunities for emigration and various employments, the pannaiyals too began to feel the irksomeness of tied service and longed for a certain amount of freedom.[71]

It is important to note that this difficulty appeared just when agriculture was exhibiting a change toward greater intensity of labour and more painstaking care and industry from labourers and farm managers were required for a better crop yield, as mentioned in the previous section. The emigration of labourers and the increasing difficulty large landholders encountered in securing labourers and in making them work hard are likely to have further weakened any advantage large farms might have possessed over small family farms. Evidence indicates that some large farmers may have decreased their farm size, leasing out a part of their land under the varam (share-cropping) system.

In South Arcot, this practice of subletting was reported to be on the increase in the Tirukkoyilur division owing to the difficulty experienced by landowners in procuring labourers, who preferred to cultivate lands themselves to working for wages.[72] In 1914, subletting was on the

---

in a fuller exploitation of lands, at least in the period before the 1920s (Baker, *An Indian Rural Economy*, p.149). It is also important to note, however, that the excessive expansion of groundnut cultivation was reported to reduce the supply of fodder and consequently to result in a low standard of cultivation in the 1920s, though a fuller consideration of the problem is beyond the scope of this paper; *Season and Crop Report for Fasli 1334 (1924-25)*, p.16.

[70] *Season and Crop Report for Fasli 1316 (1906-1907)*, p.4; *Season and Crop Report for Fasli 1317 (1907-1908)*, p.4. See also, Baker, *An Indian Rural Economy*, p.152.

[71] Sivaswamy, *Tenant*, Part I, p.64.

[72] 'Notes to G.O. Nos.3594-95, Revenue, 9th December 1914', p.20. Sivaswamy, *Tenant*, Part I, p.64.

whole increasing in Tinnevelly, where 'the spread of education may account for this in part by exciting a repugnance to field labour on the one hand, and on the other, a preference in the actual cultivation for the status of a tenant rather than that of a cooly'.[73] It was stated in 1923 that 'subletting is on the increase in the Chingleput district as it is said the Adi-Dravidas who were hitherto working as farm servants now own lands themselves and are now becoming more and more independent of the landowning classes'.[74]

As Sivaswamy has written, the varam was an evolution from the older custom of permanent farm service. The latter went into disuse with the breakup of the sense of mutual obligation and its replacement by the wage relationship.[75]

Emigration of upper-caste members to urban jobs

The third factor which promoted the deterioration of the traditional pattern of landownership was the emigration of upper-caste members to urban jobs. The upper-caste landowning communities, especially Brahmans, developed their interest in urban employment, and a considerable number of people migrated to urban areas to take jobs and obtain good education for their sons, consequently losing their commitment to agricultural management.

The case of Gangaikondan village in Tinnevelly district illustrates the changes brought about by the increasing interest of the Brahman community in urban jobs. As revealed in a survey by the University of Madras in 1936, the number of Brahman families in the village decreased from 100 to 75 between 1916 and 1936.

An aged Brahmin gentleman told me that some fifty years back they were 120 families strong, possessed 100 ploughs between them, owned 800

[73] Ibid., p.22.
[74] *Report on the Settlement of the Land Revenue of the Districts in the Madras Presidency for Fasli 1332 (1922-23)* (Madras, 1924), p.10. In Nellore district also, subletting was reported to be on the increase as many members of the depressed classes refused to work as farm servants; ibid., *Fasli 1332 (1921-22)*, p.10.
[75] Sivaswamy, *Tenant*, part 2, p.48. See also, Baker, *An Indian Rural Economy*, pp.178-81.

acres of nanja [irrigated] land out of the 1,000 acres, 3,000 out of 3,500 acres of black soil and 3,000 out of 3,500 acres of red soil in the village with a further right to a poramboke area of 3,500 acres for purposes of grazing cattle—all between 106 share-holders. But their total possessions in the village today have dwindled to 250 acres of nanja land, 100 acres of black soil land and 1,000 acres of red soil land of which only 100 are cultivable. This is because the Brahmins ceased to take personal interest in land. They were content with merely letting the land on lease or varam on terms more or less dictated by the tenant. From active agriculturists they have degenerated into idle rent-receivers.... The increasing resort to higher, English education, has been another drain for Brahmin families. They were not only the first to take to it, today they are almost the one community in the village who have sought it wholesale. Although this may not fully account for their decline in numbers in the village, it is certainly the cause of emigration of a considerable number of Brahmin families and their lack of interest in land. It looks as though the Brahmin thrives best in towns and the rural soil is uncongenial to his genius.[76]

What is indicated by this case is that (a) though the Brahmans had been active agriculturists some decades before, they ceased to take a personal interest in land, stopped their direct farm management and (b) began to let their land out to tenants; (c) this change led them to lose a part of their landownership; in addition (d) the increasing desire for higher education was one of the causes of their emigration to urban areas and their decreased interest in land.

This process of detachment from agricultural management can be seen as part of the trend outlined earlier in this paper: Tenancy became more competitive as the farm-managers' direct participation in and devotion to cultivation became an important factor in obtaining a better yield in the intensive cultivation method.

In North Arcot district, it was reported in 1904 that subletting was prevalent where Brahman landowners had other engagements to attend to, as Vakils, officials or traders for example.[77] In Tanjore district also, the practice of subletting was wide spread, especially in Brahman villages and where lands were passed out of the hands of the cultivating

[76] Thomas and Ramakrishnan (eds.), *Some South Indian Villages: A resurvey*, p.61; See also, Baker, *An Indian Rural Economy*, pp.178-81.
[77] 'Notes to G.O. Nos.3594-95, Revenue, 9th December 1914', p.20.

class into the possession of pleaders, officials and merchants;[78] and in Trichinopoly subletting was reported to be resorted to by absentee landlords and by those who followed professions other than agriculture.[79]

The emergence of low-caste small farmers and the decrease in the number of large Brahman landowners

Some depressed-caste members were elevated from the status of mere agricultural labourers to that of tenant cultivators. A confidential report in 1912 on the tenants in North Arcot district indicated the existence of tenants who came from the labourer class. According to this report, there were two kinds of tenants; those who were pattadars themselves in a village, cultivating other lands on the varam (sharecropping) system, and those who were more or less labourers, living in the village without any patta.[80]

The report stated, 'the Tanjore porakkudis [sharecroppers], as I know, consist of Pariahs [depressed caste] and Pallis, the latter more commonly known by the name of Padayachis. Both are low in their social status....'[81] Since in 1883 Tanjore Manual Pariahs were reported to be employed chiefly as agricultural labourers and Pallars (depressed caste) were stated to be praedial labourers,[82] it may not be too bold to conclude that some from the depressed castes had been elevated from the status of labourers to that of tenants.

A report from Trichinopoly district more explicitly stated that the sharecroppers came from the agricultural labourer class:

The position of a *mattu-varam* man [a sharecropper who cultivates land with his own oxen] may not be quite [so] satisfactory as that of the *varam* tenant but it evidently is an improvement on that of the agricultural labourers from whom he has sprung and to whom he has to revert in case a landlord does not care to entrust him with the cultivation of the land.[83]

[78] Ibid.
[79] Ibid., p.21.
[80] G.O., No.3594 (Confidential), Revenue, 9 December 1914, p.32.
[81] 'Notes to G.O. Nos.3594-95, Revenue, 9 December 1914', p.69.
[82] *Manual of Tanjore District*, pp.203-4.
[83] 'Notes to G.O. Nos.3594-95, Revenue, 9 December 1914', p.75.

The wet lands in Kuritalai taluk were let on varam (sharecropping) to tenants from the Pallar caste.[84]

In Madura district also, as mentioned already, agricultural labourers who returned from Ceylon and Malaya bid for lands as tenants. Some of the Perumalai Kallars of Usilampattinadu settled as tenants around Madura.[85]

[84] Sivaswamy, *Tenant*, part 2, p.44.

[85] Ibid., p.50. A transformation of a large section of the farm servant into tenant cultivators in Nellore district was suggested by M. Atchi Reddy, 'The commercialization of agriculture in Nellore District 1850-1916: effects on wages, employment and tenancy,' in K.N. Raj, Neeladri Bhattacharya, Sumit Guha and Sakti Padhi (eds.), *Essays on the Commercialization of Indian Agriculture* (Delhi, 1985), p.180. Satyanarayana asserts that the peasant-bourgeoisie in colonial Andhra attempted to increase their occupational holdings by leasing-in the land of others (*Andhra Peasants*, p.112). One piece of evidence to which he refers is a description from Baliga's study, which stated that the pattadar 'whose lands are *sufficient* for all their requirements' took additional lands on lease to make up an economically viable holding (B.S. Baliga, *Studies in Madras Administration*, vol. 2, Madras, 1960, p.130; emphasis added). Though the name of the government document on which Baliga based his description is not given, he was certainly citing sentences from a 1914 Government Order ('Note to G.O. Nos. 3594-5, Revenue, 9 December 1914'). What the original document said, however, was: 'The pattadar whose patta lands are *insufficient* for all his requirements' took additional land on lease (ibid., p.102; emphasis added). Baliga made a crucial mistake when he cited the original government report; he wrote 'sufficient' instead of the original 'insufficient'. N.G. Ranga's 1925 survey of Guntur district does not simply point to the trend for big ryots to lease-in land: while noting that, 'in the last four years', big ryots 'have begun to rent others' lands', it also reports that 'in most places, most of the ryots find it more profitable to lease out their lands after such holdings exceed the limit of ten acres than to cultivate them' (N.G. Ranga, *Economic Organisation of Indian Villages*, vol. 1, Deltaic Villages, Bezwada, 1926, pp.62-3). The result of the 1930 Economic Enquiry Committee also reveals that while in villages in dry zones of Kistna district and in upland taluks of East Godavari district, most rented land was occupied by bigger ryots, landless people and smaller ryots were the main lessee classes in irrigated villages in delta zones, leasing-in more than half of the rented area there (*Report of the Economic Enquiry Committee*, Vol. 1, Madras, 1930, pp.33-5). V.V. Sayana's study, to which Satyanarayana referred, also emphasised that the bulk of the tenants were small farmers and landless people and that cases of large landlords' leasing-in were observable only in some villages (V.V. Sayana, *The Agrarian Problems of Madras Province*, Madras, 1949, p.198). Thus, while, as correctly indicated by Satyanarayana, there were some big ryots who leased-in others' land in some areas of colonial Andhra, it may be misleading to exaggerate the trend

The acquisition of land by members of depressed castes in South India has been suggested by some researchers already.[86] Our analysis of the village settlement registers in a taluk in Trichinopoly district showed evidence of land transfer from higher-caste members to members of the depressed and other lower castes between 1895 and 1925.[87] Other contemporary evidence, including government records, attests to land acquisition by the lower castes.

A report published in 1893 stated as follows:

Taking the labouring classes as a whole, the improvement in their condition in recent years is manifested, not in any clearly visible rise in the standard of living of the lowest grades or in the comforts that they enjoy, but in the fact of a considerable portion of the labourers, who, under the old conditions, would have remained in the lowest grade, having been drafted into the next higher grade, while a portion of the latter has gone into the grade which is next higher, and so on. Thus, a percentage of labourers of the pannial class, as will be seen from Mr.Clerk's account, has gone into the grade of porakudies, and a considerable percentage of porakudies has gone into the class of tenants, paying definite rents in cash or kind, while a portion of the latter has acquired landed property and become puttadars... The statistics...show that, in the three districts of Tanjore, Chingleput and South Arcot, in which these classes [the Pariahs and the Pallars] are found in large numbers, a considerable portion possesses landed property.[88]

In Tanjore, it was stated, 'There can be no doubt that the position of the porakudis has very considerably improved, several of them having become landholders'.[89] The purchase of landed property by lower caste members was also reported from Chingleput district: 'Judging by

---

in the large landowning class to increase operational holdings, considering that the large landlords were the main lessors of tenancy land. Rather, the general trend was for the smaller peasants and the landless peasants to increase their operational holdings.

[86] Kumar, 'Landownership and inequality', p.260; Baker, *An Indian Rural Economy*, p.152, pp.182-3; Tsukasa Mizushima, 'Changes, chances and choices: the perspective of Indian villagers', *Socio-Cultural Change in Villages in Tiruchirapalli District, Tamil Nadu. India*, part 2-1, (Tokyo, 1983), pp.75-86. For the cases in Western India, see Neil Charlesworth, *Peasants and Imperial Rule* (Cambridge, 1985), p.224.

[87] Yanagisawa, 'Mixed trends'.

[88] *Progress*, p.152.

[89] Ibid., p.149.

general impressions I should say that the lower classes of agriculturists, that is, the Pallis and Panchamas [depressed castes], are steadily acquiring land in Chingleput...there were many instances of large landholders selling good land at fair prices to members of the lower castes'.[90] The Manual of Tinnevelly district indicated, 'the opening of the market to all classes alike, and giving the low castes as good a title to hold lands obtained by purchase as the high, has thrown no inconsiderable part of the land out of the high caste monopoly in which it was held, and into the hands of the thrifty merchant, artisan or laborer'.[91] 'The result is the higher classes, who were sole landholders before, have now to give up their land little by little, whereas the poor laboring classes have acquired land by dint of their economical savings'.[92]

Further evidence was provided by the District Registrar of Tinnevelly as follows:

[T]he daily increasing independence of the cultivator, his boldness to refuse to give the landlord anything more than his actual due, using his time and labor to more profitable things, his savings, &c., have enabled him to buy cattle of his own to meet the expenses of cultivation from his own pocket without depending on the mercy of his usurious landlords, who, saved of these services, is paid a much less share of the produce. The said cultivators have gone on further. They began to advance sundry sums to their landlords, and have bought, in most cases, small bits of land of their own, which they cultivate themselves, and obtain all the produce without a sharer.[93]

The Deputy Collector of Coimbatore district, who made personal enquiries into the condition of the labouring classes, concluded that 'those who have once formed the landless class, the petty traders, the artizans and the weavers, who have chosen to work in the fields and elsewhere, have now acquired landed property to some extent'.[94]

It is noteworthy that, at least in some areas, the small farmers who emerged from the labouring class seem to have pursued intensive

---

[90] G.O., No.2765, Revenue, 16 September 1912, p.4.
[91] *Manual of Tinnevelly District*, pp.30-31.
[92] *Progress*, p.ccxx.
[93] Ibid., pp.ccxxii-ccxxiii.
[94] Ibid., p.151, n.70.

cultivation, growing valuable crops using well irrigation. The Statistical Appendix for Trichinopoly District published in 1931 stated,

The area under wells has also increased and a large number of erstwhile agricultural labourers have become petty landholders having invested their savings in small extents of land and, by sinking wells in them, converting them into garden lands which yield such valuable crops as chillies, tobacco and sugarcane.[95]

A report from Tinnevelly district indicates the same trend. Referring to the enormous increase in the number of Shanan pattas as one of the most noticeable features of the resettlement conducted around 1910, it noted that 'in many villages the [Shanan] labourer is steadily converting himself into a peasant-owner'. Cultivation with the aid of wells was their specialty, and 'if water exists under the ground a Shánán will find it, and will quickly convert into a luxuriant garden a patch of poor soil which, in the time of its previous owner, had been a dreary waste'.[96] The Shanans were reported to stand out as the most enterprising of garden cultivators in all parts of the district.[97] The evidence indicates that some of the new wells were dug by small farmers coming from the labourer class.

The cultivation of well-irrigated garden land is, at least in some areas, likely to have been an important economic basis on which former labourers based their economic independence. The resettlement report on Trichinopoly district states as follows:

There is no hard and fast line between the petty ryot and the agricultural labourer and many of the former seek to eke out an existence by themselves and their families working as casual labourers during the busy season. Many of this class are investing their savings in land, and by sinking wells are enabled to take up land which by dint of hard labour can be improved until it produces good crops. It is to garden land that the petty ryot must in most cases turn. It is seldom possible for him to get large areas of dry land while the rents for wet land are usually so high that he might just as well continue

---

[95] *Statistical Appendix for Trichinopoly District* (Madras, 1931), p.84.
[96] *Gazetteer of Tinnevelly District*, p.128.
[97] Ibid., p.157.

to be a labourer. On garden land a man's industry governs his outturn.[98]

Even though a considerable number of former agricultural labourers acquired or leased land, many of the depressed-caste families remained landless and worked as full-time agricultural labourers. Our analysis of the village settlement registers in Lalgudi taluk for 1925 shows that the depressed-caste families owning any extent of land never exceeded 40 percent of total families from this caste.[99] While some among the remaining families, that is, those having no landed property, may have been tenant cultivators, many must have remained mere agricultural labourers without any operational holdings.

In addition, many small farmers who emerged from the agricultural labourer class either by purchasing or by leasing land still had to supplement their incomes by working as coolie labourers. In the 14 wet villages in Lalgudi taluk in 1925, most of the land plots owned by depressed castes were less than one acre, and scarcely ever exceeded two acres.[100] As the income they earned from their own farms was, in most cases, too small to maintain their families, they and their family members had no choice but to hire themselves out as daily labourers on other farms to earn supplementary income, which may have been larger than the income from their own farms.

The small tenant farmers also hired themselves out as agricultural coolies. A survey of tenants in 27 villages in Tinnevelly district noted that 'out of a total number of 3,139 tenants, 55 per cent were found to be coolies'.[101]

On the other hand, though the total number of agricultural labourers did not decrease,[102] the permanent labourers such as pannaiyals are likely to have been gradually displaced by daily labourers. The small farmers of depressed castes hired themselves out as daily labourers instead of being hired as permanent workers, as they now needed to spare their time for the cultivation of their own farms. The development

[98] P.B.R., No.86, 26 November 1923, p.27, para.22.
[99] Yanagisawa, 'Mixed Trends', p.416.
[100] Yanagisawa, *Minamiindo Shakaikiezaishi Kenkyu*, p.188, Table 7.9.
[101] G.O., No.3594 (Confidential), Revenue, 9 December 1914, p.115.
[102] Table 9, compiled from Census data, indicates change in the share of the agricultural labourers in the agricultural working force.

of a sense of independence among these classes also promoted this trend. The settlement report of Tanjore district noted in 1921: 'The tendency will probably be toward a further rise [of grain wage] owing to the growing fluidity of labour caused by the facilities for emigration and to some extent also to the replacing of pannaiyals by free labourers as the result of the present movement for the acquisition of house-sites for Panchamas'.[103]

In Vadamalaipuram village in Ramnad district, there were 53 permanent labourers and about 100 casual labourers in 1916, whereas in 1958 there were 140 resident agricultural labourers, of which only three were attached workers; the rest were all casual workers.[104]

On the other hand, our village settlement data from Lalgudi taluk shows a decline in the number of large Brahman landowners, which is likely to indicate a decrease in the number of large farms managed by them.

While no statistical data on the operated holdings is available, we noted a reduction in the number of the large Brahman landowners in villages in Lalgudi taluk. In the 14 wet villages in Lalgudi taluk, as shown in Tables 1-B, 2-B and 3-B, the number of large Brahman pattadars owning 50 acres or more continuously decreased between 1865 and 1925. The trend of larger landholdings decreasing to the larger extent was noticed in the intermediate and dry zone villages. If we can assume, as suggested by Sumit Guha,[105] that trends in landownership may be used as an indicator of trends in operated holdings, the change in Brahman landholding suggests that in both wet and intermediate zones, the number of and the acreage under the large Brahman farms was decreasing in the period, reflecting the general tendency for the managing unit to decrease.

The decline in large Brahman landholding seems to have been accompanied by a change in family size. The number of Brahman landholders increased from 370 in 1865 to 866 in 1895. The increase in

[103] P.B.R., No.28, 12 February 1921, p.49.

[104] Athreya, *Vadamalaipuram*, p.93.

[105] Sumit Guha, 'Some aspects of rural economy in the Deccan' in K.N. Raj *et al.* (eds.), *Essays on the Commercialization of Indian Agriculture*, p.238.

small landowners of less than one acre may be mostly nominal, being caused by a technical change in the method of describing joint-pattadars' names in the settlement registers. As each of the joint pattadars' names was not mentioned in 1865 settlement registers, many probable Brahman joint pattadars, unless they had a single patta, cannot be reckoned among the Brahman landholders in our data-processing, whereas each of the joint pattadars' names was given in the 1895 registers, resulting in a large nominal increase in Brahman landholders' numbers.[106]

However, this does not explain the increase in medium-sized landholders holding between 1 and 50 acres, because most owners in this group owned a single patta and, therefore, each of their names surely appeared in the register even in 1865. The rate of population increase between 1865 and 1895 was only about 9 percent, which was not large enough to explain the large increase in the number of Brahman landowners in this size group. A report which we cited in the previous section seems to offer an explanation, hinting at a possible correlation between the intensification of cultivation and the change in family size: 'My enquiries tend to show that, under the stress of necessity and the additional incentives to individual exertion promoted by the breakup of the joint family system, greater care is now bestowed on cultivation of lands in Tanjore district than in times past; and this is to some extent the case in other districts also'.[107]

In this connection, a change suggested by population data in these villages is worth notice. Table 8, population and numbers of houses in some villages in this taluk, shows a decrease in the number of persons per house between 1871 and 1925. Though the definition of 'house' in the statistics remains obscure, they seem to indicate a decrease in family size in this period.

The decline in family size is likely to have been associated with a decline in the size of the farm and its managing unit. It may not be too

---

[106] Yanagisawa, *Minamiindo Shakaikeizaishi Kenkyu*, ch.7.
[107] *Progress* p.56. Dharma Kumar has already indicated the breakup of large landholdings and related it to the breakup of the joint families (Kumar, 'Agrarian Relations: South India', p.234).

bold to presume that the breakup of the joint family may have influenced this increase in the number of Brahman landholders, though the data we have is too limited to generalise the conclusion, and our interpretation of the relationship between family size, change in cultivation method and increase in Brahman pattadars remains speculative.

As we discussed in the previous section, the decrease in the total acreage in Brahman landholdings in Lalgudi taluk became apparent after 1895. The transfer of land owned by Brahmans to other communities was evidenced by the 1936 survey of Gangaikondan village cited in the previous section. Our analysis of the transfer of plots of land owned by a large Brahman landlord reveals that many plots had been transferred to members of other communities between 1895 and 1925.[108]

It may be plausible to conclude that large landownership by Brahmans decreased in two ways, first through the breakup of joint families and second through the transfer of land held by large landowners to other communities, and that this change was accompanied by a reduction in farm size.

Conclusion

This chapter attempted to present a new interpretation of some aspects of South Indian agriculture, though it still remains to be attested by further evidence. Main findings are as follows.

Under the general trend in agriculture toward intensive cultivation, the relative advantage of the large farm disappeared; instead, the care and attention to be paid by farmers to cultivation acquired greater importance.

At the same time, the Brahman community, the largest landowning community in wet villages, became increasingly interested in urban jobs and higher education; some of them migrated to urban areas, decreasing their concern with cultivating land.

Due to the increase of emigration of agricultural labourers to the overseas estates and the development of a sense of independence among them, the large farmers increasingly experienced difficulty in securing

[108] Yanagisawa, 'Mixed Trends'.

labourers and making them work as hard as before.

As a result of these changes, some of the high-caste landowners who had formerly managed their farms by employing labourers leased out a part or all of their land to tenants who would devote themselves to cultivating the leased-in land with the help of their family members. Thus, 'the number of tenants are increasing while the number of farm servants are decreasing'.[109] Some landowners even sold part or all of their lands.

The advantage a small family farm possessed over the larger farm has been already asserted by Sumit Guha, who has stated that in the Deccan economy, the large farm had no advantage over the small, while the latter could superexploit its own family labour.[110] Our discussion indicates that the advantage enjoyed by the small peasant farm may have been a result of the historical change in agriculture. What seems particularly important was that one of the factors having promoted this change was the trend in agriculture toward intensive cultivation.

[109] Sivaswamy, 'Agricultural leases in ryotwari areas', in *Tenant*, part 2 p.48.

[110] Sumit Guha, 'Rural economy in the Deccan', p.240. The importance of family labour in raising the yield in paddy cultivation has been suggested by Alan Heston and Dharma Kumar, 'The persistence of land fragmentation in peasant agriculture: an analysis of South Asian cases', *Explorations in Economic History* 20, 2 (1983), p.207. D. W. Attwood, who studied the changes in landholdings in a village in Maharashtra state between 1920 and 1970, observed a tendency of some of the poor to get richer instead of getting poorer in terms of landholdings and found one reason to explain this in the fact that small holders have been shown to invest more labour and supervision per acre, generating higher yields than those obtained on larger farms with the non-mechanised technology; 'Why some of the poor get richer: economic change and mobility in rural Western India', *Current Anthropology* 20, 3 (1979).

Table 1A: *Distribution of area by size of holding: 14 wet villages in Lalgudi taluk, 1865* (acres)

Size Group (Acreage)	1 <0.5	2 <1	3 <2	4 <3	5 <5	6 <10	7 <15	8 <25	9 <50	10 <100	11 =>100	12 Total	(%)
Brahman	17	22	106	105	154	319	201	397	782	1,223	388	3,713	(38.5)
Non-Brahman	19	46	181	153	327	670	373	357	617	344	865	3,957	(41.0)
Chetty	0	1	3	7	3	19	0	15	0	54	0	103	(1.1)
Muttiriyan	1	3	28	9	12	37	40	0	0	0	0	131	(1.4)
Nadan (Nadavan)	8	15	39	54	81	241	83	188	124	200	282	1,317	(13.7)
Pillai	3	8	31	24	93	100	80	40	160	0	460	1,000	(10.4)
Reddi	0	1	0	0	0	0	0	20	0	0	0	20	(0.2)
Udaiyan	4	5	17	10	64	86	85	18	61	0	0	352	(3.7)
Others	3	13	63	49	74	187	85	76	272	90	123	1,034	(10.7)
Depressed castes	0	0	1	3	0	0	0	0	0	0	0	4	(0.0)
Occupational titles	1	9	16	3	24	42	12	0	0	0	0	105	(1.1)
Muslim	0	1	8	12	0	9	0	0	0	0	268	298	(3.1)
Christian	1	1	8	7	5	28	0	0	0	0	0	49	(0.5)
Caste Unknown-female	2	1	6	12	3	26	37	0	40	0	0	126	(1.3)
Temple	1	1	15	16	26	59	23	0	74	0	107	322	(3.3)
Others	22	42	63	39	88	117	147	194	96	259	0	1,066	(11.1)
TOTAL	63	123	404	350	627	1,270	793	948	1,609	1,826	1,628	9,640	(100)

Table 1B: *Distribution of landholders by size of holding: 14 wet villages in Lalgudi taluk, 1865*

Size Group (Acreage)	1 <0.5	2 <1	3 <2	4 <3	5 <5	6 <10	7 <15	8 <25	9 <50	10 <100	11 =>100	12 Total
Brahman	61	29	70	43	41	46	16	22	23	17	2	370
Non-Brahman	80	61	126	61	83	96	31	19	18	5	5	585
Chetty	0	1	2	3	1	3	0	1	0	1	0	12
Muttiriyan	4	4	19	4	3	5	3	0	0	0	0	42
Nadan (Nadavan)	35	22	28	21	21	34	7	10	4	3	2	187
Pillai	9	10	22	10	23	15	7	2	5	0	2	150
Reddi	1	1	0	0	0	0	0	1	0	0	0	3
Udaiyan	18	7	11	4	16	13	7	1	2	0	0	79
Others	13	16	44	19	19	26	7	4	7	1	1	112
Depressed castes	0	0	1	1	0	0	0	0	0	0	0	2
Occupational titles	3	11	11	1	7	6	1	0	0	0	0	40
Muslim	0	2	5	5	0	1	0	0	0	0	1	14
Christian	5	1	5	3	1	3	0	0	0	0	0	18
Caste Unknown-female	5	2	4	5	1	3	3	0	1	0	0	24
Temple	4	2	10	6	7	9	2	0	2	0	1	43
Others	74	60	45	17	24	16	12	10	3	3	0	265
TOTAL	233	168	277	142	164	180	65	51	47	25	9	1,361

Table 2A: Distribution of area by size of holding : 14 wet villages in Lalgudi taluk, 1895 (acres)

Size Group (Acreage)	1 <0.5	2 <1	3 <2	4 <3	5 <5	6 <10	7 <15	8 <25	9 <50	10 <100	11 =>100	12 Total	(%)
Brahman	61	71	177	182	269	439	332	669	986	439	172	3,795	(40.0)
Non-Brahman	66	157	315	278	512	702	578	454	177	262	555	4,065	(42.9)
Chetty	1	7	6	8	13	17	12	33	44	0	0	142	(1.5)
Muttiriyan	14	30	63	15	53	111	45	21	0	0	0	356	(3.8)
Nadan (Nadavan)	16	27	55	107	143	158	136	132	60	179	360	1,371	(14.5)
Pillai	10	26	56	14	88	98	160	90	47	83	195	867	(9.1)
Reddi	1	0	2	3	3	11	13	23	0	0	0	56	(0.6)
Udaiyan	5	19	44	37	72	121	82	52	0	0	0	432	(4.6)
Others	19	48	89	94	140	186	130	103	26	0	0	841	(8.9)
Depressed castes	2	10	10	8	10	0	0	0	0	0	0	39	(0.4)
Occupational titles	2	4	15	14	19	54	0	0	0	0	0	107	(1.1)
Muslim	2	3	15	0	24	47	23	0	0	0	227	341	(3.6)
Christian	3	5	16	0	8	5	10	0	0	0	0	48	(0.5)
Caste Unknown-female	20	43	52	50	74	121	24	25	30	0	0	439	(4.6)
Temple	10	18	32	26	28	34	71	36	69	0	0	323	(3.4)
Others	29	33	48	51	41	71	39	14	1	1	0	322	(3.4)
TOTAL	195	344	680	609	985	1,473	1,077	1,198	1,263	702	954	9,479	(100)

Table 2B: Distribution of landholders by size of holding: 14 wet villages in Lalgudi taluk, 1895

Size Group (Acreage)	1 <0.5	2 <1	3 <2	4 <3	5 <5	6 <10	7 <15	8 <25	9 <50	10 <100	11 =>100	12 Total
Brahman	347	98	122	75	67	59	28	35	28	6	1	866
Non-Brahman	382	214	223	114	130	105	48	24	5	4	2	1,251
Chetty	5	10	4	3	4	2	1	2	1	0	0	32
Muttiriyan	81	41	44	7	13	15	4	1	0	0	0	206
Nadan (Nadavan)	90	38	39	44	37	24	11	7	2	3	1	296
Pillai	49	35	41	6	23	15	12	5	1	1	1	189
Reddi	5	0	1	1	1	2	1	1	0	0	0	12
Udaiyan	34	24	30	14	17	19	7	3	0	0	0	148
Others	118	66	64	39	35	28	12	5	1	0	0	368
Depressed castes	9	13	8	3	3	0	0	0	0	0	0	36
Occupational titles	12	4	10	5	5	8	0	0	0	0	0	44
Muslim	16	4	9	0	6	6	2	0	0	0	1	44
Christian	15	7	10	0	2	1	1	0	0	0	0	36
Caste Unknown-female	93	63	38	20	19	18	2	1	1	0	0	263
Temple	37	24	23	11	7	4	6	2	2	0	0	116
Others	143	49	35	21	11	10	3	1	0	0	0	265
TOTAL	1,054	476	478	249	250	211	90	63	36	10	4	2,921

Table 3A: *Distribution of area by size of holding: 14 wet villages in Lalgudi taluk, 1925* (acres)

Size Group (Acreage)	1 <0.5	2 <1	3 <2	4 <3	5 <5	6 <10	7 <15	8 <25	9 <50	10 <100	11 =>100	12 Total	(%)
Brahman	50	74	187	161	324	565	375	522	388	237	0	2,884	(30.5)
Non-Brahman	155	285	550	357	506	698	271	227	212	279	789	4,328	(45.8)
Chetty	3	11	44	18	21	55	0	64	36	69	0	322	(3.4)
Muttiriyan	47	80	109	60	94	92	22	18	0	0	0	522	(5.5)
Nadan (Nadavan)	22	35	69	36	82	51	43	15	33	0	101	487	(5.2)
Pillai	17	31	83	45	41	121	73	35	49	79	474	1,047	(11.1)
Reddi	1	6	8	11	22	37	0	0	0	0	214	392	(4.1)
Udaiyan	16	34	77	71	122	210	82	15	0	76	0	703	(7.4)
Others	49	88	160	116	124	132	51	80	0	55	0	855	(9.0)
Depressed castes	26	42	49	13	0	8	0	0	0	0	0	138	(1.5)
Occupational titles	4	8	18	2	13	11	0	0	0	0	0	56	(0.5)
Muslim	10	22	32	8	8	6	11	0	0	0	109	206	(2.2)
Christian	21	28	40	9	12	21	0	0	0	0	0	132	(1.4)
Caste Unknown-female	49	113	149	142	126	147	13	16	0	0	0	755	(8.0)
Temple	14	29	44	44	31	62	36	35	144	52	229	720	(7.6)
Others	42	41	69	35	31	16	1	0	1	0	0	231	(2.4)
TOTAL	371	642	1,138	771	1,051	1,534	707	800	745	568	1,127	9,450	(100)

Table 3B: Distribution of landholders by size of holding: 14 wet villages in Lalgudi taluk, 1925

Size Group (Acreage)	1 <0.5	2 <1	3 <2	4 <3	5 <5	6 <10	7 <15	8 <25	9 <50	10 <100	11 =>100	12 Total
Brahman	359	102	136	67	83	77	31	28	11	4	0	898
Non-Brahman	953	385	388	150	132	101	23	12	6	4	5	2,159
Chetty	20	14	30	8	6	8	0	3	1	1	0	91
Muttiriyan	309	109	76	26	24	14	2	1	0	0	0	561
Nadan (Nadavan)	166	46	48	15	22	7	4	1	1	0	1	311
Pillai	94	41	58	19	11	17	6	2	1	1	2	252
Reddi	4	7	6	5	5	6	0	0	3	0	2	38
Udaiyan	82	49	57	29	31	29	7	1	0	1	0	286
Others	278	119	113	48	33	20	4	4	0	1	0	620
Depressed castes	110	56	35	6	0	1	0	0	0	0	0	208
Occupational titles	24	9	13	1	3	2	0	0	0	0	0	52
Muslim	51	31	22	3	2	1	1	0	0	0	0	112
Christian	114	37	30	4	3	3	0	0	0	0	0	191
Caste Unknown-female	275	158	113	58	33	20	1	1	0	0	0	659
Temple	59	39	32	18	8	9	3	2	4	1	1	176
Others	275	60	48	14	8	2	0	0	0	0	1	407
TOTAL	2,220	877	817	321	272	216	59	43	21	9	7	4,862

Table 4: *Cases of land transfer to depressed castes in a village:*
1895 - 1925

Patta owned by depressed castes in 1925		Owners in 1895	
Patta number	Acres	Acres	Title, community, name of pattadars
No. 5	1.03	0.41 0.22 0.40	Jointly held by 3 Pillais (Vellala) Muppan (depressed caste), Periyanna Holders unknown
No. 58	1.40	1.05 0.35	Jointly held by 2 Ayyars (Brahman) Jointly held by 2 Ayyars & 1 Ayyangar
No.147	0.10	0.10	Ayyangar Seshadri Ayyangar
No.158	3.83	0.92 0.92 0.92 1.08	Pandaram Ku Muttusami Pandaram Pandaram Muttu Pandaram Female Minakshiammal Jointly held by 2 Pandarams & 1 female
No.167	0.55	0.28 0.28	Ayyar Venkita Rangayyayyar Jointly held by 8 Ayyars, 2 Ayyangars
No.194	2.07	0.46 1.15 0.46	Ayyar Venkita Rangayyayyar Ayyangar Seshadri Ayyangar Jointly held by 7 Ayyars & 2 Ayyangars
No.203	1.37	0.86 0.51	Pillai Matturaivana Pillai Holders unknown
No.214	0.10	0.10	Ayyangar Seshadri Ayyangar
No.282	4.45	4.36 0.09	Ayyar Venkita Rangayyayyar Ayyangar Seshadri Ayyangar
No.313	0.10	0.10	Ayyangar Seshadri Ayyangar
No.344	1.03	1.03	Maruda Viran
No.363	0.09	0.09	Ayyangar Seshadri Ayyangar
No.391	0.93	0.47 0.47	Ayyangar Krishna Ayyangar Jointly held by 3 Ayyars
No.T28	5.25	5.25	Ayyar Subbaratinamay Ayyar

Table 5: *Village census data for 60 villages in Lalgudi taluk:*
average per village for 1894 and 1925

Zones		Wet	Intermediate	Dry	Total
Number of villages		18	20	22	60
Total cultivated	1894	594.2	735.6	1428.4	927.7
area (acres)	1925	591.1	739.3	1585.4	981.4
Wet land	1894	553.9	518.0	180.2	414.3
(acres)	1925	556.8	543.0	205.4	432.6
Area double-cropped	1894	352.3	273.5	108.1	235.9
(acres)*	1925	364.7	304.6	153.1	266.4
Area cultivated	1894	877.8	686.9	202.3	566.5
with paddy (acres)*	1925	920.4	795.7	294.4	649.3
Area cultivated	1894	24.6	73.1	150.3	86.8
with ragi (acres)*	1925	3.7	30.6	158.3	69.4
Area cultivated with	1894	8.6	82.9	822.3	331.7
other cereals (acres)*	1925	2.2	42.6	697.9	270.8
Area cultivated with	1894	15.9	34.6	19.0	23.3
sugarcane (acres)*	1925	3.8	13.8	9.5	9.2
Area cultivated with	1894	0.0	3.7	37.4	14.9
cotton (acres)*	1925	0.0	0.6	64.9	24.0
Area cultivated with	1894	2.2	6.4	12.2	7.3
groundnut (acres)*	1925	0.2	13.1	109.6	44.6
Number of wells	1894	0.0	2.4	47.7	18.3
	1925	0.0	3.7	88.1	33.5
Number of cattle	1894	516.1	645.2	852.2	689.0
	1925	622.7	735.2	975.7	786.5
Number of carts	1894	25.9	28.3	28.5	27.7
	1925	38.9	43.5	55.2	46.4
Number of ploughs	1894	107.4	113.5	172.0	133.7
	1925	129.8	145.0	256.1	179.9

Source : *Re-settlement Registers for Villages in Lalgudi Taluk, Trichinopoly District* (c. 1895); *Re-settlement Registers for Villages in Lalgudi Taluk, Trichinopoly District* (c.1926).
Note : * Area cropped for 1894 is the average for five years ending 1894, whereas that for 1925 is for a single year, 1925.

Table 6: *Irrigated area in Tamil districts* (thousand acres)

	1891-92	1902-03	1912-13	1922-23
Government canals	1,349	1,389	1,442	1,436
Private canals	4	7	27	23
Tanks	1,358	1,407	1,970	2,063
Wells	785	796	989	1,206
Other sources	74	93	149	139
Total	3,570	3,690	4,577	4,867

Source: *Season and Crop Reports*.
Note: Tamil districts: Chingleput, North Arcot, South Arcot, Salem, Coimbatore, Trichinopoly, Tanjore, Madura, Ramnad and Tinnevelly.

Table 7: *Area cropped more than once* (thousand acres)

	1891-2	1902-3	1912-13	1922-3	1933-4	1944-5	1946-7
Chingleput	58	153	168	233	213	237	271
South Arcot	175	286	162	267	255	272	242
North Arcot	153	247	203	287	255	287	327
Salem	190	235	202	269	234	269	263
Coimbatore	213	310	276	304	324	348	334
Trichinopoly	85	107	193	171	149	219	183
Tanjore	93	111	98	121	172	341	289
Madura	98	140	171	176	171	231	196
Ramnad	-	-	49	74	56	64	65
Tinnevelly	200	216	232	219	239	214	206
Total	1,265	1,805	1,754	2,121	2,068	2,482	2,376

Source: *Administration Report of the Madras Presidency during the year 1891-92* (Madras, 1892); *Season and Crop Reports*.
Note: A division of Tinnevelly district and that of Madura district were amalgamated into a new district, Ramnad, in 1910.

Table 8: *Change in population in villages in Lalgudi taluk*

Year	1871 Persons (houses)	1895 Persons (houses)	1925 Persons (houses)
14 wet villages	14,260 (2,214)	15,555 (2,651)	16,585 (3,445)
Persons per house	6.44	5.87	4.81
6 intermediate villages	6,660 (1,042)	6,507 (1,096)	7,222 (1,469)
Persons per house	6.39	5.94	4.92
5 dry villages	7,129 (1,178)	8,158 (1,628)	9,025 (1,937)
Persons per house	6.05	5.01	4.66

Source: Figures for 1895 and 1925 are from *Resettlement Registers of Villages in Lalgudi Taluk, Trichinopoly District*. Figures for 1871: The village names corresponding to those of 1895 are identified by the description on the change in village boundaries in the 1895 village resettlement register for each village. The 1871 figures for these identified villages are compiled from the *Census Statement of Population of 1871 in each Village of the Trichinopoly District* (Madras, 1874).

Table 9: *The share of agricultural labourers in the agricultural workforce in the Madras Presidency* (thousands)

	Adult Male	Population supported by male workers		Male workers			
	1871	1881	1891	1901	1911	1921	1931
Workforce	7,021	6,637	11,667	8,445	9,099	9,473	9,082
Cultivators etc.	4,938	4,465	8,617	6,092	6,445	6,680	5,714
Agric. labourers	2,083	2,172	3,050	2,353	2,654	2,793	3,368
Proportion (%) of agric. labourers	29.7	32.7	26.1	27.9	29.2	29.5	37.1

Source: *Census of India*.
Note: 'Cultivators etc.' includes the following census categories: 'Cultivators', 'Land proprietors', 'Landholders', 'Planters', 'Landlords', 'Farmers', 'Permanent leaseholders', 'Farm bailiffs', 'Pattadars, Ryots', 'Agriculturists', 'Tenants', 'Growers of special products', etc. 'Agricultural labourers' denotes such census categories as 'Agricultural labourers', 'Ploughmen', 'Farm servants', 'Field labourers', 'Labourers and workmen otherwise unspecified' and 'General labourers'.

Chapter 4.2

A COMPARISON WITH THE JAPANESE EXPERIENCE

Haruka Yanagisawa

Introduction

The changes in South India outlined above seem to have their parallels in agrarian change in seventeenth- to nineteenth-century Japan. Common features include the intensification of agricultural production, the acquisition of land by erstwhile agricultural labourer classes and the consequent increase of small farmers, and the decrease in the number of permanent bonded labourers. Moreover, both cases indicate that the general trend in agricultural progress in Japan and in South India has been toward a smaller rather than a larger farm. Following the general discussion in Chapter 1.2, this chapter attempts to make a more specific comparison between Japan and South India, and, by so doing, highlights the characteristics of agrarian changes in South India from a comparative perspective.

Similarities

In the early Tokugawa years (1603-1868) and the period immediately preceding it, there were many large farms, mainly cultivated by permanent servile labourers called *shoju, genin, fudai-genin,* and so on. These labourers were expected to serve for their masters throughout their lifetime, and were in turn supported by their masters, often living in their masters' houses.[1] In addition to these *fudai-genin*, annual labourers were hired and tenants worked as casual labourers in their landlords' farms at peak seasons. Such a large farm was managed not by a nuclear family but either by an extended family, as has been the popular view, or jointly by stem families, as was asserted by Osamu Saito.[2]

---

[1] As to the 'Fudai genin' in Kansai area, see Rintaro Imai and Akihiro Yagi, *Hoken Shakai no Noson Kozo* (The Structure of Rural Villages in Feudal Society) (Tokyo, 1955), part 2, ch. 2, section 1.

[2] Osamu Saito, 'Daikaikon, jinko, shono keizai' (The great reclamation, pop-

In this period, the seasonal fluctuation of labour input was so large that a large workforce was needed to cultivate the land, especially at the time of transplanting and harvesting, and for collecting the green manure from the forest. The plough was drawn by animal power, which only a big farmer could afford to purchase and maintain. The forest was important for feeding the draft animals also.[3]

Several important changes occurred in the method and technique of cultivation in the late medieval and the Tokugawa periods, particularly in the seventeenth and eighteenth centuries in the developed areas; and these new techniques spread over the backward districts in Japan.[4] First, the introduction of the hoe enabled even those small peasant farmers who did not own draft animals to cultivate their land. It could be obtained fairly cheaply, and thus became the most important farm instrument. A reduction in the number of draft animals was reported from Nobi district, where the number of draft animals (cows, oxen and horses) in 838 villages declined from 17,825 to 8,104 during the period between 1660 and 1810.[5] This change, Akira Hayami assumes, forced farmers to work harder and for longer hours than they had in the previous period.

Second, the introduction of soy-bean fertiliser, which could be cultivated in part of the dry fields, decreased the large demand for labour for the collecting of green manure in the forest, thereby reducing the importance of the forest. In the later period, such fertilisers as dried sardines became available in the markets as a result of the development of commodity transactions. This change also supported small peasant farming. Improvements in the art of soil enrichment involved not only the introduction of a variety of fertilisers, but also an increase in the amount applied and the meticulous care given to the timing of applica-

ulation and peasant economy), in Akira Hayami and Matao Miyamoto (eds.), *Nihon Keizaishi*, vol.1 (Tokyo, 1988).
[3] Teisaku Hayama, 'Shono noho no seiritsu to shono gijutsu no tenkai' (The development of small peasant cultivation method and technique), in Junnosuke Sasaki (ed.), *Zairai Gijutsu no Hatten to Kindai Shakai* (Tokyo, 1983).
[4] Hayama, 'Shono noho'.
[5] Akira Hayami, *Nihon niokeru Keizaishakai no Tenkai* (Development of Economic Society in Japan) (Tokyo, 1973), p.93.

tions to ensure optimal effecitveness.[6]

A better method of threshing was also devised. The thresher, which was initially used to separate grain from the ear, consisted of two sticks. The invention of a thresher called *senbakoki* (thousand teeth) drastically reduced the amount of labour required in threshing; and the threshing opportunities for widows, who had formerly been hired for threshing with sticks, were said to virtually disappear.[7]

The development of early-, middle- and late-ripening rice varieties with different dates of maturation enabled peasants to even out the periods of heavy labour demand, especially periods of planting and harvesting, and enabled them to introduce more extensive double-cropping. A red Indica variety of rice, which was introduced into Japan in the late medieval period, could be cultivated in less fertile land. It was supposed to have been grown by small poor farmers in newly-reclaimed lands, and thus to have promoted the process whereby the poor peasant became independent farmers.[8]

As a result of such improvements, areas of double-cropping increased during the period. Hironori Yagi's research on a Kyushu district indicated that, in addition to these two factors, the improvement in irrigation facilities and the necessity for the small farmer to reduce the labour demand at the time of transplanting were also factors that promoted double-cropping.[9]

The trend toward fragmentation of a farmer's landholding and the scattered location of the plots on a farm were not necessarily a regression in agriculture. The fact that a farm consisted of several plots under different irrigation conditions, receiving water at different times, was rather advantageous for those small farmers who could disperse the risk of farm management and mitigate the concentration of labour demand in the peak season by cultivating different varieties of rice, with differ-

[6] Chie Nakane and Shinzaburo Oishi (eds.), *Tokugawa Japan* (Tokyo, 1990), p.71.

[7] Ibid., p.69.

[8] Hironori Yagi, *Suiden Nogyo no Hatten Ronri* (The Logic of the Development of Irrigated Rice Agriculture) (Tokyo, 1983), p.90; Shuichi Miyakawa, 'Daitomai to teishituchi kaihatu' (Great China rice and reclamation of damp lowlands), in *Ine no Ajiashi* (Tokyo, 1987), vol.3, p.280.

[9] Yagi, *Suiden Nogyo*, pp.134-44.

ent times of planting and harvesting among the scattered plots.[10]

The new agricultural techniques and the new system of production were definitely more labour-intensive in character and required much more careful management than before. Fewer labourers were needed to cultivate land, and even a small farmer could manage a farm successfully. The situation was suited to a small farmer who depended on his own family members for his main labour force, as they tended to cultivate land with meticulous care. In this system, the core of the labour force was the nuclear family. The husband and wife were the principal workers, but the elderly and the children did what they could. The goal of the household labour force was to make effective use of family workers without hiring outside help.[11] This new system of agriculture is known as the 'family farm' or 'peasant farm' system.

The change in agricultural techniques accompanied the gradual establishment of the peasant farm system.[12] The change proceeded in two ways: first, there was a tendency for servile labourers to become independent small peasants; second, a small nuclear family increasingly became an independent farming unit. Along with this change, the size of an average farm decreased.[13] It is generally recognised that the number of small farmers increased in the eighteenth century, and the importance of large farms hiring permanent servile labourers decreased considerably, though such farms remained until the latter half of the nineteenth century, at least in some backward areas in Japan.

The trend is well-illustrated by the work of Hironori Yagi on the historical change in a fertile area of a district in Kyushu. In this area, the tendency for servile labourers to become small peasants started in the late medieval period. In a village examined by Yagi, there were 25 cultivators (*sakunin*) in 1656. The minimum operational holding necessary for an independent farm was about five acres at that time.

In 1792, 136 years later, the number of cultivators had risen to 42, with a big increase in the number of small cultivators operating less

[10] Hayama, 'Shono noho'.
[11] Nakane and Oishi (eds.), *Tokugawa Japan*, pp.66-7.
[12] Kanji Ishii, *Nihon Keizaishi* (Economic History of Japan), (second edition; Tokyo, 1991), pp.53-6.
[13] Hayami and Miyamoto (eds.), *Nihon Keizaishi*, vol.1, p.49.

than 2.5 acres. The number of plots grew from 217 to 373 during this 136 years, reflecting the fragmentation of farm land into separate plots, and the consequent decrease in the average size of plots.[14] A decrease in the number of persons who were registered as dwellers in a living unit was also reported. In a village in Yamashiro district, the number of households increased from 57 to 102 between 1673 and 1715, with the average number of persons per household decreasing from 7.56 to 4.08, while in an area in Shinshu district, the population per household decreased from 7.87 to 4.31 between the periods 1671-1700 and 1851-1870.[15]

An important change occurred in the type of agricultural labourer hired by the large farmers. According to research work done in the Kansai area, the most common type of agricultural labourer in the early seventeenth century was a *fudai genin*, who was obliged to serve his master throughout his life (as generally were his children after him, for generations), living in his master's house. From the middle of that century, the number of such life-long servile labourers decreased, and instead there was an increase in the number of *chonenki-hokonin*, who served a master for eight to ten years, generally from childhood until the beginning of adulthood. After completing the term of service, this labourer was supposed to return to his family's small farm, where he would work as part of the family labour force.[16] This change has been understood to indicate the emergence of many small farmers from the former agricultural labourer class, whose farms were too small to engage all of their family members in the work of cultivation, and to provide them with subsistence, so that parents were obliged to send their sons to work as *chonenki-hokonin* for other big farmers. After about 1740, the most common type of agricultural labourer in this area

[14] Yagi, *Suiden Nogyo*, pp.126-8, 132-4.

[15] Ishii, *Nihon Keizaishi*, p.55.

[16] Ryuzo Yamazaki, 'Settsu niokeru nogyo koyo rodo no hatten' (Development of hired agricultural labourers in Settsu Area), in Takamasa Ichikawa, Nobuo Watanabe and Toshio Furushima (eds.), *Hoken Shakai Kaitaiki no Koyo Rodu* (Tokyo, 1969), p.202. Dharma Kumar has compared the pannaiyal and padiyal in South India with *fudai* and *genin* of seventeenth-century Japan. Dharma Kumar, *Land and Caste in South India* (Cambridge, 1965), p.190.

was the worker hired on a one-year contract.[17] The number of daily labourers also increased at this time.[18] Thus the term of labour contract gradually shortened in the Tokugawa period.[19] Imai and Yagi revealed a decrease in the number of *genin* in one village, from 24.7 per cent of its total population in 1659 to 15.4 per cent in 1691.[20] In addition to the general tendency for servile labourers to become peasants, the development of labour demand from outside the rural society, particularly in urban areas, contributed to the disappearance of the long-term labour contract of the old type.[21]

We can conclude that in both South India and Japan, the development in agricultural techniques was generally toward intensive agriculture, which was more efficiently managed by the small peasant families and which promoted the emergence of small farmers with small landholdings, and thus diminished the importance of the large farms which hired permanent labourers on a long-term basis. In the case of paddy cultivation, which was the main crop both in wet areas of South India and in Japan, the direction of agricultural progress was toward the family farm system. The decline in .farm size and the 'fragmentation' of landholdings did not necessarily mean a deterioration of agricultural production, but rather reflected the progress of agrarian society in the paddy-cultivating districts.

Though we do not wish to be as bold as Satoru Nakamura who stated that progress in agriculture in the modern world, in both Asia and Western countries, has been toward the small peasant farm system,[22] nor do we believe that progress in agriculture necessarily implies a change toward a larger farm cultivated by hired wage-labourers. It is plausible that at least in the wet rice-producing areas in East Asia, including Japan and China,[23] and in the wet districts in South India, the

[17] Imai and Yagi, *Hoken Shakai no Noson Kozo*, pp.152-3.

[18] Ibid., p.157.

[19] Hayami, *Nihon niokeru Keizaishakai*, p.99.

[20] Imai and Yagi, *Hoken Shakai no Noson Kozo*, p.132.

[21] Hayami, *Nihon niokeru Keizaishakai*, p.99.

[22] Satoru Nakamura, 'Kindai sekai niokeru nogyo keiei, tochi shoyu to tochi kaikaku' (Agricultural management, landownership and land reform in the modern world), no.1, *Keizai Ronso* 143, 1 (1989), p.30.

[23] As to the case of Chinese agriculture, see Francesca Bray, 'Rice

main trend was toward a smaller farm. In contrast to the industrial sector, the advantage of the economies of scale does not always work in the agricultural sector in these areas.

Differences

The most notable difference between the two cases is the fate of agricultural labourers. In Japan, the number of agricultural labourers rapidly decreased in the latter half of the nineteenth century and they actually disappeared by the beginning of the twentieth century. Satoru Nakamura noted that the number of yearly-contract agricultural labourers began to decrease in the middle of the eighteenth century. Their number is assumed to have been about one million around 1890, and only 380,000 in 1920.[24]

In contrast to the Japanese case, hired labourers have remained a large group in the agricultural population in South India even in the twentieth century, though the dominant form seems to have changed from that of permanently-attached labourer to daily 'coolie' labourer. In other words, the family farm system has not been established as a dominant form of agrarian management in South India, as it was in Japan, though the trend toward the family farm system was noted in South Indian agriculture between 1865 and 1925.

Several factors contributed to this difference. First, though South Indian agriculture exhibited signs of change toward intensive agriculture during the period from 1865 to 1925, it lacked the kind of improvements which might have reduced the labour demand in peak seasons, such as harvesting and planting. These developments took place in Japan over a long period of at least three centuries.

Second, the development of rural industry played an important role in the decrease in labour supply in Japan. In the latter half of the Tokugawa period, there was a widespread development of small-scale rural industry. An important portion of this developed in the form of by-employment for cultivating families. A survey of the sericulture

economies: the rise and fall of China's communes in East Asian perspective', in Jan Breman and Sudipto Mundle (eds.), *Rural Transformation in Asia* (Delhi, 1991).
[24] Nakamura, 'Kindai sekai niokeru nogyo keiei', p.30, pp.29-30.

industry in one area indicates that, at the beginning of the nineteenth century, the spread of silkworm culture among the lower class of farmers reduced the labour supply to the agricultural daily labour market from these small farmers' families.[25]

Though there is debate as to whether or not the industrial population in India declined in the colonial period, it cannot be denied that the development of subsidiary industry among the farmers in rural areas in nineteenth-century South India did not occur in the way that it did in Japan. The caste system may have been one of the factors which prevented agricultural caste members from participating in industrial production, though the significance of this requires further examination. The colonial condition also set an important limit on the expansion of subsidiary industry. The hand-spinning industry, which had been an important supplier of subsidiary jobs for female members of farmers' families, significantly declined, if not died out, in the nineteenth century. In the latter half of that century, in some villages in South India, the depressed castes were engaged in weaving coarse cloths. However the evidence in the first half of the twentieth century shows that coarse cloth weaving by depressed castes was seriously hit by the competition from the mill industry.[26]

Third, what contributed most to the actual disappearance of agricultural labourers in Japan was the expansion of labour demand in the industrial sector after the Meiji period, particularly in the 1910s and 1920s. The increasing demand for labourers, caused by rapid industrial development, influenced the wage level of agricultural labourers, and caused large farmers to abandon the hiring of agricultural labourers.[27] The importance of the serious limit set on the development of industry under the colonial conditions in India cannot be too strongly emphasised. The major demand for labourers came only from plantations;

[25] Hiroshi Sinbo and Osamu Saito (eds.), *Nihon Keizaishi* (Economic History of Japan), vol.2 (Tokyo, 1989), p.54.

[26] Haruka Yanagisawa, 'The handloom industry and its market structure: the case of the Madras Presidency in the first half of the twentieth century', *Indian Economic and Social History Review* 30, 1 (1993), p.12.

[27] Yoji Shimizu, 'Chuno hyojunka keiko to noumin keiei' (A trend toward middle peasants and their farm management)', in Shigeaki Shiina (ed.), *Famiri Famu no Hikakushiteki Kenkyu* (Tokyo, 1987).

labour demand from the industrial sector in India was too limited to cause a general rise in the agricultural wage level in South India and to reduce the number of labourers in the agricultural sector.

Colonial rule further impeded the decrease in the number of agricultural labourers in South India. As was mentioned in the previous chapter, the process of stratification among the non-Brahman communities proceeded in Trichinopoly district between 1895 and 1925. While some rich non-Brahmans increased their landholding by exploiting new economic opportunities, a considerable number of small farmers may have lost all or a part of their land, and sunk to the status of smaller farmers, tenant farmers or agricultural labourers. Many of them were forced to wholly or partly hire themselves out as daily labourers. Though some erstwhile agricultural labourer-class (depressed-caste) people may have decreased their supply of labour to the agricultural labour market, this possible decrease in labour supply may have been offset by a new inflow of daily labourers from the declining farmer families. Thus the total supply of agricultural labour does not seem to have decreased even though its composition changed.

Fifth, colonial policy in South India, unlike the policy implemented in Japan in the late sixteenth and early seventeenth centuries, also worked against the emergence of a peasant farm system. Hideyoshi's national cadastral survey attempted to identify a single taxpayer for each plot of land and to officially deny multi-layer ownership. Many small private lords with servile labourers were reduced to peasant status. The status differentiation between the free peasant and the unfree peasant was eliminated. The Tokugawa government supported the independence of the small peasant farm, trying to prevent the appearance of large-scale production and landless agricultural labour.[28]

By contrast, in the Raiyatwari settlement in the Madras Presidency, when the government determined the taxpayer (pattadar) for each plot of land, those large landholders (*mirasidars*) who owned servile labourers or leased out their land to tenants were not precluded from the taxpayers. Actually, such large landowners were often discernible, together with small farmers, among the pattadars, particularly in wet

[28] See ch.1.2 of this volume.

villages.[29] Though slavery as a legal system was abolished in 1843, there was actually no prohibition against the hiring of debt-bonded labourers.[30] In spite of a suggestion made by a Collector that a policy be adopted to support the under-tenants,[31] no positive policy was adopted by the Madras government at least up until 1892, when Tremenheere made a strong recommendation to ameliorate the condition of the depressed castes.[32] However even Tremenheere's recommendations were not fully adopted by the government, except for a few measures, of which gradual development of a land-assignment policy to the depressed castes was the most important. The stance of government policy as a whole continued to be pro-*mirasidar* at least up to the First World War. The land aquisition rule (*darkhasht* rule), in which priority in cultivating unoccupied land in a village was given to the pattadars who already owned land in the village, impeded the acquisition of land by non-pattadar depressed-caste members. This principle remained unchanged at least up to 1918.[33]

Conclusion

Our comparison of South Indian and Japanese agricultural changes seems to indicate an important similarity in agrarian progress between the two regions. We notice a change in agricultural methods toward intensive cultivation, an emergence of small family farms from the former agricultural labourer class, and a gradual deterioration of large farms cultivated by non-family servile labourers. It is plausible that at least in the wet rice-producing areas in East Asia (including Japan and China) and in South India, the main trend in agriculture was toward smaller farms, though the evidence produced in this chapter is too limited to fully attest our conclusion. The pattern of the change thus

[29] W.H. Bayley and W. Hudleston (eds.), *Papers on Mirasi Right* (Madras, 1862), p.417.

[30] Dharma Kumar, *Land and Caste*, p.74.

[31] G.O., No.590, Revenue, 13 April 1875; G.O., Nos.437-8, Revenue, 29 April 1882, pp.387-91, 401-8.

[32] G.O., Nos.1010-1010A, Revenue, 30 September 1892.

[33] Haruka Yanagisawa, *Minamiindo Shakai Keizaishi Kenkyu* (Studies in the Socio-economic History of South India: the emancipation of low-caste folks and the change in the agricultural society) (Tokyo, 1991), ch. 9.

identified can be considered as an outgrowth of the internal force of change inherent in agriculture in these areas.

It is also very important to note that the aspects of similarity should not be overemphasised, as shown in the fact that a large section of the agricultural population remained agricultural labourers in South India. The family farming system has not been established as a dominant form of agrarian management in South India, as was the case in Japan, though the trend toward the family farming system was present in South Indian agriculture. The differences can be attributed to many factors, of which at least some seem to have been closely connected with the colonial conditions in India; the decline of some indigenous industries, the serious limit set on industrial development, the growth of stratification among non-Brahman communities and the resultant decline of some farmers to the status of tenants and agricultural labourers,[34] and the basic stance of government policy. Thus, though it is difficult to judge the impact of colonial rule on the growth of agriculture as a whole, it can not be denied that colonial rule and the structural change to society resulting from the colonial conditions, placed some important obstacles to the inherent growth of South Indian agriculture.

[34] On this aspect of change, see Haruka Yanagisawa, 'Mixed trends in landholding in Lalgudi Taluk: 1895-1925', *Indian Economic and Social History Review* 26, 4 (1989).

Chapter 5

REGIONAL PATTERN OF LAND TRANSFER IN LATE COLONIAL BENGAL[1]

Nariaki Nakazato

The degree of stratification of the peasantry forms one of the major criteria for demarcating agrarian regions of an economy. A classic attempt was made for agrarian Bengal by A. Ghosh in 1950.[2] More recently, Partha Chatterjee and Sugata Bose paid special attention to this problem when they tried to formulate typologies of Bengal agrarian society.[3] Being a zamindari area, however, this province poses a particular technical difficulty for economic historians. Barring the last years of the colonial rule, no reliable all-Bengal data are available on the size-distribution of peasant holdings at any point of time, to say nothing of the time series. This means that it is impossible in the case of Bengal to trace the historical course of peasant stratification on the basis of such relevant statistical figures. If one wants to look at the

[1] I have benefited much from suggestions and comments I received from participants both during and after the conference. Special thanks are due to Professor Binay Chaudhuri, Dr. Clive Dewey, Dr. Peter Robb, Professor Utsa Patnaik and Dr. Sumit Guha. I cannot adequately thank Dr. Peter Robb for volunteering to correct my English. Abbreviations used in this paper are as follows:

RLRA: Government of Bengal, *Report on the Land Revenue Administration of the Presidency of Bengal.*
RLRC: Government of Bengal, *Report of the Land Revenue Commission*, 6 vols. (Alipore, 1940-41).
RRD: *Report on the Administration of the Registration Department of Bengal.* These were published under various titles such as *Report on the Administration of the Registration Department of Bengal*, and *Statistical Returns with a Brief Note of the Registration Department in Bengal.*

[2] A. Ghosh, 'Economic classification of agricultural regions in Bengal', *Sankhya* 10 (1950), pp.109-18.
[3] Partha Chatterjee, 'Agrarian structure in pre-partition Bengal,' in A. Sen *et al.* (eds.), *Three Studies on the Agrarian Structure in Bengal 1850-1947* (Calcutta, 1982), pp.196-204; Partha Chatterjee, *Bengal 1920-1947: the land question* (Calcutta, 1984), ch.12; Sugata Bose, *Agrarian Bengal: economy, social structure and politics 1919-1947* (Cambridge, 1986; Indian ed., 1987). See also Willem van Schendel, *Three Deltas: accumulation and poverty in rural Burma, Bengal and south India* (New Delhi, 1991), pp.108-16.

problem of stratification in a historical perspective, one has to have recourse to some substitute. The present paper is an attempt to make a rough estimate of the historical change in the velocity and regional variations of peasant stratification in Bengal by processing the registration statistics. These statistics, compiled by the Registration Department of the Bengal Government since the 1870s, recorded among other things the annual number of voluntary sales of occupancy holdings by registered deeds. It is provisionally assumed here that the sales of raiyati holdings were one of the major factors causing stratification among the peasantry. To be sure, however, land transfer does not necessarily bring about stratification.[4] In the course of discussion an attempt will be made to compare the results obtained from the study of land transfer with the figures on the size-distribution of raiyati holdings.

Before going on to the examination of registration statistics we must consider the limitation and bias of our study. First, the registration statistics deal only with the voluntary sales that were formally registered at registration offices. The rate of registration was estimated to be rather low in the early years; and, in addition to the voluntary sales, there were compulsory sales by order of the civil court. The Registrar-General remarked in 1892 that the rate of registration was well under 30 per cent.[5] Considering the trend in the following years, his view appears to have been a considerable underestimate. There is no reason to doubt that the principle of compulsory registration of all kinds of land sales introduced in 1886 gradually permeated among the common peasants. Moreover, the number of registration offices in Bengal more than doubled from 202 to 417 between 1881 and 1920, which must have facilitated the registration of smaller transactions. It may be reasonably presumed that the rate of registration picked up steadily and that a fairly large proportion of the sales of raiyati holdings came to be registered in the twentieth century. Execution sales, on the other hand, amounted to about 30 per cent of the total number of registered

---

[4] Teodor Shanin, *The Awkward Class: political sociology of peasantry in a developing society: Russia 1910-1925* (Oxford, 1972), ch.3; Neil Charlesworth, 'The Russian stratification debate and India', *Modern Asian Studies* 13, 1 (1979).

[5] *RRD*, 1892-3.

voluntary sales in 1881, and to 12 per cent in 1904.[6] So far as the twentieth century was concerned, therefore, the number of execution sales was not so large as materially to affect an interpretation based on the study of the movement of voluntary sales. Certainly, the registration statistics should be treated more as an indicator of the trend than as actual figures. When we use them, however, with the above limitations in mind, they will sufficiently answer our purpose of finding variations among different regions of Bengal.

Secondly, it may be argued that partition of raiyati holdings owing to inheritance and other reasons will add to the number of raiyati holdings through time, and that a fair proportion of the increase in sales might be accounted for by such rise in the total number of raiyati holdings. However, this does not seem to have been an important factor in our time. The revisional survey and settlement operations in the Bakarganj and Faridpur districts revealed that the raiyati holdings had recorded only a moderate rise in number during the three to four decades after the first operations. They increased by 27.4 per cent in Bakarganj and 24.2 per cent in Faridpur.[7]

Thirdly, we have to be clear as to which aspect of the stratification process our data are concerned with. It appears to me that two basic processes should be distinguished in the agrarian history of modern Bengal. One refers to stratification of the peasantry through land transfer. With the emergence of a sale and mortgage market in peasant holdings, a new way of concentrating raiyati holdings, that is, land

---

[6] *RRD*, 1881-2 and 1904; 'Statement of classes of purchasers of ryots' holdings and small under-tenures under execution sales' in *Report of the Rent Law Commission with the Draft of a Bill to Consolidate and Amend the Law of Landlord and Tenant, etc.*, 2 vols. (Calcutta, 1883), vol.2, pp.384-92; Government of Bengal, Revenue Department, Land Revenue Branch, May 1906, Proceedings 36-8.

[7] Motaharul Huq, *Final Report on the Revisional Survey and Settlement Operations in the District of Bakarganj 1940-42 and 1945-52 and Khulna Sundarbans 1947-50* (Dacca, 1957), pp.54-5; J.C. Jack, *Final Report on the Survey and Settlement Operations in the Faridpur District, 1904 to 1914* (Calcutta, 1916), pp.vi-vii; and Promode Ranjan Das Gupta, *Faridpur Revisional Settlement Final Report 1940-45*, 2 parts (Dacca, 1954), pt. 2, p.36. Furthermore, the Settlement Officer of Faridpur remarked that the increase in his district was explained 'by further subletting by proprietors and tenure-holders as well as by further subdivisions'.

speculation, was opened up to the rural rich. The *jotedars* of today, for example, are supposed to have built up their large holdings exactly in such a manner, by accumulating holdings of their defaulting debtors or by hunting for holdings put up for sale in the land market.

At least in Bengal, however, such a state of things does not seem to have taken a full-fledged form before the late nineteenth century. It is interesting to note, for instance, that the Bengal Rent Act of 1859 did not have a provision on transferability of raiyati holdings, whereas there was a heated discussion on this very problem when the Bengal Tenancy Act was enacted in 1885. To paraphrase this change in terms of rent recovery, under the Rent Act of 1859 the landlord was expected to recover arrears of rent by selling off standing crops or the moveable property of the defaulting raiyat, while under the Tenancy Act of 1885 he could put up the raiyati holding for sale. To cite another interesting case in point, it is reported that the Collector of Bogra learnt in the early 1870s that a private sale of an occupancy holding had taken place in a corner of his district. He was so interested in this case that he took the trouble to go and see the vendor personally and enquire into the circumstances.[8] These cases suggest that a raiyati holding came to be regarded as saleable by a considerable section of rural society between 1859 and 1885, most probably in the 1870s. Two developments, economic and social, appear to have played an important role in bringing about this important change. First, growth of a new type of lucrative commercial agriculture centring around jute, together with the lowering in real terms of the rent rate owing to a steady inflationary trend since the mid-1850s, much enhanced the economic value of a raiyati holding as a commodity. Second, the powerful peasant movements that covered the whole of eastern Bengal in the 1870s struggled to consolidate peasants' rights, including the right to sell their holding at will. With the enactment of the Bengal Tenancy Act in 1885 the colonial government conceded their demands. The Bengal government gave, although partially, legal sanction and protection to occupancy rights, and thereby turned a surplus-producing peasant's

[8] *Selections from Divisional and District Annual Administration Reports, 1872-73, with the Government Resolutions on them* (Calcutta, 1874), p.184.

holding into a secure property that could be exchanged in the market. It would be no exaggeration to say that the Act gave an institutional foundation for the growth of a peasants' land market in subsequent years. Thus, the stratification of the peasantry through land transfer must be considered a relatively recent phenomenon in Bengal which gained firm root in agrarian society only in the later decades of the nineteenth century.[9]

But rural Bengal was a stratified society even before the late nineteenth century. Thus a second type of stratification took place in the traditional setup of society. This postulated neither the commoditisation of peasant holdings nor the growth of a market for them. It is well known that rural Bengal had a variety of privileged peasants in the eighteenth and nineteenth centuries. They were village headmen and superior peasants. The former were variously called *mandals, mukhyas, mukaddams, pradhans, pramaniks, matabars, bosneahs*, and so on, depending upon the locality, while the latter were known under such diverse names as *aimadars* in Midnapore, *jotedars* in Rangpur and other districts, *gantidars* in Jessore and the 24-Parganas, *chakdars* in the Sundarbans, *thikadars* in the 24-Parganas, and *haoladars* in Bakarganj.[10] Much remains to be explored as yet about their origin, but even in the present limited state of our knowledge it may safely be said that the source of their power lay in their privileged status in the system of state revenue administration or

[9] For a detailed discussion on this point, see my *Agrarian System in Eastern Bengal c.1870-1910* (Calcutta, 1994), ch.7. B. B. Chaudhuri has also pointed out that the transfer of raiyati holdings through voluntary sales was rare in Bengal until the mid-nineteenth century, citing an interesting case where even an execution sale under the court order caused much stir among the local people of Bogra in 1828: B. Chaudhuri, 'Agrarian relations: eastern India', in Dharma Kumar (ed.), the *Cambridge Economic History of India*, vol.2, *c.1757-c.1970* (Cambridge, 1982; Indian reprint, Delhi, 1984), p.152. Sirajul Islam, however, holds that voluntary sales were common from the beginning of the nineteenth century. See his *Rent and Raiyat: Society and economy of eastern Bengal 1859-1928* (Dhaka, 1989), p.7, fn.3.

[10] For instance, see C. D. Field, *Landholding, and the Relation of Landlord and Tenant, in Various Countries* (Calcutta, 1883), pp.705 ff.; *Report of the Rent Law Commission*, vol.1, 'The Report', paras. 13-7; Ratnalekha Ray, *Change in Bengal Agrarian Society c.1760-1850* (New Delhi, 1979), ch.3. See also Taniguchi's essay in this volume.

zamindari rent-collection or both. Their status in the administrative setup provided them with ample opportunities to mobilise economic and social resources at the cost of their fellow villagers. The origin of the *gantidars* of Jessore, for example, can be traced to rent farming and contracts for reclamation, and not to land speculation.[11]

When studying the process of agrarian change in a society like Bengal where the market for raiyati holdings came into being at a comparatively late stage, it will be useful for us to make a clear distinction between the two types of stratification. However, the present paper deals only with the change through time and the regional characteristics of the first or new type of stratification. The specific features of the old type as well as the relationships and interactions between the two, are left for further enquiry.[12]

II

The registered voluntary sales of occupancy holdings in Bengal[13] showed a remarkable increase from the 1880s to the early 1900s.[14] As is clearly shown in Figure 1, sales of the other classes of landed interests, whether zamindari estates or intermediate tenures, were no match for them.[15] It should be noted at the same time that this increase

[11] Nariaki Nakazato, 'Superior peasants of central Bengal and their land management in the late nineteenth century', *Journal of the Japanese Association for South Asian Studies* 2 (1991) [in English]. See also Rajat Kanta Ray, 'The retreat of the jotedars?', *Indian Economic and Social History Review* 25-2 (1988). Incidentally, the *gantidars* of Jessore operated their holdings mainly on the basis of *korfadars* (under-raiyats) and servants. The *barga* system was rare in the late nineteenth century.

[12] As regards the interactions of the two processes, Ratnalekha Ray appears to hold that the superior peasants like *jotedars* were not affected by large-scale land transfer (Ray, *Bengal Agrarian Society*, p.289). However, we cannot preclude the possibility of the inner composition as well as the social character of the *jotedar* class having undergone a considerable change due to active transactions in *jotedari* rights on the land market. Actually, much spadework remains to be done on this point.

[13] In this paper Bengal means all the districts in Bengal proper excepting Malda. Malda had to be excluded, because it belonged to the Bhagalpur division of Bihar from 1876 to 1905.

[14] A comprehensive study of the problems of stratification in Bengal and Bihar will be found in Binay Bhushan Chaudhuri, 'The process of depeasantization in Bengal and Bihar, 1885-1947', *Indian Historical Review* 2, 1 (1975).

[15] Figures and maps are appended to this chapter.

was accompanied by an equally sharp rise in the number of mortgages. During the 24 years from 1881 to 1904 the sales of occupancy holdings grew more than seven times from 25,448 to 184,233, while the mortgages with a value of less than Rs.100, which seem to have mostly consisted of occupancy holdings,[16] rose 5.4 times from 32,011 to 174,303.[17] The parallel movement was a sure sign that the sales of occupancy holdings were closely related to rural indebtedness. And during the 24 years between 1881 and 1904 the sales of occupancy holdings totalled 2,399,328. This figure was equivalent to 17.9 per cent of the total number of occupancy holdings as ascertained by the survey and settlement operations (about 13,370,000).

The total area transferred by such sales during the 22 years from 1882 to 1903 came to 3,246,730 acres,[18] or about 14.5 per cent of the total acreage under *khas*, or direct, possession of occupancy raiyats.[19] The area sold during the five years to 1903 was 1,103,729 acres, or 4.8 per cent of the raiyati land, annual rate of transfer reckoned at 0.96 per cent. It is not irrelevant here to recall that an enquiry conducted by the Floud Commission revealed that the proportion of the raiyati area transferred during the twelve years before 1939 came to 6.9 per cent, or 0.58 per cent a year.[20] To be sure, these figures are not directly comparable. Nonetheless, seeing that both of the statistics were undoubtedly understated,[21] we may feel justified in saying that occupancy holdings had been changing hands in Bengal at an annual rate between 0.5 and 1 per cent since the turn of the century.

[16] *RRD*, 1889-90, para.32.

[17] *RRD*, 1881, 1904.

[18] See Table 2 below. The figure is exclusive of Bakarganj, Jalpaiguri and Darjeeling. No returns on acreage were submitted from Bakarganj, while reports from Jalpaiguri and Darjeeling were irregular.

[19] In this paper the figures for the areas of the raiyati lands are taken from Table 6 in *RLRC*, vol.2, p.107. The acreage in the possession of the occupancy raiyats (hereafter simply as 'raiyati land') has been computed by deducting the area held under *khas* by *mukarari* raiyats from the area held under *khas* by raiyats mentioned in the table. The *mukarari* raiyats are privileged raiyats who paid rents at fixed rate or fixed amount of rent. Furthermore, the figures quoted in this report do not necessarily agree with those mentioned in the Final Reports of the Survey and Settlement Operations.

[20] *RLRC*, vol. 2, pp.120-1.

[21] On the accuracy of the registration statistics during the 1930s, see below.

The above figures indicate that the market in occupancy holdings was quite active. Who then bought those holdings? Table 1 shows the breakdown of the purchasers. It will be seen that the raiyats themselves constituted the largest group, accounting for about 70 per cent of the total. What is more, they seem to have gradually raised their share with the expansion of the land market. By contrast, all the other groups, whether moneylenders, zamindars or 'others', were involved in only a small part of the transactions. Moneylenders and zamindars accounted for 10 and 7 per cent respectively, and 'others' formed 13 per cent.

When we turn our attention from all-Bengal statistics to districtwise figures, we see that there were considerable regional differences. If we reckon the proportion of the yearly mean of the registered sales from 1881 to 1904 to 1,000 acres of raiyati land, we find the results widely dispersed from the highest figure of 12.2 recorded in Noakhali to the lowest one of 0.56 in Rajshahi.[22] As is clearly shown in Map 1, an L-shaped zone consisting of Bakarganj, Khulna, the 24-Parganas, Nadia and Rajshahi, all of which registered considerably lower rates of transfer, split this province into two parts. Of the two, the western region, excepting the frontier district of Bankura, showed a higher rate. The eastern and northern districts, on the other hand, tended to record a relatively small number of transactions, although there were significant exceptions: Tippera and Noakhali, which recorded by far the highest rates of all the districts of Bengal.

Map 2 shows the total area transferred between 1882 and 1903 as a percentage of the raiyati land. More or less the same pattern as observed in Map 1 emerges from this map, excepting that three districts of northern Bengal, Rangpur in particular, showed fairly high percentages. The high rates in these three districts are most probably attributable to the fact that occupancy holdings fetched rather low prices there as compared with the other districts of eastern and western Bengal.[23] The raiyats had to sell a larger area to fulfil their economic needs.

[22] See Map 1.
[23] For instance, occupancy holdings sold at Rs. 43 per acre on an average in Rangpur, when they sold at Rs. 77 in Dacca. Particulars are given in Table 2.

It has been noted that the majority of the purchasers of occupancy holdings consisted of raiyats. However, here too, marked regional variation existed. Broadly, the proportion of peasant purchasers was higher in the north-east than in the south-west (see Map 3). In most of the districts lying on the left bank of the Ganges, their percentages were above the all-Bengal average of 70 per cent. On the other side of the bank, by contrast, the raiyats' share was considerably smaller, even falling below 50 per cent in Nadia, the 24-Parganas, Hooghly and Howrah. When one looks into the social composition of purchasers in the districts where raiyat purchasers formed less than 60 per cent, it will be seen that the zamindars' share was comparatively small with the two notable exceptions of Bakarganj and the 24-Parganas. In these two districts the zamindars formed 26 and 21 per cent respectively. In Midnapore and Bankura the moneylenders were fairly active in the land market. They constituted 29 per cent in the former and 24 per cent in the latter. They were also active in such districts as Nadia (22 per cent), Hooghly (17 per cent), and Rajshahi (16 per cent). In most of the districts, however, it was people classed as 'others' by the Registration Department, who played an important role, second only to raiyats, in land speculation. In the Registration Statistics the term 'others' did not necessarily denote unspecified persons, but, as the Registrar-General once remarked, in many cases it referred to the 'members of the rising class of native advocates, pleaders, Judges, Magistrates, doctors, &c.'[24] To put it another way, the word 'others' meant the urban middle classes in many cases. Thus it was no wonder that 'other' purchasers bought a large number of occupancy holdings in the districts adjacent to metropolitan Calcutta such as Howrah (47 per cent), Nadia (38 per cent), Hooghly (32 per cent) and the 24-Parganas (28 per cent). In point of fact they even outnumbered the raiyat purchasers in the first two districts.

The market in occupancy holdings in Bengal before the beginning of the twentieth century was marked by considerable regional differences. The most significant features from the point of view of the stratification lay in the fact that even the understated registration statistics showed

[24] *RRD*, 1884-85, p.12.

that a fairly large number of transactions had taken place, causing transfer of a sizeable acreage, except in the districts belonging to central parts of Bengal. They also showed that there were two distinct patterns with regard to the social composition of the purchasers. The purchasers mainly consisted of raiyats in the majority of districts. However it is important that non-cultivating classes such as zamindars, moneylenders and urban middle classes also bought a large number of holdings in several important districts.[25]

We do not know whether this pattern persisted in later years. The registration statistics for western Bengal (Presidency and Burdwan divisions) indicate that a similar picture continued from 1905 to 1910, but we do not have a corresponding set of figures for the rest of the districts in the northern and eastern parts of Bengal, namely, the three divisions of Rajshahi, Dacca and Chittagong.[26] If the uneven distribution of non-cultivating purchasers continued to form a prominent feature of the land market, its cumulative effects upon the course and characteristics of stratification in respective regions must have been of far-reaching significance.

III

The market in the occupancy holdings kept expanding at an accelerated pace in the first decade of the twentieth century. The annual average number of registered sales stood at 248,000 in 1910-12 as against the five-year average of 179,000 between 1901 and 1904.[27] The increase between the two periods comes to 39 per cent. The next decade saw a slower growth. The number reached 314,000 by 1923, showing an increase of 27 per cent over the 1910-12 average.[28]

The 1930s were marked by a slump in the land market, the annual

[25] If we take into account the execution sales, in which non-cultivators should have formed a large proportion, their percentage could be still higher.

[26] The newly-created Government of Eastern Bengal and Assam stopped publishing the statistical table on the sales of peasant holdings.

[27] Government of Bengal, Revenue Department, Land Revenue Branch, Sept. 1914, Proceedings 36-43. As the numbers of sales were not returned from Bakarganj and Mymensingh, the figures for 1904, namely 3,269 and 15,594 respectively, have been taken for these two districts.

[28] M. Azizul Huque, *The Man behind the Plough* (Calcutta, 1939), p.311.

average number of transactions registered between 1930 and 1937 falling by one-third to 208,000.[29] This was partly due to the long depression, which wiped out a considerable number of potential land purchasers from the countryside, and partly to government measures for protecting indebted peasants, such as the Bengal Moneylenders Act of 1933 and the Bengal Agricultural Debtors Act of 1936, which inevitably set the rich peasants and moneylenders on guard.[30] The slump in transfers was abruptly brought to an end by two significant events at the end of the 1930s. The Bengal Tenancy Act was so amended in 1938 as to lift the last of the legal restrictions on the sales of raiyati holdings. The landlords' fees, which a zamindar used to charge a purchaser for having his name recorded on the zamindar's rent-roll, were abolished once and for all. And the boom caused by the outbreak of the Second World War in 1939, on the one hand, eased the desperate economic situation in Bengal to some extent, while on the other it adversely affected the lower classes of people because it was accompanied by rapid inflation. The result was a sharp rise in the number of registered sales which had already shown signs of gradual recovery since 1934. The number jumped to 500,000 in 1939,[31] almost doubled to 937,823 by 1942-43, and climbed to the staggering height of 1,491,469 in 1943-44, when three million people in Bengal perished in the devastating famine. The scars of the famine still cast a long shadow over rural Bengal in 1944-45. In this year 1,230,252 sales of raiyati holdings were registered. The total number of transfers that took place during the four years from 1941 to 1944 thus comes to as many as 4,371,179, or an average of 1,092,795 a year.[32]

The regional variations underwent a marked change in the eventful years of the twentieth century. When we compare the annual average number of sales of occupancy holdings for 1901-1904 with that for

[29] *RLRA*, 1930-1937.

[30] Karunamoy Mukerji, *The Problems of Land Transfer (A Study of the Problems of Land Alienation in Bengal)* (Santiniketan, 1957), ch.3; Bose, *Agrarian Bengal*, ch.4.

[31] Ibid., Table 5.1 on p.151.

[32] *RLRA*, 1940-41; *Statistical Report on the Land Revenue Administration of the Presidency of Bengal*, 1941-42 to 1944-45. See also Bose, *Agrarian Bengal*, p.152.

1910-1912, we see that in all but one district they had increased.[33] The only exception was the district of Jessore in central Bengal, which witnessed a decline of 20 per cent. The districts on the west bank of the Bhagirathi-Hooghly system saw a moderate rise of 10 to 20 per cent. On the other hand, in almost all the districts in eastern and northern Bengal the number was raised by more than 50 per cent, with the notable exceptions of Tippera and Noakhali, where strikingly high rates of transfer had already been attained. The uneven changes readjusted regional pattern. Map 4 clearly shows that the west and the north and east of Bengal were now, by and large, squarely balanced in terms of the frequency of land transfer measured by the number of sales per 1,000 acres of raiyati land, although the central districts, showing the lowest rates of transfer, continued to form a dividing corridor.

The regional characteristics received further modifications in the 1930s. The annual average number of sales per one thousand acres from 1930 to 1937 is shown in Map 5. We learn from this map that the regional balance between the east and the west acquired in 1910-12 began to tilt towards eastern Bengal in the 1930s. As pointed out above, the overall trend was for the sales to drop sharply from the 1910-12 level. However, some districts were conspicuous in registering a moderate increase over 1910-12. They included Dacca, Faridpur, Pabna, Rajshahi, Nadia, Khulna, 24-Parganas and Hooghly, six of which were within eastern or central Bengal. By contrast, all the districts on the west bank of the river Hooghly recorded a considerable decrease, the single and important exception being Hooghly. It was these movements towards opposites that overturned the regional balance.

The stagnant state of the land market was dramatically reversed in the 1940s. All the districts without exception saw a sudden upswing in the number of sales, ranging from 2.5 times to more than 9 times the average of the 1930s. The interesting fact is that the districts which saw an increase exceeding the all-Bengal average not only broadly agreed with the area most severely hit by the famine but also conformed neatly with the front line with the Japanese army in Burma.[34] Those districts

[33] Particulars are given in Table 2.
[34] As regards the regional differences in the impact of the famine, see Paul

were Jalpaiguri, Rangpur, Mymensingh, Dacca, Tippera, Noakhali, Chittagong, Bakarganj, Khulna, 24-Parganas, Jessore and Faridpur. As is evident from Map 6, the high land-transfer area had now decidedly shifted from the west of the Hooghly to the eastern districts located around the confluences of the Ganges, the Brahmaputra and the Meghna.

IV

According to the estimate made in section 2 above, 0.5 to 1 per cent of the raiyati land had been transferred annually by registered sales since the beginning of the twentieth century. And the acreage of the occupancy holdings put up for sale from 1882 to 1903 amounted to 14.5 per cent of the raiyati land. It may be reasonably presumed on the basis of these figures that the proportion of the occupancy holdings transferred to 1947 fell between 36 and 58 per cent in terms of area. It should be stressed here that this is a conservative estimate arrived at by taking the official statistical data for the years of lower rate of transfer at face value, and, above all, by totally ignoring the alarming scale of land sales in the 1940s.

We can form a somewhat more accurate estimate of the cumulative number of sales. Crude as it is, estimate attempted in Table 3 below shows that the total number of sales during the 64 years from 1881 to 1944 adds up to 16,194,000. It is interesting that this figure happens to be very close to the number of 'raiyati tenancies' in Bengal according to the Land Revenue Commission, that is, 16,400,000.[35] Needless to say, this does not mean all the occupancy holdings had been sold during this period. Nor does it mean that all the peasant households became landless, for it was a common practice among the Bengal peasants to sell only a portion of a holding as the occasion arose.[36]

R. Greenough, *Prosperity and Misery in Modern Bengal: the famine of 1943-1944* (New York, 1982), p.144.

[35] *RLRC*, vol.1, p.83. This figure includes Malda.

[36] Chaudhuri, 'Depeasantization', pp.139-40. Furthermore, this is one of the reasons why we have not attempted to reckon the ratio of the number of sales to that of occupancy holdings when studying the movement in the land market. Moreover, the Bengal raiyat usually had a few holdings or more in different places of his village, and it does not appear to have been rare that he put up for

Moreover, there would have been a lot of resales of the same holding. Still it is important to note that the land sales in late colonial Bengal were of such an overwhelming volume as could have, at least in theory, affected every raiyati holding in greater or lesser degree.

We have seen that there were wide regional differences in the extent of land transfer during our period. Map 7 below shows the proportion of the annual average number of the sales of occupancy holdings per thousand acres of raiyati land in 1881-1904, 1910-12, 1930-37 and 1941-44. On the basis of the frequency of land transfer in the whole period that we have discussed, Bengal may be broadly divided into the three regions: (A) Murshidabad, Birbhum, Burdwan, Hooghly, Howrah, and Midnapur; (B) Dinajpur, Rajshahi, Nadia, Jessore, 24-Parganas, Khulna and Bakarganj; and (C) Rangpur, Bogra, Pabna, Mymensingh, Dacca, Faridpur, Tippera, Noakhali and Chittagong.

Region B consists of seven districts which showed a low rate of transfer, that is, below 7.5 sales per thousand acres on an annual average. As if wedged into the middle of Bengal from the south to the north, this region divides Bengal into the west (region A) and the east (region C). Regions A and C both contain medium-transfer districts (7.5 to 15 sales per thousand acres) as well as high-transfer districts (above 15). However, region C, with 13.3 sales per thousand acres, shows a slightly higher rate of transfer than region A (11.2 sales).

Geographically, region A corresponds to what is called the Rarh land, which has a long historical tradition.[37] Here the zamindars were powerful, the rent rates were high, tight caste restrictions were enforced, and agricultural labourers formed a fairly large part of the population. On the other hand, region C is a fertile upper delta created by active alluvial action of the three great rivers of Bengal. It comprises all the five leading jute-growing districts of the 1930s, namely, Mymensingh, Dacca, Rangpur, Tippera and Faridpur.[38] Blessed with

---

sale a number of holdings at a time.

[37] The below is just an impressionistic sketch. For a detailed description of regional features, see Bose, *Agrarian Bengal*, ch.1 and his recent work *Peasant Labour and Colonial Capital: rural Bengal since 1770* (New Cambridge History of India III: 2) (Cambridge, 1993), ch.1.

[38] The five districts accounted for about two-thirds of the total jute acreage in Bengal in 1936-37 (Huq, *The Man*, p.59).

natural resources and mainly inhabited by Muslim peasants who were relatively free from the caste system, so Partha Chatterjee and Sugata Bose hold, the peasantry of this region was less stratified in the beginning of the twentieth century. From an ecological point of view, region B is a mixed area which may be subdivided into two zones. The districts in the south (24-Parganas, Khulna and Bakarganj) comprise the vast virgin land, the Sundarban forests, along the coast of the Bay of Bengal, where reclamation started in earnest only in the early nineteenth century. The districts in the north, by contrast, form an ageing backward area. Nadia and Jessore, which were located on the unhealthy moribund delta, and Rajshahi, which had an extensive, sickly, swampy tract along the Ganges, were the only three districts in Bengal that recorded a decline in population between 1881 and 1931.[39] Yet these two zones have one feature in common: both were prominent for the existence of various types of superior peasants. Besides Dinajpur with its *jotedars*, of whom Buchanan-Hamilton left a famous account, Jessore and the 24-Parganas were known for their powerful *gantidars*, the Sundarbans for *chakdars*, Bakarganj for *haoladars*, and the 24-Parganas for *thikadars*.

The most noteworthy feature in the regional pattern of land transfer in our period would be the rapid growth of the sales of occupancy holdings in eastern Bengal (region C), and the sharp contrast this area makes with the northern zone of region B. It is interesting to note that the land market was much more active in the relatively prosperous districts of eastern Bengal than in the region suffering from declining economic activities. This pattern came to assume definite shape as early as in 1910-12, when neither the sharp drop in jute prices nor the acceleration in population growth, which told against the economy of eastern Bengal, had made an appearance on the horizon. It goes without saying that economic strain causes land sales. A flood of sales after the years of economic distress in the 1930s and during the Bengal famine of 1943 most eloquently demonstrates this point. It does not seem, however, that poverty alone can explain the whole process of the growth of the land-market in Bengal. It seems that the phenomenal rise

[39] *Census of India, 1931,* vol.5, pt.2, p.4.

in land sales was connected not only with economic distress, but also with the intensification of economic activities among the peasantry that went together with commercialisation of agriculture. In fact registration officers were well aware of this relationship. For instance, trying to explain the increase in sales and mortgages in east Bengal in 1905-07, an officer remarked: 'the fall in the price of jute, too, in the present season has probably hit those raiyats hard who borrowed money in order to extend their cultivation of that product...but it appears to be extremely doubtful whether poverty is in the main responsible for the increase'.[40] Most probably production loans for the purpose of extending commercial agriculture formed a considerable part of mortgages in Bengal, and when the price of commercial crops like jute collapsed, defaulting peasants were forced to part with their mortgaged holdings. It would be by taking this sort of wider perspective, rather than by simply stressing the hardships of the cultivators, that we can arrive at a better understanding of the whole process from the emergence of a peasants' land market on a sound institutional foundation in the 1870-80s to the sudden outburst of sales in the 1940s. Concerning the emergence of the land market, we have already suggested a hypothesis which stresses the importance of positive factors existent in the late nineteenth century as well as the legislative initiative by the colonial state. Growth of the land market was one of the consequences of the limited development of Bengal's rural economy within the framework of colonialism.

V

We may conclude this short essay by enquiring into how land transfer was related to the stratification of the peasantry in Bengal. Due to the paucity of reliable data, it is rather hard to say with certainty to what extent the Bengal peasants were stratified. Here let us confine ourselves to examining the results of the Ishaque Survey. This was an agrarian survey conducted in October 1945 by the Department of Agriculture under the direction of H.S.M. Ishaque, and covers 5,284 families in 77

[40] *Report on the Administration of the Registration Department in Eastern Bengal and Assam for the Years 1905 to 1907*, para.7.

randomly-selected villages in 77 subdivisions of the 25 districts.[41] It should be kept in mind, however, that Ishaque's figures are biased in that the survey was carried out immediately after the famine. Table 4 has been prepared by processing his figures according to our scheme of regions.

The table clearly shows that landholding in rural Bengal had a highly polarised structure in 1945. Ishaque divided the rural families into five classes by the size of land under their *khas*, or direct, possession. The last row of Table 4 shows that in Bengal as a whole families with a holding exceeding 5 acres (class E families), held as much as 62.4 per cent of the total acreage, despite their accounting only for 14.2 per cent of the total number of families. On the other hand, class A refers to families which were either landless, or in possession of nothing but their homestead land. They amounted to 36.4 per cent in number, but owned only 1.8 per cent of the total land. Families classed as B, C, and D owned less than 1 acre, 1 to 3 acres, and 3 to 5 acres respectively. The table shows that more than one out of three rural families were virtually landless, and that three in four held less than 3 acres. Yet, even in the double-cropped area in one of the most fertile districts, 2.5 acres were said to be required to maintain an average family in reasonable comfort. The size of an economic holding was estimated at 5 acres for northern and eastern districts, and at 8 acres for the other regions.[42]

It will be difficult to give a full explanation for polarisation of this intensity without taking account of the land transfer. There is evidence to show that the large-scale land transfer in the last phase of the colonial rule resulted in a sharp increase in the number of poorer peasants. The percentage of the smaller peasants who held less than 3 acres stood at 57.2 per cent when the Land Revenue Commission conducted an extensive survey around 1939,[43] whereas, as has been shown in Table 4, the figure for 1945 came to 76.2 per cent. The proportion increased by as much as one-third during these six years, when raiyati holdings were being put up for sale on an unprecedented

[41] H.S.M. Ishaque, *Agricultural Statistics by Plot to Plot Enumeration in Bengal 1944-45*, 3 parts (Alipore, 1946), part 1, pp.46-56, 121-33.
[42] *RLRC*, vol.1, pp.85-6.
[43] *RLRC*, vol. 2, pp.114-5.

scale.

There were regional variations to a certain extent within this general tendency towards steep polarisation. It will easily be seen from Table 4 that the most polarised area was region A. Here more than half of the families were landless or almost landless, while two-thirds of the total acreage was owned by less than 10 per cent of the population. Region B constituted the least polarised tract, although there was but a small difference between this region and region C. In region B the families in possession of 5 acres or more (class E) formed a fairly large proportion in number (19.3 per cent), while in region C we find very small peasant farms below 3 acres (class B and class C) accounting for 49.7 per cent in number and 30.2 per cent in area.

When we compare these findings with the results of our study on land transfer, we find that the relatively low degree of polarisation and the smaller volume of land transfer go well together in region B. In region A the relationship between stratification and land sales becomes a little blurred. This area stands out from the other two regions in showing strikingly high degree of polarisation, but ranks second to region C in the frequency of land sales. The relationship becomes further blurred in Region C where the highest rate of land transfer coexists with a low degree of stratification much as in region B. In short, the frequency of land transfer does not necessarily correspond to the degree of polarisation at the regional level. It seems that in Bengal, increased rates of land transfer did not always bring about steeper polarisation; and, conversely, smaller rates did not always point to the persistence of a less polarised peasant society. The effects of land transfer seem to have permeated rural society through the mediation of various regional factors.

What then caused such discrepancy at the regional level? It seems there are at least two questions to be considered. In the first place, region C poses an interesting question. This region of expanding economic activities recorded a larger number of sales of occupancy holdings, but the peasantry there was not so stratified as could be expected from the scale of land transfer. In particular, the districts of Tippera and Noakhali offer a case in point. These districts witnessed a very high frequency of land transfer throughout the period under study. Nonetheless, the degree of stratification in both districts was among the lowest of all in Bengal.[44] This may be *partly* explained by assuming that brisk economic activities stimulated social mobility in both directions: not only downward but also upward mobility, and that the upward mobility had the effect of taking the edge off the downward stratification process. As has been shown above, at least until 1904 the overwhelming majority of the purchasers of occupancy holdings in eastern Bengal consisted of raiyats themselves. In other words, land transactions there showed a stronger tendency for occupancy holdings to circulate among the peasant class. This might have provided conditions favourable for the multidirectional mobility.[45]

In the second place, and conversely, regions A and B recorded relatively high degrees of stratification compared with their rates of land transfer. Perhaps this should be seen in the proper historical perspective. For one thing, the high stratification recorded in region A is mainly due to the existence of a large population of landless or virtually landless peasants. This had more to do with the traditional social hierarchy combined with the rigid caste system peculiar to this region than with land transfer. Moreover, various privileged peasants had already emerged in the traditional setup of Bengal rural society even before stratification developed further through land transfer. In

[44] Ghosh also includes Tippera and Noakhali among the least stratified districts. See Ghosh, 'Classification', p. 113.

[45] On this point, see also Willem van Schendel, *Peasant Mobility: the odds of life in rural Bangladesh* (New Delhi, 1982), Part 1, where van Schendel has put forward a cyclical model regarding eastern Bengal on the basis of his own case study and Shanin's model; and Amit Bhaduri, Hussain Zillur Rahman and Ann-Lisbet Arn, 'Persistence and polarisation: a study in the dynamics of agrarian contradiction', *Journal of Peasant Studies* 13, 3 (1986) which aroused a series of debates on the same journal. My standpoint is to admit a certain cyclical or multi-directional tendency in the general trend of stratification through land transfer. It appears to me that colonial Bengal lacked basic institutional conditions suitable for the cyclical model. It neither had the village community of Russia nor the strong rural kin group of the Panjab. On the relationship between land transfer and stratification in Panjab, see Clive Dewey, 'Some consequences of military expenditure in British India: the case of the Upper Sind Sagar Doab, 1849-1947', in Clive Dewey (ed.), *Arrested Development in India: the historical dimension* (New Delhi, 1988), pp. 112-17.

short, when we examine statistical data regarding stratification in the late colonial period, we are always looking at the outcome of two types of stratification which overlap each other. And it is quite probable that privileged peasants had more chance of surviving the vicissitudes in regions where land market remained inactive. The relatively high degrees of stratification registered in regions A and B might be attributable to the survival of the privileged peasants and of the older social order represented by them.

Thus the effects of the large-scale transfer of occupancy holdings were differently felt in the three regions of Bengal. But the fact remains that Bengal peasants were highly stratified at the end of the colonial rule and that land transfer was one of the major causes of their stratification.

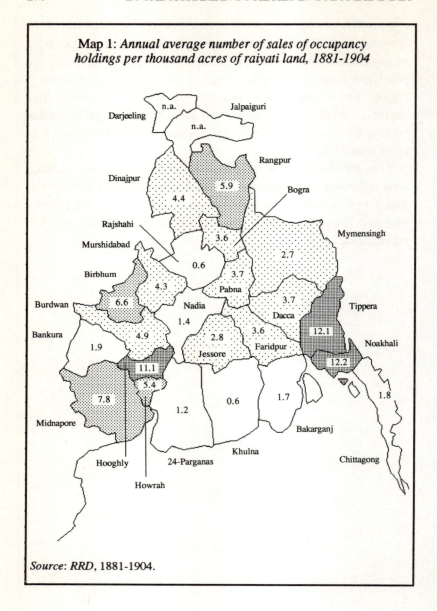

Map 1: *Annual average number of sales of occupancy holdings per thousand acres of raiyati land, 1881-1904*

Source: *RRD*, 1881-1904.

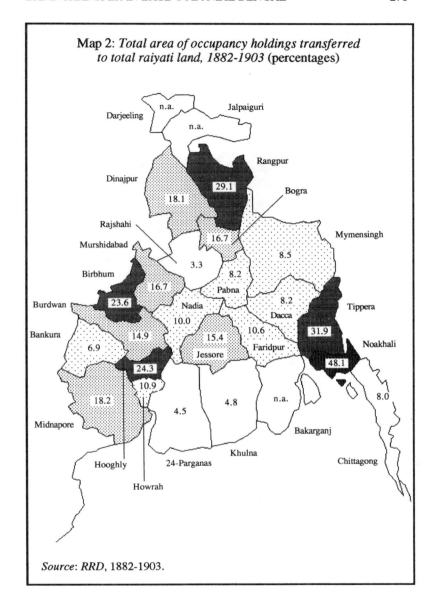

Map 2: *Total area of occupancy holdings transferred to total raiyati land, 1882-1903* (percentages)

Source: *RRD*, 1882-1903.

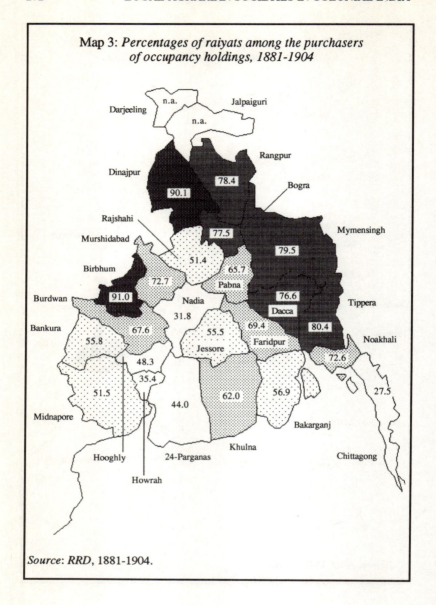

Map 3: *Percentages of raiyats among the purchasers of occupancy holdings, 1881-1904*

Source: *RRD*, 1881-1904.

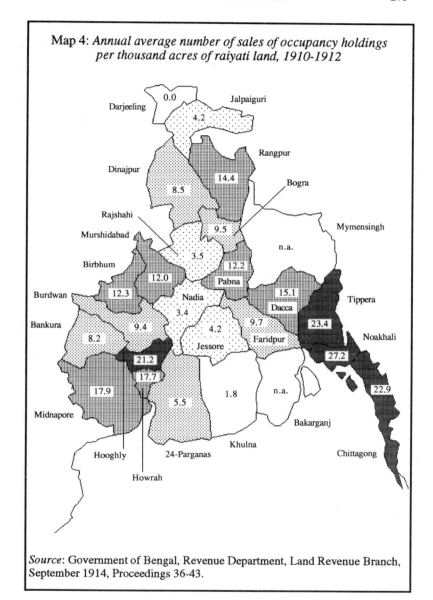

Map 4: *Annual average number of sales of occupancy holdings per thousand acres of raiyati land, 1910-1912*

Source: Government of Bengal, Revenue Department, Land Revenue Branch, September 1914, Proceedings 36-43.

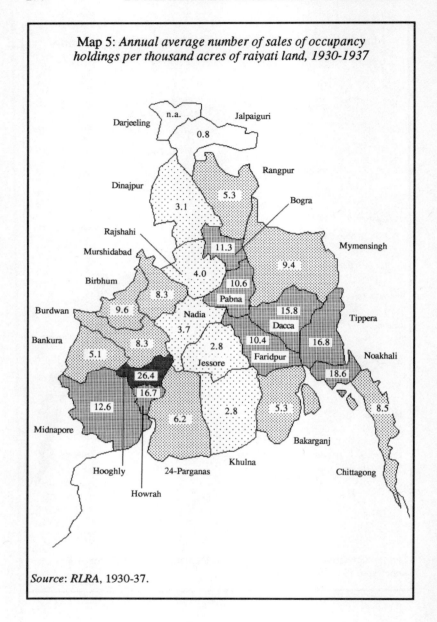

Map 5: *Annual average number of sales of occupancy holdings per thousand acres of raiyati land, 1930-1937*

Source: *RLRA*, 1930-37.

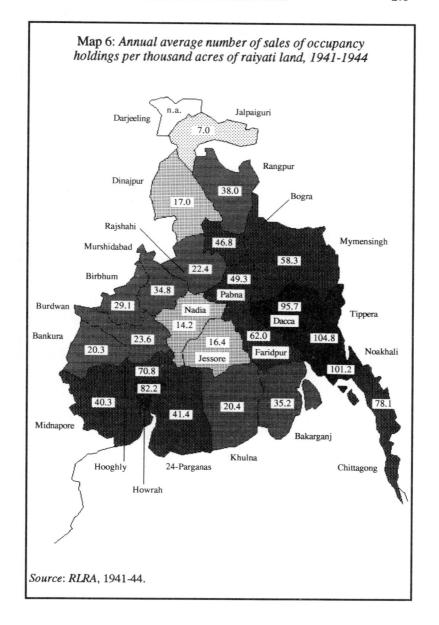

Map 6: *Annual average number of sales of occupancy holdings per thousand acres of raiyati land, 1941-1944*

Source: *RLRA*, 1941-44.

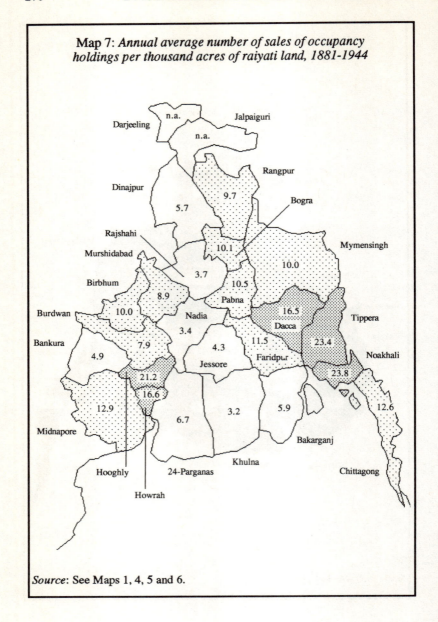

Map 7: *Annual average number of sales of occupancy holdings per thousand acres of raiyati land, 1881-1944*

Source: See Maps 1, 4, 5 and 6.

Table 1: *Purchasers of occupancy holdings in Bengal, 1881-1904 (percentages)*

	M	Z1	Z2	R	O
1881-1885	12	1	4	66	16
1886-1890	11	2	5	69	13
1891-1895	11	2	6	70	11
1896-1900	10	2	5	71	12
1900-1904	9	1	5	71	13
1881-1904	10	2	5	70	13

Source: Computed from *RRD.*. *Note*: M = mahajans, traders, or money-lenders; Z_1 = landlord of the holding transferred; Z_2 = zamindar or landlord of holding other than that transferred; R = raiyats; and O = others, including unspecified.

Table 2: *Sales of occupancy holdings by registered deeds*

	Annual average number of sales							
	1881-1904	1891-5	1901-4	1910-12	1930-7	1941-4	Acres	Rs.
Burdwan	5,856	6,270	9,208	11,166	9,795	27,921	177,002	72.3
Bankura	1,056	655	3,067	4,559	2,843	11,316	38,279	42.8
Birbhum	4,600	4,696	7,722	8,541	6,650	20,199	163,878	55.2
Midnapr.	14,309	14,431	28,001	32,888	23,105	74,163	334,070	68.1
Hooghly	2,765	2,478	4,565	5,268	6,557	17,548	60,236	81.8
Howrah	1,095	713	2,929	3,552	3,349	16,524	21,962	104.7
24-Parg.	1,216	882	2,408	5,702	6,472	43,326	46,963	65.8
Nadia	1,364	1,461	2,680	3,219	3,458	13,409	94,764	24.7
Jessore	2,486	1,897	4,640	3,788	2,463	14,687	137,398	21.1
Khulna	551	411	1,062	1,509	2,354	17,295	40,741	22.2
Murshdbd.	3,673	3,749	7,454	10,175	6,982	29,399	140,930	45.0
Dinajpur	6,918	8,229	11,090	13,294	4,815	26,659	284,421	43.2
Rajshahi	681	620	1,379	4,300	4,898	27,208	40,395	33.1
Rangpur	7,648	7,864	10,253	18,781	6,941	49,557	379,673	43.1
Bogra	2,454	2,845	4,278	6,545	7,777	32,161	114,857	44.1
Pabna	3,221	4,034	4,854	10,629	9,214	42,924	71,603	81.6
Jalpaiguri	n.a.	n.a.	1,879	2,537	504	4,164	n.a.	n.a.
Dacca	5,192	5,760	9,302	21,365	22,396	135,217	115,451	76.8
Faridpur	3,879	4,637	6,723	10,416	11,240	66,740	114,094	50.6
Bakargnj.	2,195	2,798	2,997	n.a.	6,853	45,981	n.a.	n.a.
Mymngh.	7,678	8,085	14,605	n.a.	27,012	167,931	246,040	57.3
Tippera	15,120	14,135	25,896	29,272	21,056	130,969	398,676	64.7
Chittagng	891	357	2,840	11,254	4,166	38,433	39,539	33.4
Noakhali	4,713	5,093	8,841	10,488	7,196	39,069	185,759	33.2
TOTAL	99,561	102,100	178,673	229,248	208,096	1,092,800	3,246,731	53.5

Table continues

Table 2 continued

Source: *RRD*; *RLRA*; and Government of Bengal, Revenue Department, Land Revenue Branch, September 1914, Proceedings 36-43. *Notes*: The final two columns show the total transferred area in acres and the average price per acre paid in rupees, between 1882 and 1903. Darjeeling figures are unavailable.

Table 3: *Sales of occupancy holdings in Bengal, 1881-1944*

Year	Number	*Derivation*
1881-1904	2,399,000	*RRD*, 1881-1904
1905-1909	893,000	Yearly mean for 1901-4 multiplied by 5
1910-1919	2,481,000	Yearly mean for 1910-12 multiplied by 10
1920-1929	3,140,000	The number for 1923 multiplied by 10
1930-1937	1,665,000	*RLRA*, 1930-37
1938-1940	1,245,000	Bose, *Agrarian Bengal*, p.151
1941-1944	4,371,000	*RLRA*, 1941-44.
Total	16,194,000	

Source: *RRD*; *RLRA*: Government of Bengal, Revenue Department, Land Revenue Branch, September 1914, Proceedings 36-43; Huque, *The Man*, p. 311; Bose, *Agrarian Bengal*, p.151.
Note: Figures for Jalpaiguri from 1885 to 1897 and Darjeeling in most years were not returned to the Registration Department, and are excluded from the totals here. When reckoning the yearly mean for 1910-12, the figures for Mymensingh and Bakarganj have been taken from those of 1904.

Table 4: *Distribution of khas land in Bengal in 1945*

Region	Class A		Class B		Class C		Class D		Class E	
	No.	Area	No.	Area	No.	Area	No.	Area	No.	Area
A	53.4	2.9	13.9	4.3	17.0	15.9	5.8	11.5	9.9	65.4
B	31.2	1.7	16.3	3.2	21.0	12.8	12.3	15.2	19.3	67.2
C	30.1	1.6	22.5	6.1	27.2	24.1	8.8	15.4	11.4	52.8
Bengal	36.4	1.8	17.7	4.2	22.2	16.9	9.6	14.7	14.2	62.4

Source: Computed from *Ishaque Survey*, part 1, pp.121-33.
Note: The classes of families are as follows: A = landless or having no *khas* land other than homestead lands; B = holding less than 1 acre *khas* land; C = holding 1 to 3 acres; D = holding 3 to 5 acres; E = holding over 5 acres.

Figure 1: *Number of sales of various landed interests in Bengal, 1881-1904*

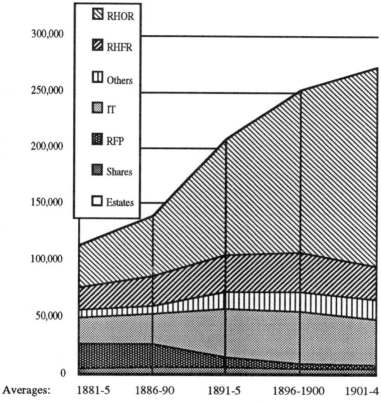

Source: Compiled from *RRD*. *Key*: RHOR, raiyati holdings (occupancy right); RHFR, fixed-rate raiyati holdings; IT, intermediate tenures of all kinds; RFP, revenue-free properties; Shares, shares in estates; Estates, entire estates.
Notes: The graph shows the quinquennial or quadrennial average of sales. Being negligible, the number of entire estates cannot be shown on the figure. Bengal is here defined as all the districts of Bengal proper except Malda.

Chapter 6.1

FAMINES, EPIDEMICS AND MORTALITY IN NORTHERN
INDIA, 1870-1921[1]

Kohei Wakimura

Introduction

This chapter examines the causes of high mortality in Northern India during the period from 1870 to 1921, and discusses the relative importance of famines and epidemics and the close relationship between them.[2] It suggests ways in which famines contributed to the prevalence of epidemic malaria, and argues for its vital importance in the high mortality rate and population change.

The importance of a high mortality rate in this period is obvious when seen in the context of the long-term demographic trend of India as a whole. The annual rate of population growth in the period from 1871 to 1921 was about 0.37 per cent, while that for the period from 1921 to 1941 was 1.22 per cent.[3] Although no definitive statement can be made for the population trend before 1871 when the first census was collected, it is possible that the mortality rate may have been higher in this period than in the previous one.

On the other hand, if we are nevertheless to assume that the rate of population growth before 1871 was closer to the low level in our period than to the high rate after 1921, we then see a clear upward shift of the rate of population growth around 1921. Does the demographic transition theory account for this? What kind of socio-economic circumstances were responsible for this change? Much groundwork needs to

[1] I am grateful to those who gave me comments on the earlier versions of this chapter, especially Clive Dewey, Sumit Guha, Yukihiko Kiyokawa, Dharma Kumar and the editors. [The wording of this chapter has been revised by Kaoru Sugihara and Peter Robb.]
[2] In this chapter mortality rate means crude mortality rate.
[3] L. Visaria, and P. Visaria, 'Population (1754-1947)', in D. Kumar (ed.), *Cambridge Economic History of India, Vol.2: 1757-1970* (Cambridge, 1982), p.490.

be fully undertaken to answer these questions.[4] This study tries to answer part of these questions with respect to the United Provinces (U.P.),[5] one of the provinces most affected by famines and epidemics in this period (see Figure 1).

Let us first examine the available statistics. There are very few original sources of demographic data on the pre-census period. S. Commander implied that, as far as the Doab region of the U.P. was concerned, population growth rates during the first half of the nineteenth century were higher than those of the second half.[6] The question remains, however, as to whether or not the population growth during the period from 1871 to 1921 was slower than that of the pre-census period. The growth rate of the U.P. was lower during the period from 1871 to 1941, compared to that of British India as a whole (see Table 1). Many divisions experienced population decreases after 1901. It can be assumed that the main reason for decreases in this period was the frequent occurrence of such epidemics as malaria, plague, cholera and influenza.

There were a few cases of population diminution from 1872 to 1901. Some cases may be explained by the impact of famines. In the 1870s, for example, the population of Agra and Rohilkhand divisions fell in absolute terms (by 6.7 per cent and 2.5 per cent respectively). These

[4] Concerning these two questions, see the following studies: M.B. McAlpin, 'Famines, epidemics and population growth: the case of India', in R. I. Roberg and T. K. Rabb (eds.), *Hunger and History: the impact of changing food production and consumption patterns on society* (Cambridge, 1985); I. Klein, 'Population growth and mortality in British India, part 1: The climacteric of death', *Indian Economic and Social History Review* 26, 4 (1989); I. Klein, 'Population growth and mortality in British India, part 2: The demographic revolution', ibid. 27, 1 (1990); S. Guha, 'Mortality decline in early twentieth-century India: a preliminary enquiry', ibid. 28, 4 (1991); C.Z. Guilmoto, 'Toward a new demographic equilibrium: the inception of demographic transition in South India', ibid. 29, 3 (1992).

[5] The U.P. stands for the United Provinces of Agra and Oudh (currently called Uttar Pradesh).

[6] S. Commander, 'The mechanics of demographic and economic growth in Uttar Pradesh, 1800-1900', in T. Dyson (ed.), *India's Historical Demography: studies in famine, disease and society* (London, 1989), pp.51-2. Dyson also felt that the mortality rate during the period from 1830 to 1891 was lower than that in the period from 1891 to 1920, with regard to British India as a whole. See his 'Indian historical demography: developments and prospects', in ibid., pp.8-11.

divisions were the regions most affected by the 1877-78 famine. On the other hand, in eastern U.P., that is, Benares and Gorakhpur divisions, the population increased dramatically (by 18.2 per cent and 21.7 per cent respectively). It is clear that the famine did not seriously affect these divisions, although improvements in the accuracy of the census account for much of the increase.[7]

Mortality rates clearly reflect the effects of famines and epidemics (see Figure 2). The Report of the Sanitary Commissioner (R.S.C.) was published annually in each Province.[8] The U.P. government published this report from 1868 on an annual basis. It includes vital statistics that are useful in understanding changes in mortality rates and identifying causes of death. But it is well known that this information was originally taken from the reports of illiterate *chaukidar* at the lowest level. Therefore the mortality rate was considerably underestimated. Even in 1946 it was probably underestimated by about 30 per cent.[9]

Vital statistics of the R.S.C. include information on the causes of death. There is no doubt that *chaukidar* had very little medical knowledge, and their classification of the causes of death is very dubious.[10] The information on cholera, smallpox and plague is relatively creditable because they were easily indentifiable diseases,[11] but the definition of death by fever is problematic. Many unidentifiable diseases were classified as fever. And fever was the largest cause of death, the percentage of deaths from fever being about 70 to 85 per cent in the U.P. There is a strong possibility that this fever mortality rate was exaggerated.[12] It is difficult to say what proportion of fever deaths actually came from malaria, but one estimate puts it as one-third.[13] This proportion must have increased in times of epidemic outbreaks.

[7] *Report of the Census of NWP and Oudh,* 1881, p.7.
[8] *First Annual Report of the Sanitary Commissioner of the North-Western Provinces, 1868* (Allahabad, 1869); hereafter *RSC* (with year).
[9] The Health Survey and Development Committee, *Report of the Health Survey and Development Committee,* vol.1 (1946), p.154.
[10] K. Davis, *The Population of India and Pakistan* (Princeton, 1951), p.42.
[11] Ibid., p.43.
[12] I. Stone, *Canal Irrigation in British India: perspectives on technological change in a peasant economy* (Cambridge, 1984), pp.146-8.
[13] Davis, *The Population of India and Pakistan,* p.53.

There appears to be crucial defects in the mortality statistics for the period before the late 1870s, as the system was not yet well-established. But mortality rates after 1877 are generally reliable. The period from 1879 to 1948, can be divided into three periods. Figure 2 suggests that compared with the first period (1879-1900) or the second period (1901-1920), the average mortality rate in the third period (1921-1948) was much lower.[14] It will be interesting to explore the reasons why the level of mortality rate declined after the 1920s, particularly because we do not find any remarkable changes in such aspects as living standards, nutritional and health or medical conditions during this period.[15] But here we will concentrate on the cause of high mortality in the period up to 1920.

The mortality rate peaked in 1879, 1894, 1897, 1905, 1908, 1911, and 1918. These fluctuations almost definitely reflect famines or epidemics. The mortality peaks in the years 1879, 1897 and 1908 can be explained by both famines and epidemic malaria. The year 1894 saw a malaria epidemic only, while 1905 and 1911 experienced plague epidemics. But 1918 was an exceptional year, being in the middle of an influenza epidemic, and with famine conditions also existing in this year.

Figure 2 also shows the long-term movement of the fertility rate during the period from 1879 to 1948. Because the fertility rate in the first period is assumed to be underestimated due to the inadequacy of a reporting system at the village level, it is only possible to compare the second period (1901-1920) with the third period (1921-1948). The average level of the fertility rate during the second period was much higher than that of the third period.[16] The reason might be that the frequency of famines and epidemics pushed up the average fertility rate during the second period. An implication is that insurance to guard against the risk of mortality crisis increased the fertility rate.

[14] This difference of average between two periods is statistically significant.
[15] See McAlpin, 'Famines', Klein, 'Population growth and mortality', and Guha, 'Mortality decline'.
[16] This difference is also statistically significant. I am very grateful to Professor Yukihiko Kiyokawa for the calculations mentioned here and in note 14 above.

In fact, the fertility rate declined in the years immediately after famine. It might be suggested that famine conditions caused the decline of the conception rate, which resulted in the decline of the fertility rate after a while. It is reasonable to assume that the time-lag of decline between the conception rate and the fertility rate is about 9 to 10 months.[17] Unfortunately, we do not have monthly data of the fertility rate, and cannot confirm the presence of this time-lag with regard to the U.P.[18] However, the fertility rate tended to increase sharply a few years after the famine. This phenomenon can be called the rebound of the fertility rate. This rebound of the fertility rate must have compensated for the previous decline to a large extent.

Finally, there is ground for thinking that famines did not directly cause so many deaths from starvation. An examination of the movements of sown area, agricultural prices and mortality rate suggests that the sown area and agricultural prices moved in opposite directions (see Figure 3). Needless to say, they are assumed to be inversely correlated. But the same Figure shows that no such correlation existed between agricultural prices and mortality rate. A sharp increase in agricultural prices did not always accompany an increase in the mortality rate. It was epidemics that pushed up the mortality rate, sometimes even without a food crisis. Some epidemics, such as plague, cholera, smallpox, influenza and malaria, had serious effects on the mortality rate.

This study will show that, while famine was an important factor in explaining the demographic trends, the main determinant of death during the famine was epidemics, particulary epidemic malaria. There are two kinds of malaria. One is endemic malaria, that was always present and moderate. The other is epidemic malaria which occurred suddenly causing widespread death.[19] According to the R.S.C., three

[17] G. Hugo, 'The demographic impact of famine', in B. Currey and G. Hugo (eds.), *Famine as a Geographical Phenomenon* (Dordrecht, 1984), pp.19-20.

[18] Using the monthly data, T. Dyson showed this relation graphically in his study on the famines in Madras Presidency, Central Provinces and Bombay Presidency. See T. Dyson, 'On the demography of South Asian famines, part 1', *Population Studies*, 45 (1991).

[19] On the distinction between endemic malaria and epidemic malaria, see A.

years out of the above-mentioned seven high-mortality years were connected with famine-cum-epidemic malaria. In the following two sections we examine the relationship between famine and epidemic malaria in some detail, primarily on the basis of descriptive evidence contained in the R.S.C. It will be argued that when famine met epidemic malaria, the mortality rate increased substantially, while this did not occur in times of food crisis without the prevalence of epidemic. The fourth section examines the relationship between famine and other epidemics. The final section locates our findings in a general historiographical context.

Famine and epidemic malaria: the case of 1877-78

Let us first examine the case of 1877-78. From 1876 to 1877, huge areas of southern and western India faced very severe drought. In the north too, the south-west monsoon was not adequate; in 1877 rainfall was about half the average in the greater part of U.P.[20] Thus the 1877-78 famine affected both southern and northern India. The Government of India acknowledged that the death toll reached about 5.25 million.[21]

Clearly drought was the biggest and most immediate reason for the famine in the U.P. But there were other factors as well. During the second half of the 1870s, the export of wheat from this region to Europe increased substantially. This trade began with the opening of the Suez Canal in 1869 and the development of railways. Thus, although food-grain production was about average from 1874 to 1876, foodgrain stocks had been considerably reduced before the drought occurred.[22] In addition, many coarse grains flowed from this region to the Madras and Bombay Presidencies in the years before the drought, because these provinces had already faced very severe famines.

From the middle of 1877 foodgrain prices doubled within the space

---

Learmonth, *Disease Ecology* (Oxford, 1988), pp.195-7.

[20] Only 20.3 inches (516 mm.) of rain fell in 1877, although the average annual rainfall between 1871 and 1876 was 41.7 inches (1,060 mm). *RSC 1877*, p.4.

[21] *Report of the Indian Famine Commission*, part 1, *Famine Relief*, British Parliamentary Papers (1880), cd. 2591, p.28.

[22] Ibid, p.191.

of six months. The quantity of wheat purchasable by one rupee decreased from about 21 *seers* in June 1877 to about 12 *seers* in December 1877, for example.[23] The price increase from 1877 to 1878 was the steepest during the period from 1870 to 1920.[24] The number of food riots in August and September of 1877 exceeded 150,[25] and the number of those who were sent to prison increased.[26] The situation deteriorated when sufficient rain did not arrive until the end of September. It was expected that the *rabi* sowing would be reduced. At the end of September, the U.P. government began relief work and established poor-houses.[27] With the arrival of rain by early October, the *rabi* sowing started, and the employment situation for agricultural labourers improved. However, the real pressure of famine intensified after this period. The changes in the monthly death toll showed a different pattern from that of normal years. The number of deaths gradually rose after November. Abnormally high levels were recorded in January and February of 1878.[28] This is very unusual as mortality rates were normally very low in these months.

The R.S.C. did not provide any information on deaths from famine or starvation, but a series of mortality enquiries was undertaken by the U.P. government at the beginning of 1879.[29] They covered 196 villages in Agra, Oudh and Rohilkhand, the three areas which were most severely affected by the famine. The total population of these 196 villages declined from 36,588 in 1872 to 35,674 in 1879. This diminution was

[23] *RSC 1878*, p.9.

[24] T. Matsui, *Kitaindo Nosanbutsu Kakaku no Shiteki Kenkyu* (An Historical Study of Agricultural Prices in North India), vol.1 (Tokyo, 1977), p.165.

[25] W.C. Bennett, *Report on Scarcity and Relief Operations in the North-Western Provinces and Oudh during 1877-78 and 1879* (1880), p.3.

[26] Proceedings of the Government of the N. W. Provinces and Oudh in the Revenue Department, no.6, February 1879 (India Office Records, London), p.17.

[27] W.C. Bennett, *Report*, pp.11-12.

[28] Proceedings , p.12 (see note 26 above).

[29] Proceedings of the Government of the N. W. Provinces and Oudh in the Revenue Department, no.27, June 1879 (India Office Records, London). This file contains three reports on the Agra area by D.T. Roberts, on the Oudh area by W.C. Bennett, and on the Rohilkhand area by D.G. Pitcher. Hereafter these reports are referred to as Roberts, Bennett and Pitcher respectively.

due not only to deaths, but also to migration, since many people left their villages and temporarily migrated to other regions during the famine. But some of the 196 villages saw a population increase. It seems that they were located within irrigated areas in the Agra, Etah and Mainpuri districts.[30]

What were the causes of death? And what is the relation between famines and epidemics? The Oudh inquiry reports that deaths from starvation were greater (51 per cent) than deaths from smallpox and fever (20 per cent and 13 per cent respectively).[31] A classification of causes of death in some villages of the Agra area is shown in Table 3. This table is useful in the sense that it includes items which were omitted by the R.S.C., that is, deaths from starvation. But there were not as many of these as might be expected, and in this respect we can assume that the Oudh inquiry was more accurate. In Agra deaths from starvation may have been under-reported because of the shame felt by some families. There is also a strong possibility that some deaths from starvation were returned as deaths from fever, bowel complaints or other causes.[32] It is likely that deaths from diarrhoea and dysentery were included under bowel complaints or other causes, despite diarrhoea and dysentery being very closely related to starvation. A report on the 1877-78 famine in Madras showed that there were increases in deaths from bowel complaints and other causes in 1877. W.R. Cornish suggested that these increases were almost totally due to starvation.[33] In the U.P. also, deaths from bowel complaints and deaths from other causes increased in 1878. Therefore, these increases can be interpreted as being closely connected with starvation. Nonetheless, in the case of Agra, fever was the largest cause of death. In particular, malaria was prevalent in the Agra and Muttra districts during the second half of 1878.[34] We will examine this point later on.

Caste classification of death was also included in the Oudh inquiry (see Table 4). Most of the deaths from starvation were centred in the

[30] Roberts, paras.13-15.
[31] Bennett, Form 1.
[32] Ibid., paras.53-56.
[33] *Review of the Madras Famine 1876-78* (Madras, 1881), p.129.
[34] *RSC 1878*, pp.63-72.

five lower castes, such as Kori, Chamar, Pasi, Lodha and Ahir. They were rural poor classes.[35] Chamar and Pasi were mainly agricultural labourers. Lodha and Ahir were mostly small tenants and Kori were weavers. On the other hand Brahmans and Thakurs were relatively unlikely to die from starvation, as they were land-owning castes in this region. Agricultural labourers made up 68 per cent of deaths from famine. The cultivators of less than five *bighas* accounted for 22 per cent, and those with more than five *bighas* for 10 per cent. The rural poor were clearly most at risk of death from famine or starvation.[36]

Finally, although quantitative data is not available, it is interesting to note that accounts of the Rohilkhand inquiry show that the mortality rate in a relatively homogeneous village with few castes was lower than that in a multi-caste village, because relief activities were more generous in the former. Similarly, it is interesting that mortality rates were low in villages where the landlords were known not to be oppressive.[37]

If we look at crude death rates in 1877, 1878 and 1879 (19.67, 35.62, and 44.81 respectively), we find that the highest level was in 1879—not a year of famine. According to deaths registered in the R.S.C., excess deaths during the famine period, November 1877 to December 1878, were about 0.73 million people. As the registered numbers must have been much less than the actual ones, real excess deaths were estimated at 1.27 million.[38] If we estimate the real excess deaths for 1879 in the same way, they would constitute about 1.86 million people, being therefore much greater than the real excess deaths during the famine period. What happened in 1879? According to the 1879 R.S.C.:

The record for 1878 was unusually large, as a result of general distress approaching to famine in some portions of the province, the unusual death incidence, from that cause, being increased by the prevalence of an extremely

[35] Bennett, paras.13-19.
[36] Ibid., paras.20-26
[37] Pitcher, paras.32, 42, 75, and 91.
[38] Excess deaths are calculated by subtracting the average number of deaths from the number of deaths during the famine year. *Report of Indian Famine Commission*, part 3, *Famine Histories* (1885), cd.3086, p.203.

fatal form of malarial fever in the Agra and Muttra districts. But the record for 1879 is far in excess of that for 1878, and this very excessive mortality has been due to one cause only, the extended prevalence of the same form of malarial fever as affected the people of Agra and Muttra only in 1878.[39]

Thus epidemic malaria was prevalent in 1879. There is no doubt that this was the same kind of malaria as in Agra and Muttra. There were two kinds of malaria in that year. The 1879 R.S.C. says:

In the case in which fever action continued without marked interval, from the day of attack to the day of death, the common symptoms were: a commencement of illness by shivering and headache with pain in the back and limbs, soon followed by increases of temperature, with a quick pulse and hurried respiration. Cases in which the patient enjoyed intervals of freedom, from active fever distress, were marked by exactly the same commencing symptoms as described above, in the speedily fatal form of the disease. But sometimes as early as the following morning, or within 24 hours after first attack, the symptoms of diseased action would begin to decline—and after an interval, of comparative relief, varying from a few to 24 hours, the former urgent symptoms would return.[40]

Judging from the symptoms, the first kind must be *Plasmodium falciparum*. It is much more fatal and very virulent, but does not cause relapses. The second kind must be *Plasmodium vivax*. It is very widespread and less virulent, but remains in the liver, causing relapses for a long time.[41]

Malaria is an endemic disease in this region. It is well known that deaths from malaria were classified as deaths from fever in the vital statistics of the R.S.C. In an average year, between 39 per cent and 50 per cent of deaths from fever were in effect deaths from malaria.[42] But the prevalence of malaria in 1879 was much greater than in an average year. In this sense, this should be called epidemic malaria. This kind of epidemic malaria also occurred in 1894, 1897 and 1908. Why did epidemic malaria occur? First of all, rainfall should be considered. In an average year malaria is prevalent from September to December. The

[39] *RSC 1879*, p.3.
[40] Ibid., pp.51-2.
[41] A. Learmonth, *Disease Ecology*, p.193.
[42] K. Davis, *The Population of India and Pakistan*, p.53.

monsoon plays a very important role. It was recorded that the south-west monsoon brought heavy rainfall in 1879.[43] The monthly fever mortality rate began to rise from September, reaching its peak in October, and then gradually declined. This pattern was usual, but the level of mortality was high.

Secondly, there is a strong possibility that hunger contributed to epidemic malaria. According to the 1879 R.S.C.:

I think it likely that some portion of the excessive mortality, recorded during 1879, may have been due to this continuance of high prices. And especially I believe that many very poor people, who lived with difficulty during the last three years, had fallen into a low state of health which, in great measure, took away their power to recover from the attack of the fever disease prevailing so generally in the later months of the year.[44]

The high prices of foodgrains had been continuing since July 1877. It is reasonable to assume that high prices for more than two years physically weakened the poorer classes and made them less resistant to malaria.[45] Most of the districts where malaria was highly active are located around the Doab region (see Figure 4). Mortality was highest of all in Aligarh and Bulandshahr in the central Doab, next highest in the adjacent districts (Meerut, Etah, Muttra, Mainpuri and Budaun), only the last of which was outside the Doab, and high, though somewhat lower again, in another ring of districts around these ones.[46] These districts were also the most severely damaged by the famine.

Thirdly, the western and southern parts of the U.P. belong to what may be regarded as areas of epidemic malaria. As is seen in Figure 4, most of the epidemic malaria which followed famines occurred in this epidemic zone. Areas where malaria was endemic were usually humid, and malaria was constantly present. But there were other areas where

[43] *RSC 1879*, p.22.
[44] Ibid., p.13.
[45] Ibid., p.59.
[46] Ibid., p.21. The top 20 districts that experienced high mortality rates were: Aligarh (113.5), Bulandshahr (113.3), Meerut (81.9), Etah (77.8), Muttra (66.9), Mainpuri (63.4), Budaun (61.2), Muzaffarnagar (58.7), Hardoi (54.4), Farrukhabad (52.7), Agra (52.3), Kanpur (46.3), Etawah (45.8), Shahjahanpur (43.0), Unao (43.0), Saharanpur (42.4), Fatehpur (37.6), Moradabad (36.4), Jhansi (32.1) and Jalaun (31.5).

epidemics of malaria occurred periodically. These areas were mostly semi-arid. Immunity was low, and morbidity and mortality rates were very high.[47]

Finally, it should also be considered that there is a causal relationship between malaria and the extension of canal irrigation.[48] As is widely known, the Doab region was blessed with canal irrigation.[49] In the 1879 R.S.C., C. Planck wrote:

There is but one circumstance which can account for this peculiarity of disease prevalence, and that is canal irrigation. As the result of local enquiry and investigation, extending to all portions of the country influenced, I have on previous occasions recorded the opinion that the excessive prevalence of malaria in the Doab, and its neighbourhood, is due to the increase of moisture, which has entered into its land, since the introduction of this form of irrigation.[50]

The causal relationship between malaria and canal irrigation has been often pointed out. Even in the 1869 R.S.C., Planck claimed that although the introduction of canal irrigation made possible the cultivation of rice where only *bajra* and *jowar* had been hitherto planted, the irrigation made dry and healthy land very damp and unhealthy.[51] They thought that bad or damp air was the cause. The Doab was originally a semi-arid and drought-prone region. Canal irrigation was introduced in order to prevent famines.[52] Ironically, this

---

[47] A. Learmonth, *Disease Ecology*, p.195-7. More detailed medical explanation is given in I. Stone, *Canal Irrigation*, pp.151-3.

[48] This causal relationship has been discussed in I. Klein, 'Malaria and mortality in Bengal, 1840-1921', *Indian Economic and Social History Review* 9, 2 (1972), 'Death in India', *Journal of Asian Studies* 32, 4 (1973) 'Population and agriculture in northern India, 1872-1921', *Modern Asian Studies* 8, 2 (1974); Stone, *Canal Irrigation*; and E. Whitcombe, *Agrarian Conditions in Northern India, vol.1: the United Provinces under British rule, 1860-1900* (Berkeley, 1971).

[49] Thirteen districts comprise the Doab; from the north and west they are: Saharanpur, Muzaffarnagar, Meerut, Bulandshahr, Aligarh, Muttra, Etah, Agra, Mainpuri, Farrukhabad, Etawah, Kanpur and Fatehpur. See note 46 above.

[50] *RSC 1879*, p.25.

[51] *RSC 1869*, pp.88-9. At the time people did not know that malaria was transmitted by mosquitoes.

[52] A refurbished West Jumna Canal reopened in 1820, and the East Jumna Canal ten years later. The huge new Ganges Canal was fully operational by

canal irrigation amplified the damage resulting from famines in the sense that it enhanced the extent of epidemic malaria.

Famine and epidemic malaria: other cases

In this section we examine other mortality peaks to see whether the close relationship between famine and epidemic malaria was a general phenomenon. We find some evidence for the year 1869. The people suffered from famine during 1867-69. The western and south-western parts of the U.P. suffered very intense drought. Here again, famine accompanied epidemic malaria. Districts with high death rates from fever were located in areas badly hit by the famine. They were Bareilly, Saharanpur, Muzaffarnagar, Banda and Jhansi.[53] Of course, excessive rainfall was the basic cause of malaria epidemics in these cases also. But another cause was more significant. According to the 1869 R.S.C.:

Another cause of the increased mortality from fever was doubtless, I think, that the poor people of the country were badly nourished; eating, as a rule during a great part of the year, only the commoner sorts of grain, and those often in bad condition. During my journeys of the past year, I heard often of this as a cause of fever, disease and death; and, although it may be accepted as true that very few people died of starvation, it cannot be doubtful that many died from eating too little to support vigorous life, so that disease appearing found many ready victims.[54]

After 1879, as we have seen, there were six peaks in the mortality rate. Out of these, fever recorded peaks in four years (Figure 5). More

---

1857. The Agra Canal opened in 1874 (serving Agra and Muttra), the lower canal was Ganges, completed in 1878, taking over the lower part of the former Ganges Canal; that in turn was greatly extended in the north by new branches through the centre of Saharanpur, Bulandshahr, Muzaffarnagar and Aligarh. Thus, even if one accepts the mortality statistics (which may exaggerate the incidence of malaria), the connection between the disease and the canals, though undoubted, is complex: malaria was particularly prevalent in Doab towns, and though the extensions of canals between 1868 and 1878 may account for some of the worsening of the disease, the highest mortality in 1879 was in districts which had had canals for a generation. There was also more than twice the area under canal irrigation early in the twentieth century than there had been in 1879 (the worst malaria epidemic). See Stone, *Canal Irrigation*, pp.13-33, and, on malaria, pp.144-57. [Editor's note: PR.]

[53] Ibid., p.3.
[54] Ibid., p.4.

important, three peaks coincided with epidemic malaria, while 1918 was an influenza pandemic.

In 1894, famine did not play a role. The malaria epidemic occurred only because of excessive rainfall. Although the death rate from cholera amounted to 3.80 per million, the death rate from fever reached the very high level of 31.88. Particularly in the eastern region (Benares and Gorakhpur divisions), high death rates were recorded.[55] In 1897, by contrast, the high death rate was undoubtedly related to famine conditions. Already in 1895 the Bundelkhand region faced extremely deficient rainfall in both kharif and rabi seasons. Relief works started early in 1896. Again the south-west monsoon carried very little rainfall. Two years of deficient rainfall badly affected the residents of Bundelkhand. High mortality rates were recorded especially in southern and central districts of U.P.[56] Except for Agra and Bareilly these districts were severely stricken with famine. The 1897 R.S.C. emphasised:

As has been mentioned in the famine report for these Provinces, it is very difficult to dissociate famine and malarial fever as causes of mortality. Of starvation pure and simple it is probable there were but few instances, but the population of these Provinces generally was so reduced in stamina by continued deficiency of food that many people fell easy victims to the ever present and insidious malarial poison of the country.[57]

Many other districts suffered from malaria: 'Excluding the hill tracts, very few of the death-rates were low. Facilities for communication had equalised values all over the country, and there were not many localities where the pinch of high prices was not felt.'[58] Foodgrain price increases spread beyond the region seriously affected by the famine.

In the same way, the mortality rate in 1908 (52.7) was much higher than in 1907 (43.5), due to epidemic malaria. The fever mortality rate reached 41.31 in 1908. We read similar findings in the 1908 R.S.C.:

[55] *RSC 1894*, pp.20-21.

[56] *RSC 1897*, p.5. The districts were (death rates in parentheses): Hamirpur (62.07), Jalaun (59.71), Bareilly (55.77), Agra (55.55), Fatehpur (55.09), Mirzapur (52.46), Hardoi (51.96), Banda (51.83), Lucknow (51.16), and Rae Bareli (50.39).

[57] Ibid., p.5.

[58] Ibid., p.5.

The consensus of opinion expressed by Civil Surgeons and others in these provinces, which is in accord with my own observations, is that, as was the case in 1879, the recent epidemic affected all classes of the population with but few exceptions (such as Tharus in the Tarai). The rich and the poor, and Hindus and Muhammadans, all suffered alike. Undoubtedly the poorly nourished people, those whose stamina had been affected by the high prices prevailing during the preceding two years, fell easy victims to the disease while the rich recovered.[59]

The very young (infants and children) and the very old suffered most severely. The infant mortality rate was 345.1, which was much higher than the previous average of 249.2 (1901-1907).[60] Epidemic malaria was responsible for this.

In this case too, districts where the mortality rate was high lay not only in the famine region, but outside it. For example, Muttra, Budaun and Bareilly where mortality rates were high (70.31, 68.63, and 67.31) did not suffer from a very severe famine. Bulandshahr (64.29), Aligarh (57.33), Meerut (51.58), Muzaffarnagar (49.73) and Saharanpur (46.28) did not experience the famine at all. We should emphasise here that these districts belonged to the Doab. It was pointed out that water-logged land caused by canal irrigation led to an increase of the *anopheles* population.[61]

Famine and other epidemics

Here we examine the relationship between other epidemics and famines. Figures 8 to 10 show the changes in the number of deaths from smallpox, cholera, and plague. In the case of smallpox, peaks in death numbers coincided with famine years: 1878, 1897, and 1908 (see Figure 6). In northern India, smallpox was usually prevalent during the dry season from February to June before the monsoon came. An especially dry year, or a year which saw drought, was favourable to the spread of the smallpox virus.[62] Moreover, malnutrition due to famine probably increased its prevalence. Almost all of the deaths from

[59] *RSC 1908*, p.12.
[60] Ibid., p.4.
[61] Ibid., p.12.
[62] *RSC 1897*, p.27.

smallpox were those of children.[63] Figure 6 shows that the extent of epidemic smallpox was somewhat reduced after the end of the nineteenth century. It can be argued that the spread of vaccination was successful in weakening the effect of smallpox to some extent. Vaccination was done in relief works and conducted in poor houses during the 1907-08 famine.[64] Perhaps it played some role in checking the disease.

Cholera epidemics occurred frequently: in 1887, 1892, 1894, 1906, 1911, and 1918, but they did not coincide with famine in the U.P. (see Figure 7). However, the 1877-78 Madras famine was accompanied by a cholera epidemic. In this case the incidence of cholera was at its highest intensity in January and February 1877, although the food crisis was most severe in July and August. W. R. Cornish explained:

The pestilence was present with us before the food dearth began, but it was aggravated by the wandering habits of the famine sufferers, and by their aggregation in unusual numbers on works instituted for their relief and in market towns and food centres, and also by the resort to strange and unnatural articles of diet, and by the use of contaminated water.[65]

Thus, the outbreak of cholera epidemics was not so dependent on nutritional conditions, as on the deterioration of societal conditions. In this sense, as Cornish emphasised, cholera epidemics pursued a course relatively independent of famine. Indeed, the movement of the pilgrims was a more important factor for disseminating cholera. In 1894, the people who left Kumbh Fair at Allahabad helped disperse cholera.[66]

Plague affected parts of the U.P. for the first time in 1901. After that, large-scale epidemics occurred in 1905, 1907, 1911, and 1918 (see Figure 8). In particular, 1905 and 1911 were the years which substantially pushed up the mortality rate (see Figure 2). India had been free from plague for a long time until the end of the nineteenth century. It

[63] *RSC 1878*, p.16.

[64] *RSC 1908*, p.9.

[65] *Review of the Madras Famine 1876-78* (Madras, 1881), p.126. Concerning the relationship between cholera and famine, see D. Arnold, 'Cholera mortality in British India', in T. Dyson (ed.), *India's Historical Demography*, pp.267-9.

[66] *RSC 1894*, p.29. For the role of pilgrimage in dispersing cholera, see D. Arnold, 'Cholera mortality in British India', pp.271-3.

broke out in the city of Bombay in 1896 and then spread all over India. In the case of the U.P., 383,802 people died from plague in 1905.[67] In 1907 and 1911 the epidemics were similar in severity. But the outbreaks hardly coincided with famines. Nor do they seem to have depended on the nutritional condition of the people.

Finally, there was influenza, which in 1918 spread all over the world as pandemic. India was the country most severely affected . According to I. D. Mills' estimate, influenza killed an astonishing 17 million Indians in half a year. The government's estimate was between 12 and 13 million.[68] U.P. was one of the worst affected provinces. The morbidity rate was very high, and more than 50 per cent of the population suffered from the disease. The number of deaths amounted to about 1.7 million.[69]

This form of influenza was so often fatal because of 'the tendency for sufferers to succumb to pneumonia'.[70] It was the pneumonia that was deadly. An important characteristic of the disease was that there were many victims aged between 20 and 40, while old people are usually the most numerous victims for ordinary influenza. A further characteristic was that many women died as a result of the pandemic.[71] A strong link can be found between pneumonia and malnutrition. In 1918 the south-west monsoon did not supply much rain, and crops failed in the U.P. as well as in several other parts of India. In this sense the situation was almost like a famine. In the U.P. the yield of the principal kharif crops in the year 1918 was almost half of a normal year's.[72] I. D. Mills pointed out:

Thus, it would appear that the famine and the pandemic, in the Indian context, formed a set of mutually exacerbating catastrophes. While the synergistic effects of malnutrition and infection are well recorded at the individual level,

[67] *RSC 1905*, p.7.
[68] I.D. Mills, 'Influenza in India during 1918-19', in T. Dyson (ed.), *India's Historical Demography*, pp.223 and 228.
[69] *RSC 1918*, pp.8A-11A.
[70] Mills, 'Influenza in India', p.233.
[71] Ibid., p.238.
[72] *Season and Crop Report of the United Provinces of Agra and Oudh, 1918-19*, p.5.

this suggests that such relationships, though brought about by somewhat different mechanisms, also occur at the societal level. Whether such inter-linkages exist for other crises is a topic worthy of further research.[73]

Conclusion

The study of famines has occupied a central place in the debate on modern Indian history. Nationalists such as Romesh Chandra Dutt, Dadabhai Naoroji and William Digby, argued that famines were not just natural disasters, but occurred because of the poverty of the Indian people, which was brought about by British rule. Indian poverty was assumed to have been caused by heavy land revenue, military expenditure spent on wars that had no relation to the Indian people, and the drain of wealth that continuously flowed away from the country. On the other hand, the colonial government maintained that famines occurred as a result of severe drought which was beyond human control, and that Indian poverty was a result of excessive population or an outcome of the improvidence of the Indian people.[74]

According to Dutt, a series of famines during the second half of the nineteenth century was caused by the overassessment of land revenue and the 'drain' of wealth. In Bengal land revenue had been fixed by the Permanent Settlement of 1793, but in Bombay and Madras it was settled every thirty years. In the latter cases, even if the cultivators endeavoured to increase their agricultural production, any surplus would be taken away by the government. Thus the high revenue implied a low level of incentive. The resultant low level of production led to a high frequency of famines. Also, the wealth squeezed through land revenue flowed from India to Britain. The drain of foodgrains to foreign countries intensified the scale of famines.[75]

Other nationalists argued that the development of the railway system during the second half of the nineteenth century changed the character

[73] Mills, 'Influenza in India', p.252.
[74] D. Naoroji, *Poverty and Un-British Rule in India* (London, 1901); R.C. Dutt, *Open Letters to Lord Curzon on Famines and Land Assessments in India* (Edinburgh, 1900); R.D. Dutt, The *Economic History of India*, vol.2, *In the Victorian age* (London, 1903); W. Digby, *'Prosperous' British India* (London, 1901).
[75] R.C. Dutt, *Open Letters*; R.C. Dutt, *The Economic History of India: Vol.2.*

of famines. After the introduction of rail transport, increases in food-grain prices began to spread beyond the region severely affected by a famine. As a result, famine damage began to concentrate on the rural poor. The famine thus became a class-oriented phenomenon. The changes in the nature of famines could also be explained by the dis-integration of rural social ties. These studies identified the structural changes in rural society as a basic cause of frequent famines.[76]

Nationalists' views on famines were recently refuted by M.B. McAlpin. Criticising Dutt's arguments, she questioned whether the frequency of famines was high during the latter half of nineteenth century in comparison with the previous period.[77] There are few historical records on famines before the British rule, and we cannot be sure of the frequency. McAlpin also questioned whether the land revenue increased so substantially as to cause famines. She sought to prove, through an extensive use of quantitative evidence, that the land revenue collections per acre in fact decreased both in nominal and real terms even after two periodical settlements. She further refuted the claim that forced exports of foodgrains led to the amplification of famine damage. Emphasising the more positive aspects of foodgrain export, McAlpin argued that the commercialisation of foodgrains raised the standard of living of peasants and prevented the occurrence of famine.

In the meantime some new studies had begun to focus on epidemics rather than famines. I. Klein pointed out that during this period epidemics killed many more people than famines did. According to Klein, malaria was the worst cause of death.[78] L. Visaria and P. Visaria also emphasised the importance of epidemics for demographic trends.[79] In addition, many other studies on epidemics have emerged.[80] These

---

[76] B.M. Bhatia, *Famines in India: A study in some aspects of the economic history of India, 1860-1965* (Bombay, 1967); H.S. Srivastava, *The History of Indian Famines and Development of Famine Policy, 1858-1918* (Agra, 1968).

[77] M.B. McAlpin, *Subject to Famine: food crises and economic change in western India, 1860-1920* (Princeton, 1983).

[78] Klein, 'Malaria', 'Death in India', and 'Population and agriculture'.

[79] L. Visaria and P. Visaria, 'Population (1754-1947)'.

[80] See, *inter alia*, D. Arnold, 'Touching the body: perspectives on the Indian plague, 1896-1900', in R. Guha (ed.), *Subaltern Studies 3* (New Delhi, 1984);

studies however seem to have regarded epidemics as a phenomenon independent of famines. Certainly, as we have seen with reference to the U.P., some epidemics like cholera and plague were not the result of famine situations. On the other hand, neither classical studies nor McAlpin seriously considered epidemiological aspects of famines.

Thus, the causal relationship between famines and epidemics, or between nutritional conditions and disease conditions, is yet to be fully examined. In the original version of this paper (published in Japanese in July 1989),[81] I argued for the importance of the *links* between famines and epidemics, with special reference to epidemic malaria. As we have seen, in the case of the U.P. epidemic malaria was the most important cause of death. It usually occurred after a famine, which helped prepare the condition for it, and substantially increased the mortality rate. If we include both endemic and epidemic malaria, it becomes the largest single factor accounting for the low rate of population growth in this period. Thus the demographic regime of northern India in this period can be characterised by a very high mortality rate, a relatively high fertility rate and a slow population growth. Both very high mortality and high fertility resulted, not directly from starvation, but largely from a specific sequence of famines providing an environment in which death tolls from epidemics increased.

Recent studies on other regions support this view.[82] T. Dyson examined the famines of 1876-78 (in Madras), 1896-97 (in the Central Provinces and Bombay Presidency) and 1899-1900 (in the Central

D. Arnold, 'Smallpox and colonial medicine in nineteenth-century India', in D. Arnold (ed.), *Imperial Medicine and Indigenous Societies* (Delhi, 1989); P. Bala, *Imperialism and Medicine in Bengal: a socio-historical perspective* (New Delhi, 1991); I.J. Catanach, 'Plague and the tensions of empire: India 1896-1918' in D. Arnold (ed.), *Imperial Medicine*; J.H. Hume, 'Colonialism and sanitary medicine: the development of preventive health policy in the Punjab, 1860 to 1900', *Modern Asian Studies* 20, 4 (1986).

[81] K. Wakimura, 'Jyukyuseiki Kohan no Kitaindo ni-okeru Kikin to Ekibyo' (Famines and epidemics in northern India during the second half of the nine-teenth century), *Keizaigaku-zasshi* , 90, 2 (1989).

[82] T. Dyson, 'On the demography of South Asian famines'; S. Zurbrigg, 'Hunger and epidemic malaria in Punjab, 1868-1940', *Economic and Political Weekly* (25 January 1992); D. Guz, 'Population dynamics of famine in nine-teenth-century Punjab, 1896-7 and 1899-1900', in T. Dyson (ed.), *India's His-torical Demography*.

Provinces and Bombay Presidency). In these famines 'death peaks occurred relatively late and were comparatively confined in time'. His findings also suggest that malaria was the most important famine disease.[83] In addition, S. Zurbrigg examined nutritional aspects of malaria epidemics in Punjab and argued that death from malaria was an indirect barometer of hunger.[84]

Unfortunately, from a medical perspective, the causal relationship between malarial deaths and nutritional decline has yet to be proven.[85] In more general terms, there is a claim that the occurrence of epidemics cannot be traced to nutritional deterioration, because a micro-organism, that is, a pathogen, needs the same kinds of nutrients as those which are crucial to the host.[86] In the Indian case, both the epidemiological aspects of famine and the nutritional aspects of epidemics should continue to be considered in future studies.

[83] T. Dyson, 'On the demography of South Asian famine', p.41-2.
[84] S. Zurbrigg, 'Population dynamics of famine'.
[85] Anonymous, 'The relationship of nutrition, disease, and social conditions: a graphical presentation', in R.I. Rotberg and T.K. Rabb (eds.), *Hunger and History*, p.308.
[86] A.G. Carmichael, 'Infection, hunger and history', ibid., pp.52-3.

Table 1: *Population changes in India and the U.P. (1871-1931)*

	Population (thousands)		Index (1871=100)	
	India	U.P.	India	U.P.
1871	255,166	42,627	100	100
1881	257,380	44,876	101	105
1891	282,134	47,682	111	112
1901	285,288	48,477	112	114
1911	302,985	47,997	119	113
1921	305,679	46,510	120	109
1931	338,171	49,614	133	116

Sources: For India, K. Davis, *The Population of India and Pakistan* (Princeton, 1951), p.27; for U.P., *Census of India 1931, United Provinces of Agra and Oudh*, Part 2, Imperial Tables.

Table 2: *Population changes in the U.P. (1872-1921)*
(Province and divisions)

	1872	1881	1891	1901	1911	1921
U.P.	42,626,996	44,876,499 (5.3)	47,682,197 (6.3)	48,476,793 (1.7)	47,997,364 (-1.0)	46,510,688 (-3.1)
Meerut	3,890,312	4,105,108 (5.5)	4,267,991 (4.0)	4,761,498 (11.6)	4,626,188 (-2.8)	4,509,572 (-2.5))
Agra	4,526,458	4,225,181 (-6.7)	4,267,991 (-0.02)	4,761,498 (11.7)	4,626,188 (-4.3)	4,509,572 (-7.3)
Rohil-khand	5,252,459	5,122,586 (-2.5)	5,344,007 (4.3)	5,479,738 (2.5)	5,650,841 (3.1)	5,198,773 (-8.0)
Allaha-bad	4,801,899	4,969,135 (3.5)	5,045,086 (1.5)	5,168,175 (2.4)	4,946,504 (-4.3)	4,795,666 (-3.0)
Jhansi	2,161,755	2,249,040 (4.0)	2,299,532 (2.2)	2,106,085 (-8.4)	2,207,923 (4.8)	2,065,297 (-6.5)
Benares	4,076,677	4,819,508 (18.2)	4,980,415 (3.3)	4,706,979 (-5.5)	4,451,988 (-5.4)	4,443,898 (-0.2)
Gorakh-pur	4,810,016	5,852,386 (21.7)	6,508,526 (11.2)	6,333,012 (-2.7)	6,524,419 (3.0)	6,720,715 (3.0)
Kumaun	928,823	1,046,263 (12.6)	1,181,567 (12.9)	1,207,030 (2.2)	1,328,790 (10.1)	1,292,399 (-2.7)
Luck-now	5,315,676	5,325,692 (0.2)	5,856,652 (10.0)	5,977,177 (2.1)	5,911,642 (-1.1)	5,567,241 (-5.8)
Fyza-bad	5,905,367	6,062,140 (2.7)	6,794,272 (12.1)	6,855,991 (0.9)	6,646,362 (-3.1)	6,599,401 (0.7)
Native States	957,548	1,099,460 (14.8)	1,179,947 (7.3)	1,163,454 (1.4)	1,189,874 (2.3)	1,134,881 (-4.6)

Source: *Census of India 1921, United Provinces of Agra and Oudh, Part 2 Imperial Tables*, pp. 6-8. *Note*: Figures in parentheses are growth rates per decade.

Table 3: *Causes of death (Agra Inquiry)*

	November 1877 to June 1878		July to December 1878	
	Deaths	%	Deaths	%
Cholera	9	0.8	22	2.0
Smallpox	276	23.1	34	3.1
Fever	476	39.8	708	64.2
Bowel complaints	110	9.2	175	15.9
Starvation	221	18.5	57	5.2
Other causes	103	8.6	106	9.6
Total	1,195	100.0	1,102	100.0

Source: Roberts, Statement No.3.

Table 4: *Classification of deaths by caste*

	Numbers	%	Hunger deaths	%	Smallpox deaths	%
Brahman	63,427	13.7	53	3.0	87	12.6
Thakur	30,209	6.5	16	0.9	15	2.2
Kori	22,832	4.9	264	15.2	14	2.1
Chamar	27,132	5.8	260	14.9	51	7.4
Pasi	54,220	11.7	453	26.0	108	15.7
Ahir	49,910	10.8	194	11.1	68	9.9
Kurmi	27,427	5.9	68	3.9	70	10.1
Lodha	23,859	5.1	176	10.1	37	5.4
Others	165,046	35.6	256	14.7	240	34.8
Total	464,062	100.0	1,740	100.0	690	100.0

Source: Bennett, Form 3.

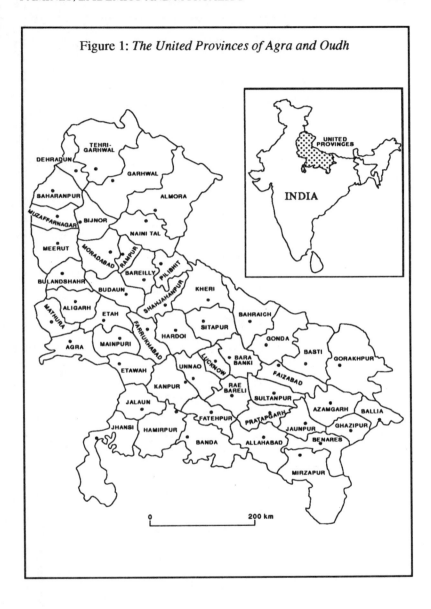

Figure 1: *The United Provinces of Agra and Oudh*

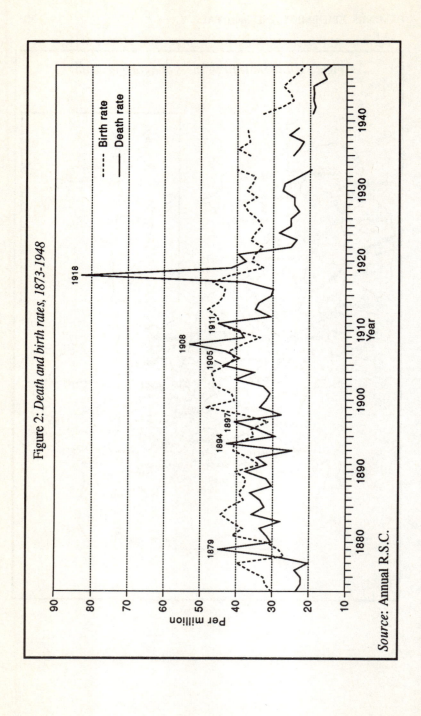

Figure 2: *Death and birth rates, 1873-1948*

Source: Annual R.S.C.

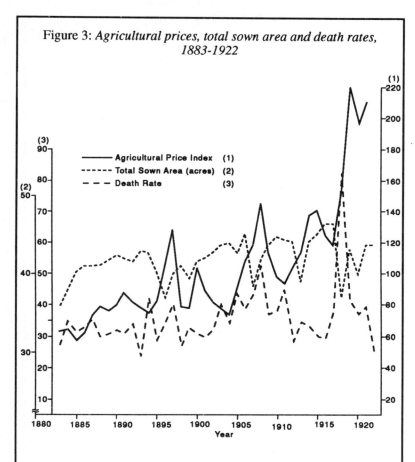

Figure 3: *Agricultural prices, total sown area and death rates, 1883-1922*

Sources: For the agricultural price index and sown area, T. Matui, *Kitaindo Nosanbutsu-kakaku no Shiteki-kenyo* (Agricultural Prices in Northern India, 1861-1921), vol.2 (Tokyo 1977), tables I-1 and A-1; for the death rates, annual R.S.C. *Note*: The agricultural price index is the index of retail prices. The base year is the average between 1901 and 1910.

Figure 4: *Ten districts with highest mortality in 1879, 1897 and 1908*

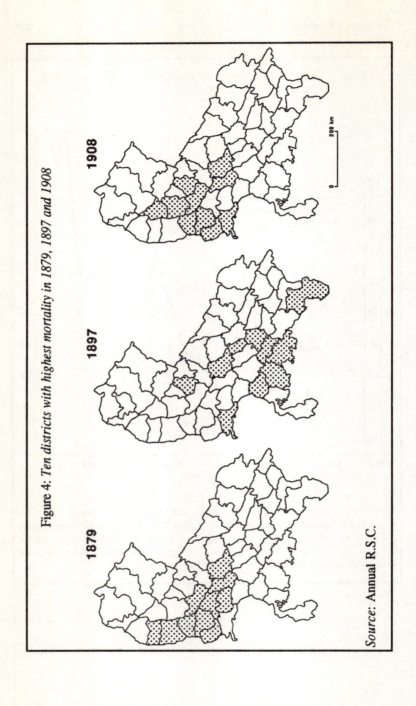

Source: Annual R.S.C.

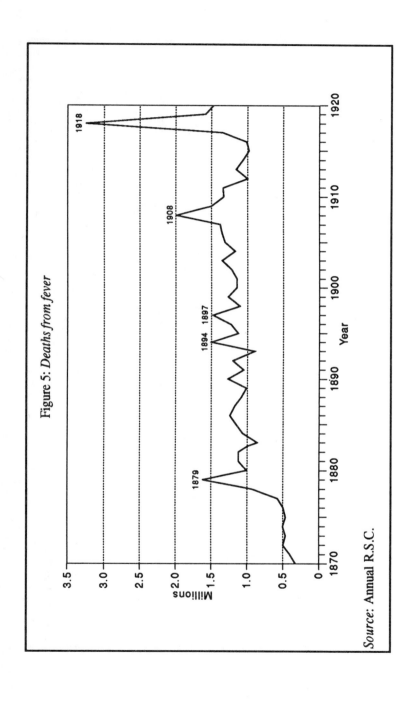

Figure 5: *Deaths from fever*

Source: Annual R.S.C.

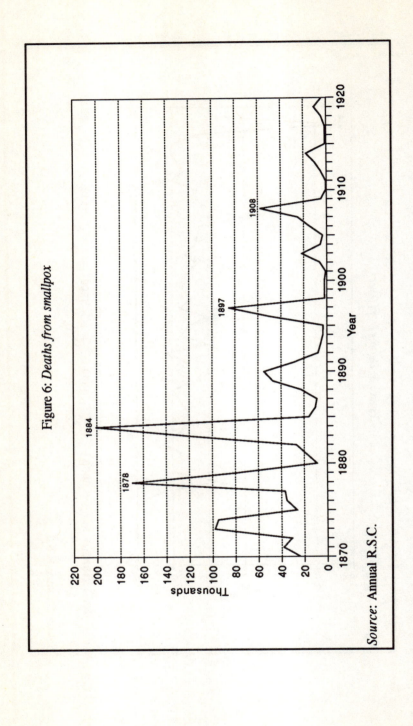

Figure 6: *Deaths from smallpox*

Source: Annual R.S.C.

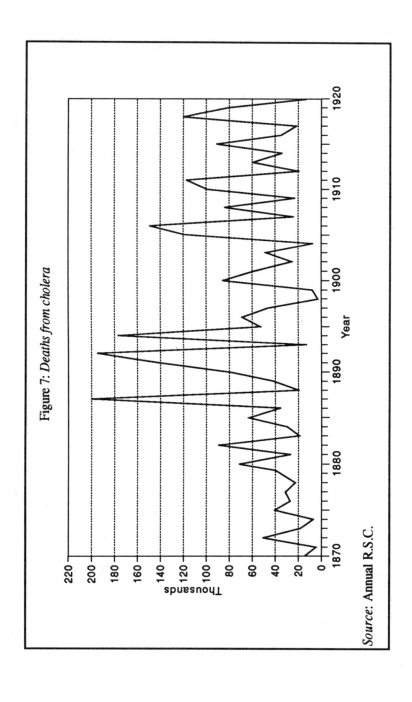

Figure 7: Deaths from cholera

Source: Annual R.S.C.

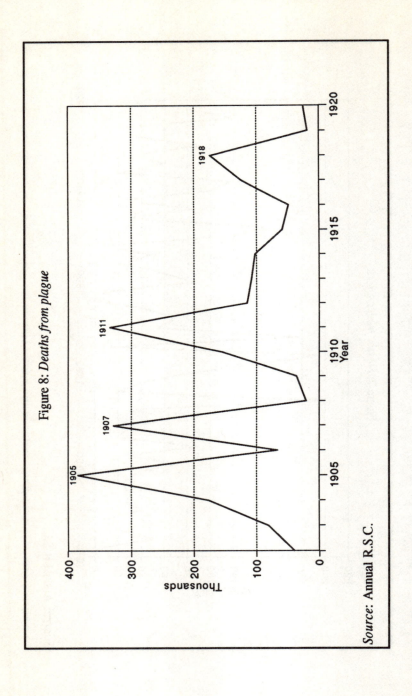

Figure 8: *Deaths from plague*

Source: Annual R.S.C.

Chapter 6.2

FAMINES AND EPIDEMICS:
A COMPARISON BETWEEN INDIA AND JAPAN

Osamu Saito

Wakimura's paper draws our attention to high death rates in India in an age of economic transformation. In the period from 1870 to 1921, despite the development of railway and canal networks and the commercialisation of agriculture, there were repeated famines and epidemics which kept the country's general level of mortality high during this period. In one northern region alone epidemics killed as many as 1,860,000 people in the famine years of 1877-79, most of whom are believed to have died of malaria. Indeed, in a half-century period from 1870, the United Provinces experienced at least four mortality crises of this magnitude for which the famine-epidemics link can be established. A pattern identified by the Wakimura paper, as well as Tim Dyson's recent article on a similar subject,[1] is that severe drought caused crop failures which prompted food prices to rise sharply. This was accompanied by a decrease in conceptions, and then, with the return of the rains, followed by malaria outbreaks. Although it is now widely recognised that pre-modern societies were prone to repeated 'crises' triggered by either famines or epidemics, or both, we are nonetheless overwhelmed by the sheer numbers of famine casualties and the severity of their demographic effects in British India. It is not entirely clear whether an immediate effect of economic transformation was that it worsened mortality conditions, or that the Malthusian check continued to operate in spite of the transformation. Either way, the following discussion should offer some relevant points of comparison.

II

Like other 'pre-modern' societies, traditional Japan too experienced a

[1] T. Dyson, 'On the demography of South Asian famines', parts 1 and 2, *Population Studies* 45, 1 (1991), pp.5-25, and 45, 2 (1991), pp.279-97.

number of recorded famines. According to one authority, there were at least 130 during the Tokugawa period (1603-1868), of which 21 were particularly serious. Of the 21 ten took place in the seventeenth century, nine in the eighteenth and two in the first half of the nineteenth century. Overall, this implies that a serious failure in harvest occurred every 12.6 years.[2] Of these 21, the Kyoho (1732), Tenmei (1782-87) and Tenpo (1833-38) famines are known as the Three Great Famines of the Tokugawa era.[3] It is well documented that rice prices soared in these famine years, while in many areas, especially in the north-east, starvation cases were reported by contemporary witnesses. Even after the Meiji Restoration, famines did not completely disappear. As late as the first half of the 1930s, villages in the north could not escape from bad harvests, leading many to near-starvation. The 'famine' mortality itself was not high, but the press gave wide coverage to those impoverished peasants who were forced to 'sell' their daughters to sweat shops or to the prostitution racket in the cities; it was a rural crisis which alarmed ultra-nationalists, thus having immense implications for the nation's political course in this period.[4]

Epidemics were not infrequent, either. Under the *sakoku* (seclusion) policy of the Tokugawa shogunate, overseas trade and communications were severely restricted by the government, so that pandemics of plague and cholera did not reach this Far Eastern land until the West knocked on the door in the 1850s. Nevertheless, epidemics such as smallpox, measles and dysentery took heavy tolls of human life, of which smallpox was 'the most terrible minister of death' in the period before the opening of the Treaty ports in 1859. According to Ann Jannetta's study of remote hill villages affiliated to the Ogenji temple in Hida, central Japan, smallpox epidemics occurred 17 times in the 81-

---

[2] Naotaro Sekiyama, *Kinsei Nihon no Jinko Kozo* (Population Structure of Early Modern Japan) (Tokyo, 1958), p.159.

[3] See, for example, Bonsen Takahashi, *Nihon Jinkoshi no Kenkyu* (Studies in the History of Japanese Population) (Tokyo, 1941), Hidetoshi Arakawa, *Kikin* (Famines) (Tokyo, 1979), and Kazuo Kikuchi, *Nihon no Rekishi Saigai: Edo koki no jiin kako-cho ni yoru jissho* (Historical Calamities in Japan: evidence from late-Edo Buddhist temple death registers) (Tokyo, 1980).

[4] Tsunao Inomata, *Kyubo no Noson* (Destitute Japanese Villages) (Tokyo, 1934), part 1.

year period from 1771, accounting for 10 per cent of all death during this period, and 26 per cent of those under the age of ten.[5]

The encounter with the West brought in epidemic diseases which had previously been virtually unknown, such as cholera, typhoid, typhus and diphtheria. Cholera was particularly rampant from the late 1850s to the turn of the century. Indeed it is recorded that there were at least seven major outbreaks of cholera between 1858 and 1902, of which the Ansei outbreak of 1858-60 is believed to have killed between 100,000 and 200,000 people in Edo (former Tokyo) alone. The 1877-79 epidemic was responsible for more than 100,000 deaths, and the 1886 one for at least 150,000 deaths over the whole country. There were other epidemics also. Dysentery broke out in 1893 and its death toll reached 200,000. A little more lethal was the 1918-19 pandemic of influenza known as 'Spanish flu', whose deaths totalled up to 257,000, with a case-fatality rate of 12.2 per thousand.[6]

Another demographic effect of famines, as revealed in the Indian case, was a substantial decline in birth rates. This effect too is evident from nineteenth-century Japanese materials. Griffith Feeney and Kiyoshi Hamano estimated annual crude birth rates for 1807-86 from early Meiji data, that is, single-year age distributions in 1886, and examined relationships with annual rice price indices. The correlations are visibly clear: when the rice price series rose sharply as happened in the famine years of 1834 or 1837, the birth rate fell markedly. Since evidence on marriage patterns in Tokugawa Japan suggests that it is unlikely that nuptiality acted as an adjusting mechanism between economic resources and fertility, Feeney and Hamano are inclined to think that infanticide, or what they call 'natural infant mortality' (that is, infant mortality exclusive of infanticide) or both might be an explanation.[7] To

[5] A.B. Jannetta, *Epidemics and Mortality in Early Modern Japan* (Princeton, 1987), pp.77 and 91.

[6] See Koseisho Imu-kyoku, *Eisei Tokei kara Mita Isei Hyakunen no Ayumi* (A Hundred Years of Japanese Medical Care seen in Health Statistics) (Tokyo, 1976), for a brief history of epidemics and public health. The best account of the 1918-19 'Spanish flu' is G.W. Rice and E. Palmer, 'Pandemic influenza in Japan, 1918-19: mortality patterns and official responses', *Journal of Japanese Studies* 19, (2) 1993, pp.389-420.

[7] G. Feeney and K. Hamano, 'Rice price fluctuations and fertility in late

be added here, however, is the possibility suggested by Tim Dyson in the Indian case, that a harvest failure had a direct effect on the birth rate by reducing the level of conceptions through mechanisms such as a reduction in coital frequency and a decrease in fecundity, before a sharp rise in mortality occurred.[8] In Okago, a mountain village in the northeast, for example, when the yearly total of all deaths was 162 or 243 per 1,000 population for the famine year of 1837, only one birth was recorded in the population register.[9] In the Ogenji area of Hida too, the number of births in 1835-39 declined by 30 per cent from the preceding five-year period, which, according to Jannetta, must have been 'a consequence of acute malnutrition'.[10] Although it is not possible in this case to separate the famine's initial effect on fecundity from the obvious effect of high mortality on the number of pregnant women who survived to give birth, it is not unlikely that both were operating. Indeed, as a contemporary vulgar rhyme—'shinuhodo yasetewa ko wa dekinu' (starving skin and bone, produce no birth)—did indicate this link too may have been at work in the Tokugawa period.[11]

Of all these instances of crises, the Tenpo famine of 1833-38 is probably most illuminating since it was apparently a case in which famine and epidemics occurred together. The immediate cause of this crisis was a series of poor harvests caused by rainy and cold summers, to which the northern regions and mountainous areas in the central regions were particularly vulnerable. The death rate peaked in 1834 and

---

Tokugawa Japan', *Journal of Japanese Studies* 16, (1) 1990, pp.1-30.

[8] Dyson, 'On the demography of South Asian famines', part 1. His 'conception index' is estimated from monthly birth rates, which are, unfortunately, unavailable for the United Provinces.

[9] Masao Takagi, 'Crisis mortality and its effects on fertility in northern Japan, 1830-45', paper presented at the Hitotsubashi Conference on Demographic Change in Economic Development, held in Kunitachi, Tokyo, Japan, 1-3 December 1991. Takagi combined the village's population registers (*shumon aratame-cho*), which tended to understate births and deaths, with temple registers (*kako-cho*), whose death records were supposed to be complete as far as people affiliated with those temples are concerned. This enabled him to give a more accurate death rate of 243 per thousand than that of 232 calculated directly from the *shumon aratame-cho*.

[10] A.B. Jannetta, 'Famine mortality in nineteenth-century Japan: the evidence from a temple death register', *Population Studies* 46, 3 (1992), pp.439, 441.

[11] Quoted in Sekiyama, *Kinsei Nihon no Jinko Kozo*, p.167, n.24.

1837, of which probably the latter was worse. There exists a number of contemporary accounts of 'disasters' in various areas, many of which suggest that people died from epidemic diseases as well as starvation. The rice price series too rose sharply in both years and a number of food riots were recorded in many places. All those accounts were impressionistic in nature, and in many cases exaggerated. However, based on a series of unusually detailed temple records in the Hida area, Ann Jannetta found that the 1837 peak in death totals overlapped with the one in the number of deaths attributed to 'starvation' as well as the one to *jieki* which literally means 'prevailing epidemic'. From 1834 to 1839 there were 105 people who died of starvation, of whom 75 per cent were recorded in the 1837 death register, while of all *jieki* deaths in the same period, 55 per cent occurred in 1837 and 33 per cent in 1838. A closer look at the monthly distribution of these deaths shows a lead-and-lag pattern more clearly: starvation deaths rose first, followed by a sharp rise in deaths from *jieki*, while the former fell first but the latter continued to claim the lives of the villagers in 1838. The timing of these two cases of death is certainly not coincidental.[12] It is therefore very likely that, as with the Indian famine, the Tenpo famine was a genuine 'subsistence crisis' exacerbated by the outbreaks of infectious diseases that followed, although what those 'prevailing epidemics' actually were is unfortunately unknown.

III

All this, on the face of it, seems to indicate that India and Japan had much in common in relation to famine, epidemics and mortality change in the past. However, a closer look at the Japanese evidence will reveal some significant differences.

First, background mortality conditions in the two countries must have been different. Life expectancy at birth, a summary measure of the mortality conditions, was significantly shorter in late nineteenth-century India than in Tokugawa Japan. According to Mari Bhat's estimates

[12] Jannetta, *Epidemics and Mortality in Early Modern Japan*, pp.178-83. See also her 'Famine mortality in nineteenth-century Japan', p.432, where she gives 95 as the number of people who died of starvation in 1834-39, of whom 76 died in 1837.

for 'All-India', it was 26.3 years for men and 27.2 years for women in the intersensal period of 1881-91, in which no mortality peaks are identified by Wakimura. In the period after the 1918-19 pandemic, 1921-31, it became a little higher, that is, 29.6 and 30.1 years respectively, but with the severe famines and epidemics occurring between the two periods, the expectation of life in 1891-1921 was even shorter than those figures.[13] On the other hand, most village studies done so far for the latter half of the Tokugawa period show much higher figures. For example, my own estimates for three villages in central Japan for the period 1751-1869 put it as 39.2 years (both sexes combined). Even Jannetta and Preston's figures for the Ogenji area in Hida, which are likely to be slightly overstated, indicate that it was a little over 32 years in the period 1776-1875.[14] In the years of famine, it is true that extraordinarily high figures were often recorded. In the Kyoho famine of 1732, villages in the Fukuoka *han* are said to have lost approximately 30 per cent of the population.[15] In the case of the Tenpo famine, 16 per cent of the population over eight years of age in the Owase-*gumi*, Kishu, died in the year 1837 to 1838; the percentage must have been even higher if children under eight had been included.[16] At the village level, Okago's 243 deaths per thousand was probably one of the highest, for it is estimated that the death rate for the villages affiliated with Ogenji, another example of the hard-hit, was approximately 90.[17] As these were all local examples, national averages must have been lower than those. More important, however, is that general health

[13] P.N. Mari Bhat, 'Mortality and fertility in India, 1881-1961: a reassessment', in Tim Dyson (ed.), *India's Historical Demography: studies in famine, disease and society* (London, 1989), p.92.

[14] Osamu Saito, 'Infant mortality in pre-transition Japan: levels and trends', paper presented at the IUSSP conference on Infant and Child Mortality in the Past, held in Montreal, Canada, 7-9 October 1992, and A.B. Jannetta and S.H. Preston, 'Two centuries of mortality change in central Japan: the evidence from a temple death register', *Population Studies* 45, 3 (1991), pp.417-36. For other estimates, see Table 1 of my 'Infant mortality in pre-transition Japan'.

[15] A. Kalland and J. Pedersen, 'Famine and population in Fukuoka Domain during the Tokugawa era', *Journal of Japanese Studies*,10, 1 (1984), pp.31-72.

[16] Akira Hayami, 'Kishu Owase-gumi no jinko susei' (Population trends of Owase-gumi in Kishu). Tokugawa Rinseishi Kenkyujo *Kenkyu Kiyo* (1968), pp.303-84.

[17] Jannetta, *Epidemics and Mortality in Early Modern Japan*, p.181.

conditions of Tokugawa villages in normal years were never the ones characterised by an expectation of life of below thirty.

Secondly, whatever the magnitude of the Tenpo crisis-mortality would have been, it is evident that not all famines and epidemics went hand-in-hand. Unfortunately, not many detailed studies of famine and epidemics have so far been done, but it is probably safe to say that the Tenpo famine was the *only* example in which both occurred together as far as the time period after 1800 is concerned. Indeed, Ann Jannetta and Samuel Preston note, based on a series of life tables reconstructed from the Ogenji death registers in Hida, that even in this poor mountainous area, 'adult death rates remained relatively stable throughout the nineteenth century, with one outstanding exception', that is, 'a severe demographic crisis' known as the Tenpo famine.[18] In other words, the link between famine and epidemics must have been rather weak for much of the period in question. Some instances of epidemic disease, such as cholera and the 'Spanish flu', claimed a huge number of deaths, but none of these cases was occasioned by a famine, nor by other forms of subsistence crisis. Indeed, except for the 1918-19 influenza, almost all the epidemic outbreaks in the period after the opening of the Treaty ports are likely to have been *urban* phenomena. They were a consequence of foreign contacts at port towns, and their diffusion was very much density-specific. In fact, according to a local doctor's records in a village only 20km west of Tokyo, less than 7 per cent of 329 deaths from 1873 to 1909 were caused by infectious diseases, of which only two were cholera deaths. Moreover, a longer series of death registers at a temple in the same village shows that none of the cholera and dysentery outbreaks in the early Meiji period, nor even the 1918-19 influenza epidemic, left a mark on the graph of annual death totals.[19] Clearly rural factors had no bearing on epidemics in Meiji Japan.

[18] Jannetta and Preston, 'Two centuries of mortality change in central Japan', pp.417-36.
[19] Osamu Saito, 'Meiji *mortality* kenkyu josetsu: Tokyo-fuka Kokubunji no shiryo o chushin ni' (Towards an analysis of Meiji mortality: with reference to the source material on Kokubunji, Tokyo Prefecture) *Keizai kenkyu*, 38, 4 (1987), pp.321-32.

IV

This, however, does not mean that Tokugawa mortality conditions were ready for a modern mortality decline, nor does it imply that a substantial decline in death rates took place as Meiji economic development proceeded.

As we have already seen, it is likely that life expectancy was not particularly short for the Tokugawa population, and it is also likely that if a comparison is made between the first half of the nineteenth century and the 1910s by omitting the period of epidemiological rampage from the 1850s to the turn of the century (for which we have no reliable statistical evidence), there must have been a decline in overall mortality. However, the decline was much less substantial than some demographers had previously assumed. Moreover, the decline was very likely to be a *net* result of increasing *and* decreasing movements of various mortality measures. In fact, while childhood mortality showed a steady decline, there must have been no substantial decline in the infant mortality rate, and young adult mortality did increase, however slightly, after the mid-Meiji period.[20] This delayed decline of overall mortality undoubtedly held *rapid* population growth *in check* during early phases of modern economic development, but for different reasons than the high Indian mortality retarding population growth in 1870-1921.

Unfortunately, what accounted for such differences is yet to be investigated. What is certain at this stage of research is that Malthusian positive checks, such as subsistence crises, played little part in keeping Japan's population growth moderate. Both urban and rural factors should be explored. On the rural side in particular, given a good performance of Tokugawa and Meiji agriculture in the form of very

[20] Osamu Saito, 'Keizai hatten wa *mortality* teika o motarashita ka? Obei to Nihon ni okeru eiyo, taii, heikin yomei' (Did economic development bring down the mortality rate?: nutrition, physique and life expectancy) *Keizai Kenkyu*, 40, 4 (1989), pp.339-56, and 'Jinko tenkan izen no Nihon ni okeru *mortality* pataan to henka' (Japanese mortality before the demographic transition: patterns and changes) *Keizai Kenkyu* 43, 3 (1992), pp.248-67. See also Osamu Saito, 'Infant mortality in pre-transitional Japan'. It is worth noting here that despite a possible overestimation bias in their estimated levels of mortality for the Hida area, Jannetta and Preston, 'Two centuries of mortality change in central Japan', also provide a quite similar pattern of change over time.

intensive farming, the observed correlation between poor harvests and decreases in the birth rate, be it accounted for by infanticide and/or 'natural' infant mortality or reduced conception rates, for the late Tokugawa period, as well as the delayed decline in infant mortality (a measure usually quite sensitive to a long-run change in living standards) throughout the late Tokugawa and Meiji periods, should be seen in a different light. It might be that even agricultural progress could have exerted adverse effects on women's health and nutritional status.[21]

[21] See, for example, a discussion advanced in Osamu Saito, 'Nogyo hatten to josei rodo: Nihon no rekishiteki keiken' (Agricultural development and female labour: Japan's historical experience) *Keizai Kenkyu* 42, 1 (1991), pp.31-42, and Osamu Saito, 'Gender, workload and agricultural progress: Japan's historical experience in perspective', manuscript for a book to be published in memory of the late Franklin Mendels (forthcoming). A new look at the question of infanticide is given in Osamu Saito, 'Infanticide, fertility and "population stagnation": the state of Tokugawa historical demography', *Japan Forum* 4, 2 (1992), pp.369-81.

Chapter 7.1

TECHNOLOGY AND LABOUR ABSORPTION IN THE INDIGENOUS INDIAN SUGAR INDUSTRY: AN ANALYSIS OF APPROPRIATE TECHNOLOGY[1]

Yukihiko Kiyokawa and Akihiko Ohno

Introduction

The nineteenth century, in particular its latter half, saw a large increase in the production of sugar. This increase was largely the result of the growth of beet sugar production. The beet sugar production began to become economically viable at the beginning of the nineteenth century, and took off during the 1830s with the strong stimuli given by the Prussian and French governments. By the 1860s, its production accounted for one quarter of world production. As early as the 1890s, its share exceeded 50 per cent, taking over the dominant position occupied by the cane sugar production.

This rapid development of beet sugar production was brought about both by highly mechanised production methods and production management techniques based on strict chemical control. A string of path-breaking technological innovations from the middle of the nineteenth century onwards made this development possible. They included the improvement and spread of the centrifuge, the development of the multiple-effect evaporator and vacuum (boiling) pan, the diffusion process and the carbonation method.[2] Thus the competitiveness of beet sugar grew and its production expanded rapidly, whilst the production of cane sugar was increasingly put under pressure. Towards the end of the nineteenth century and the beginning of the twentieth century, this modern technology began to be introduced to cane sugar production as well. In this way modern methods of

---

[1] The authors would like to express their great thanks for the valuable comments and help with improving the English expression of this chapter provided by Timothy Fox (of SOAS) and our editors.

[2] For a detailed description of the technology and for dates of patents, see N. Deer, *Cane Sugar* (2nd ed., London, 1921).

production saw a remarkable growth all over the world. In Cuba and Java, for example, the indigenous sugar industry was usually transformed in a comparatively short time to the one of large factory production based on new technology.

The Indian case was an exception to this rule. The indigenous sugar industry managed to enhance its competitiveness, and the modern sugar industry failed to establish itself firmly in India as a result. The latter's development was remarkably late, and its international competitiveness was poor. Plantation white sugar continued to be imported from Java and Mauritius.

Two kinds of indigenous sugar, *gur* and *khand*, have been produced by the cottage industry in India. The former is 'unrefined' sugar, which contains molasses (approximately 10 per cent) in addition to sucrose (approximately 70 per cent). *Gur* is generally sweeter, but is not suitable for long-term storage, since the molasses it contains tends to absorb moisture.[3] Hence, *gur* deteriorates rapidly, compared with *khand* and white sugar (defined as 'refined' sugars), which are both refined with vacuum pans and centrifuges and have had the molasses content almost totally removed.[4] We would like to emphasise that there exists a high degree of substitutability in consumption between *gur* and the refined sugar, though some social historians maintain the non-substitutability between them. Those historians point out an intuitive reasoning that *gur* is consumed mainly by lower-income class people, particularly for *puja* and in traditional life style. However, the data on the prices and

[3] In India, not just *gur* but all kinds of sugar are generally produced during the dry season between December and April. *Gur* comes on to the market soon after production and a substantial proportion is consumed before the onset of the rainy season.

[4] White sugar produced in *cane-cultivating countries* is manufactured in almost the same way as raw sugar, except that more lime clarificant is used to decolorise it into white sugar *for direct consumption*. More specifically, the production of white sugar differs mainly from that of raw sugar in the use of the carbonation method and the sulphitation method for clarification. Also, in the process of concentration and crystallisation, more sophisticated treatments are adopted than in the case of raw sugar. Historically, such white sugar is known as plantation white sugar. Raw sugar is usually produced for foreign consumption. In the country with final demand, the raw sugar is carefully refined (washed, decolorised and crystallised) into quality white sugar in refineries.

demand for *gur* and white sugar seem to support our hypothesis.[5]

Although *khand* is a refined sugar, its crystals are coated with a small amount of molasses. That is, the sucrose content of *khand* is about 97 per cent, which is slightly lower than that of white sugar produced by modern technology. However there is no doubt that *khand* and factory-refined white sugar are close substitutes.

This chapter examines whether the indigenous sugar industry in India was competitive enough from a technological viewpoint, and discusses its implications. The experience of the indigenous sugar industry in India provides a valuable lesson for the understanding of the significance of appropriate technology, since it was a typical rural industry equipped with intermediate technology, and remained competitive enough, at least in the domestic market. We would like to stress the importance of intermediate technology remaining competitive, which is our definition of appropriate technology, because indigenous technology often fails to stay competitive against modern technology, and the intermediate technology without competitive edge tends to have a negative long-run impact on the development of the economy as a whole, even if it had a strong positive effect on labour absorption in the indigenous industry.

In the next section, we support the proposition that the modern sugar industry suffered from poor market competitiveness and briefly indicate its causes. For example, the difficulties of procuring raw material (sugar-cane) and the lack of managerial ability have often been identified as factors accounting for such poor performance.[6] However, a full examination of these issues will not be detailed here, since those factors are beyond our scope and evidence. In particular, we regard the lack of managerial ability as a more important cause, but the micro data which could sufficiently confirm this proposition are unfortunately not available.

---

[5] Figure 6 indicates both products to be rather homogeneous. Note that the price of *gur* reflects its lower quality.

[6] For example, see R. G. Padhye, *Sugar Industry in Western India and Methods of Sugar Manufacture* (Bulletin no. 116, Department of Agriculture, Bombay, 1924). We are of the same view as Padhye concerning the first cause, but differ with him on the other cause.

In the following section, we show that, by contrast, the indigenous sugar industry enhanced its competitiveness through the gradual improvement of traditional sugar technology with the introduction of technological innovations. These improvements occurred through the influence of modern sugar manufacturing technology, which was introduced to India around the end of the nineteenth century. We argue that these innovations made it possible for the indigenous technology to qualify as appropriate technology. More specifically, we attempt to show the relationship between the two industries (as depicted in Figure 1) in sections II and III. That is to say, the modern sugar industry in India, which had a weaker competitive edge than those in other countries, failed to become competitive with the indigenous sugar industry, even in the domestic market.

One major significance of such appropriate technology was the absorption of labour in the rural village. Thus section IV provides an estimate of the labour absorption effect in the indigenous sugar industry, in comparison with the hypothetical absorption in the modern sugar industry.

Since the competitiveness of the indigenous sugar industry is the key to our analysis, we need to choose a period for study over which it can be considered that the survival principle under the free market mechanism was functioning. There were two important pieces of legislation relating to the Indian sugar industry. The first was the Sugar Industry Protection Act in 1932, which aimed to protect the domestic sugar industry from imported sugar. The second was the U.P. (the United Provinces of Agra and Oudh) and Bihar Sugar Factories Control Act in 1938, passed in the wake of the sugar crisis of 1937. This act introduced a set of zoning arrangements, a minimum price support policy and licensing of sugar factories so as to protect the modern sugar industry. Thus, it can be assumed that the period before 1932 was one of free competition, both internationally and domestically, and the period up to 1938 one of free competition, at least domestically, for the sugar industry in India. Hence our analysis here is confined to the period up to the latter half of the 1930s, although the post-independent period provides an even more typical example of the use of appropriate technology.[7]

The regions covered in this study are the areas surrounding the River Ganges, as seen from Figure 2. The main sugar-producing regions at the end of the 1930s were more or less limited to the northern states: U.P. accounted for the largest proportion of total production—(a) area under cane, 53 per cent; (b) production of *gur*, 48 per cent and (c) production of *khand*, 94 per cent—followed by Punjab (14 per cent; 9 per cent; and negligible), Bihar (11 per cent;   14 per cent; and 3 per cent) and Bengal (8 per cent; 14 per cent; and 1 per cent).[8] The production of *khand* was concentrated in the Rohilkhand Division in U.P. In principle, our discussion refers to the regions listed above.

It may be useful to mention here some of the main sources used in this chapter. Intensive surveys of the Indian sugar industry were not conducted till the latter half of the 1910s. After that, several important survey reports were issued by the government, since it became interested in the question of protection of the modern sugar industry. The first of those was the *Report of the Indian Sugar Committee 1920*, issued by the Sugar Committee which was established in 1917.[9] Also important are the two reports by the Tariff Board, which examined the appropriateness of the Sugar Industry Protection Act. They are the *Report of the Indian Tariff Board on the Sugar Industry 1931* and the *Report of the Indian Tariff Board on the Sugar Industry 1938*.[10] The

---

[7] For the development of appropriate technology in the contemporary indigenous sugar industry, refer to Institute of Applied Manpower Research, *Employment Study of the Khandsari and Gur Industry* (New Delhi, 1979), and United Nations Industrial Development Organisation, *Appropriate Industrial Technology for Sugar* (New York, 1980).

[8] Some cane was also produced in areas such as the Madras Presidency (now known as Tamil Nadu) and the Bombay Presidency (now Maharashtra), but these regions are not the immediate object of study here. Government of India, *Report on the Marketing of Sugar in India and Burma* (Delhi, 1943), pp.280-9.

[9] Government of India, *Report of the Indian Sugar Committee, 1920* (Simla, 1921); hereafter *Report 1920*.

[10] Government of India, *Report of the Indian Tariff Board on the Sugar Industry* (Calcutta, 1931), and Government of India, *Report of the Indian Tariff Board on the Sugar Industry* (Delhi,1938); hereafter *ITB Report 1931* and *ITB Report 1938*.

Katiha survey[11] contains the most useful information relating to *khand* production. The *Report on the Marketing of Sugar in India and Burma* is most informative as a comprehensive report on the Indian sugar industry as a whole.[12]

II. The market competitiveness of the modern sugar industry

The formation and decline of the modern sugar industry. A modern sugar factory is defined here as a sugar factory which produces refined sugar directly from sugar-cane with multiple-effect evaporators and vacuum (boiling) pans. These technologies were imported from Europe. Vacuum-pan technology was not used in the indigenous sugar industry because its use would result in the diseconomies of scale and its low cane-crushing capacity would be a constraint. Open pan technology was used instead. This distinction between vacuum pan technology and open pan technology comprised the main technological distinction between the indigenous and modern sugar industries.

The modern sugar industry started in India with the establishment of the Kanpur sugar factory in U.P. in 1884.[13] This factory was converted from an indigo factory when the indigo business went into decline. It is frequently claimed that the first factory-style sugar manufacturer was the Roza sugar factory, which was established with the participation of British capital in Shahjahanpur (the Rohilkhand Division) of U.P. in 1805.[14] The Roza factory, however, used technology similar to that of the traditional *gur* refinery, and cannot really be described as a modern sugar factory as defined in this chapter. The Kanpur sugar factory used multiple-effect evaporators and vacuum pans,[15] and was therefore not a traditional sugar factory.

[11] N.K. Katiha, *Sugar in the Fields, the Workshops and the Factories: a survey of the prospects of the Sugar Industry in Rohilkhand* (Bulletin no. 5, Bureau of Statistics and Economic Research, Allahabad, 1938).

[12] Hereafter *Marketing*.

[13] Government of India, Indian Tariff Board, *Written Evidence recorded during Enquiry on the Sugar Industry*, vol. I (Calcutta, 1932), p.161.

[14] Ibid, p.98.

[15] Both are essentially the same technology, which is called the vacuum pan technology. Conversely, the technology used in the indigenous sugar industry is the open pan system. It is the difference in the two technologies which separates the modern and indigenous sugar industries.

However, the Kanpur sugar factory was not of a type commonly known as a standard modern sugar manufacturing factory either. It was known as a modern *gur* refinery, which produced white sugar by refining *gur* produced by farmers.[16] In the Kanpur district, there were seven modern *gur* refineries of this kind in 1916, increasing to 18 in 1924.[17] The cultivation of sugar-cane was rare in Kanpur, and it was difficult to transport cane from outside, so the refining of *gur* was the only way of manufacturing sugar.

Thus a section of the modern sugar industry in India first took the form of refining *gur* produced by farmers, rather than producing sugar directly from raw sugar-cane. The sucrose in sugar-cane starts to acidify immediately after cutting, and micro-organisms cause fermentation, leading to rapid decomposition. Also, the transportation of sugar-cane over a long distance is costly and tends to cause decomposition. Therefore, the sugar manufacturer needed to be situated near the site of the raw material if he wanted to produce sugar directly from raw sugar-cane.

The refining of *gur* is a relatively inefficient process. The recovery ratio (the proportion of sugar derived from raw cane, weight for weight) for *gur* refining amounted to no more than 5.5 per cent. This figure is noticeably low compared to 9 per cent recovery ratio achieved by the modern sugar factories that were established later, which produced refined sugar directly from raw cane.[18] On average, cane grown in India contains a sucrose content of 12 per cent, which implies that the total wastage rate for *gur* refining exceeded 50 per cent. As will be mentioned in the next section, the low juice extraction ratio and the poor crystallisation of sucrose in the production of *gur* by farmers resulted in this low recovery ratio in the modern *gur* refineries.[19] As a consequence,

[16] To distinguish this from the indigenous *gur* refining industry, we shall call it the modern *gur* refining industry.

[17] S. Amin, *Sugarcane and Sugar in Gorakhpur* (Delhi, 1984), p. 96.

[18] *ITB Report 1931*, p.28.

[19] The recovery ratio for *gur* is approximately 9 per cent. Ten per cent of *gur*, however, is relatively heavy molasses and *gur* has a sucrose content that is nearly 30 per cent lower than white sugar. For the rest of the chapter we shall only use simple comparisons of weight, but it is as well to bear the above point in mind.

the modern *gur* refining industry failed to be sufficiently competitive. Even at the zenith of the *gur* refining industry there were only 26 refineries which were called pure refineries, as distinct from orthodox sugar factories equipped with auxiliary refining plants. The modern *gur* refining industry relied on the price differential between *gur* and white sugar. As the differential narrowed during the latter half of the 1920s (see Figure 3), the industry gradually went into decline.

Some of the early standard modern sugar factories, which produced white sugar directly from raw sugar-cane, using vacuum pan technology (hereafter refered to as modern sugar factories), were established in East U.P. and Bihar in the first decade of this century. The usual recovery ratio in the modern sugar industry was about 9 per cent,[20] a ratio which compares favourably with those in other cane-sugar-producing countries. However this modern sugar industry did not develop successfully in India, in spite of the reasonable efficiency of modern technology. There were only 29 such factories in 1930/31 and they failed to eliminate the modern *gur* refineries with the low recovery ratio.[21] In the 1920s, for example, out of all sugar-cane used in the production of sugar, only 5 per cent was used by those modern sugar factories (Table 1). These facts seem to suggest the lack of competitiveness of the modern sugar industry.

The stagnation of the modern white-sugar industry was reflected in its weak international competitiveness. At the beginning of the twentieth century, India turned from a net sugar exporter to a net sugar importer. Sugar came principally from Java (mostly plantation white sugar). Java had lost its market in America due to the implementation of the American sugar industry protection policy, and had started to look for new outlets in Asian countries. In 1910/11 sugar imported into India amounted to 750,000 tons (out of which 60 per cent was

[20] Even today the recovery ratio in the Indian modern sugar industry is 9-10 per cent. Thus, we can say that the refining technology of the modern sugar industry was efficient in those days too.

[21] The decline of the *gur* refining industry in the late 1920s was not brought about by competition from the domestic modern sugar industry. Rather, it was caused by the import of foreign-produced white sugar, as is mentioned later in the chapter.

Javanese), causing a sugar crisis in the early months of 1911. Although the increase in sugar imports (especially beet sugar from European countries) was brought to a halt during the First World War, imports increased rapidly once again during the 1920s. In spite of the frequent upward revisions of the import tariff (Table 2), imports amounted to nearly one million tons in 1929/30.

The white-sugar market in India had become completely dominated by imported sugar. At that time, the domestic production of white sugar amounted to not more than 100,000 tons (see Figure 4). The growth in imports of foreign sugar led to a fall in the domestic price of all types of sugar, including those of the indigenous sugar type (see Figure 3). The price for white sugar experienced the largest fall, since imported white sugar and domestically-produced white sugar were perfect substitutes, but the price differential between *gur* and white sugar was also reduced. Had this reduction of differential continued, the *gur* manufacturing industry would have lost its competitiveness as the quality of *gur* remained low.

The 1932 Sugar Industry Protection Act and its effects. Throughout the 1920s, the price of sugar continued to fall, leading to a fall in the price of sugar-cane, which in turn depressed farmers' income. This jeopardised the colonial government's attempts to stimulate the cultivation of sugar-cane, the archetypal cash crop, to secure sufficient land revenue. Consequently, a comprehensive enquiry into the protection of the sugar industry was conducted.

In June 1929, the Imperial Council of Agricultural Research was set up to examine the tariff question. It commissioned the Indian Tariff Board (ITB) to study the appropriateness of setting up protectionist tariffs, and the ITB produced a full report in 1931. On the basis of the comparison of production costs of Indian and Javanese sugars, the report recommended that a protectionist tariff (specific duty) be introduced to make the prices of imported and domestic sugars level. The Sugar Industry Protection Act was passed in 1932. The tariff imposed, when calculated on an *ad valorem* basis, was as high as 185 per cent. As a result, sugar imports decreased drastically, and by 1940 imports had all but ceased. All branches of the Indian sugar industry including the indigenous sugar industry experienced growth under this protection. Modern sugar factories were established one after another in the cane growing regions of U.P. and Bihar (Figure 5). It did not mean, however, that the modern sugar industry recovered its real competitiveness.

It is clear that, since the protection policy took the form of a uniform tariff rather than discretionary subsidies, the indigenous sugar industry as well as the modern sugar industry received protection. Thus, both sectors competed freely in the domestic market after 1932. The expansion of the domestic sugar industry led to an increase in the demand for sugar-cane, which made it possible for the government to maintain sufficient land revenue. In this respect, the colonial government had achieved its aim.[22] Although the competition from imported sugar had been removed by the Sugar Industry Protection Act, the indigenous sugar industry remained an able competitor. A study of the trends in the markets for each type of sugar after 1932 shows that in contrast to an expanding market for *gur*, domestic white sugar merely replaced imported white sugar, rather than expanded its domestic market. The production of white sugar began to stagnate in the latter years of the 1930s (see Figure 4). These trends imply that the *gur*-manufacturing industry was strong enough to compete with the modern sugar industry in the Indian market.

The production of *khand* grew immediately after the passing of the Sugar Industry Protection Act in 1932, but it declined after 1935. This decline was caused by the imposition of an excise tax in 1934 on both white sugar and *khand*. The tax amounted to between 10 to 15 per cent when calculated on a specific tax basis. Judging from the figures on sugar production after 1934, *khand* was probably less competitive than white sugar (see Figure 4). Nevertheless, the production of *khand* did undergo a recovery to some extent with the lowering of the tax rate in 1937. After independence, production was further increased with significant improvements in technology.

[22] The cultivation of cane was included in the Sugar Industry Protection Act and was one of the most important targets of protection. *ITB Report 1931*, p.39.

Hence, the lack of competitiveness of the modern white sugar industry needs to be emphasised. It failed to replace both *khand* production and the modern *gur* refining industry. The latter survived competition, despite the fact that it produced almost a perfect substitute for white sugar with inferior sugar recovery ratios. This fact seems to suggest that the factors hindering the development of the modern sugar industry lie outside the scope of technological factors in the narrow sense, since both the *khand-* and *gur*-refining industries had inferior technological systems, but nevertheless continued to co-exist with the modern sugar industry over a long period of time. An investigation and analysis of the main relevant factors is beyond the immediate scope of this chapter, so we will briefly point out a few of the factors below.

One explanation that is conventionally used for the stagnation of the modern sugar industry in India is that sugar-cane was grown by small-scale farmers. The successful management of a large modern factory with huge cane-crushing capacity, compared with indigenous manu-facturers, required that the plant lie idle for as short a time as possible, so that the productivity of capital could be increased, if only by a little. To achieve this, it was necessary that a constant supply of sugar-cane be ensured over a long period of time, in order to meet the processing capabilities of the factory. In India, however, many farmers grew cane only on one part of their holdings (on average, less than half an acre).[23] A modern factory in the early 1930s could crush up to 500 tons of cane per day on average, which suggests that the sugar-cane needed for that factory had to be procured from more than 10,000 farmers. In addition, difficulties could arise when the direct procurement of a large amount of cane was attempted in a market where there was insufficient information. These difficulties were exacerbated by the rapid decomposition of sucrose shortly after harvesting the cane. In India, middlemen (contractors or agents) entered to fill the gap between factory and sugar-cane cultivator. Consequently, the modern sugar factory had to procure sugar-cane at a higher price than the indigenous manu-facturer.

[23] G. Watt, *A Dictionary of the Economic Products of India*, vol. 6, part 2 (Calcutta, 1883), p.256.

In many sugar-producing countries, efforts were directed towards the vertical integration of raw material suppliers, an example of which is the creation of the plantation system. In Java, it took the form of planning the supply of cane from the estates over which the factory had complete control (known as *Cultuurs telsel*).[24] In India, however, the colonial government gave priority to the collection of land revenue, and this policy hindered the vertical integration. At that time, land revenue accounted for one third of the total government revenue. The forced purchase of cane-growing land was the last measure the government wanted to take, since such a measure would lead to the loss of a source of revenue. Certainly, the tiny average size of land under sugar-cane cultivation worked against the modern sugar industry, but it worked in favour of the very adaptable indigenous sugar industry.

The stagnation of the modern sugar industry cannot be explained by the above factors alone, however. It is well known that the modern sugar industry in India used a managing agent system, which made long-term management of industrial capital difficult. The Begg and Sutherland Company is a case in point. In the modern sugar industry, the introduction of technological innovations was not positively pursued, since most managing agents pursued risk-averting behaviour with a myopic view. Modern sugar factories often held redundant labour, and were not keen on cutting costs by reducing the number of workers. Therefore, they could not even begin to envisage the strengthening of direct control over cane supply as a means to increase competitiveness, since cane supply was a much more difficult problem to solve than that of over-manning. In light of other countries' experiences, it is also clear that not all problems were related to the Indian landholding structure or the behaviour of the colonial government. A serious problem arose from lack of effort on the part of management or lack of entrepreneurial spirit in modern sugar factories. There were other problems such as the lack of road networks and means of transport, and insufficient irrigation facilities. The relatively low sucrose content of Indian sugar-cane (or the backwardness in improving strains) compared with other countries was also a problem.

[24] For more detail refer to *Report 1920*.

Thus, as a result of a combination of several factors, the stagnation of the modern white sugar industry allowed the indigenous sugar industry, a typical rural industry, to grow in a gradually expanding domestic sugar market. The earlier establishment of a competitive indigenous sugar industry also made the development of the modern sugar industry more difficult. Whenever the price of *gur* rose, sugar-cane cultivators produced *gur* themselves. As a result, the modern sugar industry found itself short of sugar-cane and the scale of operations inevitably shrank. In contrast, so long as enough price differentials between *gur* and white sugar existed, the modern *gur* refining industry could survive without serious shortages of raw material (*gur*).[25] This allowed the modern *gur* refining industry to coexist with the modern sugar industry, in spite of the former's low recovery ratio. It is interesting to note that in 1919/20, 9 out of 22 and, in 1930/31, 20 out of 29 modern sugar factories possessed *gur*-refining equipment, to deal with sugar-cane shortages.[26] *Gur*-refining was a manifestation of the peculiarity of the Indian modern sugar industry. It was also a symbol of the weakness of its competitiveness.

III. *The indigenous sugar industry—technological progress and market competitiveness*

The indigenous sugar industry enhanced its competitiveness through technological innovation. In this section, we throw light on the actual innovations that are believed to have contributed to its competitiveness.

The manufacture of sugar from cane consists of three operations: (a) crushing cane for the extraction of juice, (b) clarifying and boiling (concentrating) the juice, and (c) refining, that is, removing molasses and other unwanted substances from the crystallised sucrose.[27] Needless to say, the last process is not required for the production of *gur*. The technological innovations or improvements can be classified by these three processes.

---

[25] The average working days per season were 150 for modern refineries and 300 days for *gur* refineries. *Marketing*, p.96.

[26] *ITB Report 1931*, Appendix I, p.256.

[27] For more detailed descriptions refer to *ITB Report 1931*, Appendix I, p.256, *Marketing*, and Katiha, *Sugar in the Fields*.

Cane crushing. In the crushing process a crusher or mill of the same technological sophistication was used for the manufacture of both *gur* and *khand.* This equal degree of technological sophistication reflected the fact that the choice of crushing techniques was limited by the use of bullocks as the power source.

The efficiency of a crusher is evaluated by its extraction ratio (weight of juice per weight of cane) and its crushing capacity per unit of time. The oldest type of crushing technology is the mortar and pestle. A stone or wooden cane crusher was worked by a pair of bullocks. With this method, however, the cane had to be cut into small pieces and impurities became mixed in with the juice. The operational efficiency of this technology was so low that we are told that 'working day and night a *kolhu* (crusher) will not press more than 1.5 acres of cane a month'.[28] The limited source of usable power, that is, the limited number of bullocks available, created a bottleneck. It was impossible to expand the area under cane cultivation. Wooden two-roller mills were devised, but they were so heavy that they needed to be worked by two or three pairs of bullocks. In addition, cane had to be passed through the rollers three or four times, whilst the juice extraction ratio remained at the level of 40 per cent.[29] The greatest advance in crushing technology came with the introduction of the iron-made two-roller mill in 1874. Messrs Thomson and Mylne started the manufacture of iron-made two-roller mills after being invited to India by the government, with the aim of developing the sugar industry. The mills that were common in the islands of western India were used as a model,[30] and it is said that between 1874 and 1891 some 250,000 mills were produced.[31] In 1880, the production of iron-made three-roller mills began. The juice extraction ratio of the three-roller mills ranged between 50 to 60 per cent, and the cane needed to be passed through the rollers only once

[28] I. Stone, *Canal Irrigation in British India* (Cambridge, 1984), pp.295-6. This corresponds to a cane crushing capacity of 12 to 13 kilograms per hour.
[29] M.S. Randhawa, *A History of Agriculture in India*, vol. 2 (New Delhi, 1984), p.253. In this connection, the juice extraction ratio in the modern sugar industry exceeded 80 per cent.
[30] Watt, *Dictionary*, pp.84-114.
[31] Amin, *Sugarcane*, pp.124-5.

when this type of mill was used. Thus, two to three *maund*s (75 to 112 kilograms) of cane could be crushed in one hour,[32] which meant that the three-roller mill could crush five to six times the amount crushed by the old mortar and pestle crusher. Moreover, it could be worked by one pair of bullocks. This removed the above-mentioned bottleneck for the expansion of the indigenous sugar industry.

Since the construction of iron mills was similar to that of the wooden mills, they could be manufactured using indigenous technology. Along with the increase in demand for sugar from the mid-nineteenth century onwards, factories were established to make iron-made mills all over India. By the turn of the century, old-fashioned and inferior crushers had been almost totally replaced by new, improved mills.[33] This diffusion of the more efficient cane crushers increased total crushing capacity, leading to a rapid increase in the area under sugar cane.[34]

In the 1920s, the power mill began to be introduced into the indigenous sugar industry, but the use of power mills was mainly confined to *bel*s (a boiling plant for making *rab*, that is, massecuite, from which *khand* is made), where a large amount of cane was crushed.[35] In U.P. in 1935, for example, power mills accounted for only 0.24 per cent of all crushers, and even in Rohilkhand, the centre for the production of *khand*, only 0.45 per cent (205 mills) were power mills.[36] The introduction of power mills was very limited, probably because the boiling capacity of a *gur* plant was lower than the crushing capacity of power mills. Furthermore, the extraction rtio of improved bullock-driven types of crushers was as high as that of power mills, although the crushing capacity of the latter was much greater.[37]

Thus the main technological advances in the manufacture of *gur*

[32] One *maund* is equal to 37.7 kilograms.
[33] Stone, *Canal Irrigation*, p.296.
[34] Watt, *Dictionary*, p.257.
[35] The word *bel* also signifies both the boiling pan and the place of *gur* and *rab* manufacturer. To avoid confusion, the word *bel* in this chapter refers only to the combination of furnace and boiling pan.
[36] Government of U.P., *The Livestock Census Report for the U.P.*, (Allahabad, 1935).
[37] *Marketing*, p.51.

were in the crushing technologies. As will be mentioned in the next section, it is believed that the diffusion of these innovations was almost completed by 1920. The iron-made three-roller mill could process six to seven times the amount of the old mills and its extraction ratio was 10 to 20 per cent higher. Hence its diffusion contributed greatly to the competitiveness of the *gur* manufacturing industry, since it raised the sugar recovery ratio.

The manufacture of gur. Advances in boiling technology for *gur* manufacturing were slow, however. The usual boiling pan was placed on a simple furnace, which was made by digging a pit in the ground. With this type of *bel*, irregular heat and wide fluctuations in temperature could cause inversion, caramelisation and charring. As a result, the quality of *gur* deteriorated and the sugar recovery ratio for *gur* was lowered. It was recommended that the *bel* could be improved by constructing the furnace with a chimney and grate and by the use of a flat-bottomed pan, so that the heat was equalised over its surface.[38] These recommendations were rarely adopted, however. It was also possible to avoid charring and inversion by boiling the cane juice in two or three pans on top of each other and moving the juice. This multiple-pan technique failed to become popular, and during the 1930s the single-pan technique remained dominant.[39]

Technological innovation had to be capital-saving, since the manufacture of *gur* was in the hands of small-scale farmers. One plausible method that the *gur* industry could use to increase its competitiveness was to use a combination of power mills and multiple-pan furnaces under a co-operative or special contract arrangement. In fact, this is nowadays a common form of production in Punjab, Haryana and western U.P.

The manufacture of khand. The *khand* refinery is divided into two stages: (1) the *bel*, used for making *rab* and, (2) the *khandsari*, where *rab* is refined into *khand*. From the middle of the nineteenth century, when the demand for sugar started to increase, there were technological

[38] For diagrams of the *bel*, refer to S.C. Roy, *Monograph on the Gur Industry in India* (Kanpur, 1951).
[39] Roy, *Monograph*, p.161.

advances in both parts of the *khand* refinery. There were also important changes in the organisational form of management.

(1) The *bel*. Let us first look at the technological advances at the *bel* stage. Until the mid-nineteenth century, cane-growers supplied *rab* they had produced themselves to the *khandsari* under a contractual arrangement offering cash advances.[40] The process used by farmers to prepare *rab* was more or less the same as that for *gur*. In the production of *gur* it is chiefly the colour that is important, whilst in the case of *khand*, it is the size of the sucrose crystals that is vital, since the size determines the *khand* recovery ratio. Consequently, in the production of *khand*, it is necessary to prevent sugar inversion, caramelisation and charring in order to improve the *khand* recovery ratio. Hence, more attention had to be paid to the production of *rab* in the manufacture of *khand* than that of *gur*. For this reason, two pans were normally used in the manufacture of *rab*, in contrast to the single-pan technique used for *gur*. Regional *khandsal*s (the owners of *khandsari*) even needed to send their watchmen to the cane-growers to guarantee a good quality *rab*, and they had to ensure that all the contracted cane was used in the manufacture of *rab*.[41]

The problems encountered in the manufacture of *rab* were largely solved with the appearance of the Rohilkhand *bel*, which had been introduced widely in the Rohilkhand Division during the latter half of the nineteenth century.[42] The Rohilkhand *bel* was large and had five iron-made open pans. These *bel*s were generally constructed and operated by the *khandsal*s themselves. Thus, the production of *rab* was transferred from the farmers to the *khandsal*s and at the same time the supply of high quality *rab* for use in the *khandsari*s was secured. Only the crushing process remained in the hands of the farmers, because bullocks owned by the farmers were required as its power source.[43] As

[40] Amin, *Sugarcane*, pp.57-64.
[41] Watt, *Dictionary*, p.285.
[42] For a detailed description of the *Rohilkhand bel*, refer to R. C. Srivastava, *The Open Pan System of White Sugar Manufacturing* (Calcutta, 1932). The diagrams are particularly useful.
[43] Ten to twelve crushers were installed around the *bel* and hired by the farmers. *Marketing*, p.62.

power mills began to be introduced, however, even the crushing process was taken out of farmers' hands. In this way the only task left to farmers became the supply of sugar-cane.

It is reasonable to claim that the Rohilkhand *bel* was the greatest technological advance in the manufacture of *rab*. The diffusion of the power mill also had an impact and the *khand* industry managed to increase its competitiveness as it gradually internalised the *rab* manufacturing stage.

(2) The *khandsari*. The competitiveness of the *khand* industry was also strengthened by a technological advance at the *khandsari* stage. The technological advance here is seen as a transition from the *kanchi* technique to the centrifuge technique.

The *kanchi* technique was an indigenous process and references to it can be found in writings dating back over 2,000 years. *Rab* produced by the farmers was placed in small hempen bags which were stacked and trodden on by a labourer so that the molasses was squeezed out of the *rab*. The resulting contents were then removed from the bags and placed in a room made with bricks (called a *kanchi*)[44] and then covered with a layer of aquatic weed. The water from the weed gradually washed away the remaining molasses to produce *khand*. The washing process took two or three months. Furthermore, the sugar recovery ratio in the production of the first *khand* was only 3.3 per cent. Even the second *khand* made from the second *rab* only gave an overall sugar recovery ratio of 4.4 per cent.[45]

The greatest technological advance in the *khandsari* was the introduction of the centrifuge from the modern sugar industry. The centrifuge was created in Europe in 1837 and, like the iron-made roller mill, was first introduced into India in the 1870s. As improvements and simplifications were made,[46] the simplified centrifuge became widespread in Rohilkhand, the main *khand*-producing region. Using the centrifuge, the

[44] The word *kanchi* refers to both the brick rooms used for refining at the *khandsaris* and the refining technique itself.

[45] Government of India, Indian Tariff Board, *Written Evidence*, p.92. The second khand can be made from the squeezed-out molasses from the first *rab*.

[46] Concerning the simplification of the centrifuge, refer to S. M. Hadi, *The Indian Sugar Industry* (Bophal, 1929), ch.15.

recovery ratio in the production of *khand* was 5.5 per cent, including the second *khand* (4.4 per cent excluding the second *khand*), which was high compared with the recovery ratio of 4.4 per cent achieved with the *kanchi* technique.[47] As a result, the *kanchi* method had almost died out by the end of the 1930s.

It is possible to prove that these innovations strengthened the competitiveness of the *khand* industry, by examining the cost performance of each method of production. There were three forms of organisation for producing *rab*: (a) the *rab* is made by the cane growers themselves; (b) the *rab* is made at the *khandsal*'s *bel* from cane juice by the cane growers, and (c) the *rab* is made directly from sugar-cane at the *khandsal*'s *bel*. The unit cost of *khand* when *rab* was produced under these three forms can be compared. Let the unit cost of *khand* using the type-(a) form of *rab* production be equal to 100; The unit cost using type (b) was 81.9 and that using type (c) was 76.6.[48] The impact of technological innovation is clearly shown by these figures. Furthermore, we can compare the unit cost of producing *khand* using the *khanchi* technique with that of the centrifuge technique. If the former is set at 100, the latter was reduced to 81.2.[49]

As was demonstrated above, the technology of the indigenous sugar industry was not static. Rather, various technological advances were made by the industry. It is worth noting that the *khand* industry in particular absorbed the imported technology.[50] A series of developments were possible in the *khand* industry because the modern sugar industry in India could not wield an overwhelming competitiveness over the

[47] *Marketing*, p.64.

[48] *ITB Report 1838*, p.98, calculated from Table XLI.

[49] Indian Tariff Board, *Report of the Indian Tariff Board on the Sugar Industry* (Calcutta, 1932), calculated from pp.225-6.

[50] Indigenous efforts for the development of sugar manufacturing technology should not be underrated, though we could not touch upon them for want of space. Raja Ahamsher Prakash (1856-98), for example, started a foundry at Nahan and produced the bullock-driven three-roller mills, inviting an engineer from England who was patentee of the several types of sugarcane mills. Randhawa, *Agriculture in India*. Syed Muhammad Hadi's work on the Rohilkhand *bel* named the Bopal method is well known. Several improvements from the Bopal method were said to be incorporated into the *khandsari* technology. Srivastava, *The Open Pan System*.

indigenous sugar industry, giving the latter enough time to import and develop its own technology. Furthermore, the indigenous sugar (especially *gur*) had enough competitiveness, in spite of its poor quality, since it was cheap and the proportion of the low-income groups in India was high. Thus, the very competitiveness that had been fostered in the indigenous sugar industry came to hinder the development of the modern sugar industry.

IV. *Labour absorption in the indigenous sugar industry*

The labour coefficient in the gur manufacturing industry. We will now attempt to evaluate the importance of the indigenous sugar industry, specifically from the point of view of labour absorption, since unemployment was a serious problem in Indian villages. We intend to measure labour absorption using a labour coefficient. The labour coefficient is generally defined as the units of labour input per unit of output. In this study, however, we are comparing the labour absorption in different types of sugar production using different technologies. Therefore, it is not possible to use output as a denominator, since efficiency and quality of output will naturally be affected by the type of technology or equipment in use. Here, we define the labour coefficient as the units of labour per unit of raw material, (L/R). R represents an amount of sugar-cane and L represents the amount of labour required to produce sugar from R. In using this definition, it is assumed that the indigenous sugar industry was sufficiently competitive against the modern sugar industry,[51] which has already been shown. Under this assumption the high labour (to raw materials) coefficient is off-set by the low capital to raw material coefficient. The unit of measurement for cane is tons and one working day is assumed to be eleven hours.[52] The labour coefficient is calculated by specifying the sugar-manufacturing

[51] Here we are concerned with the point, that, although there are differences in the juice extraction ratio and the use of molasses, these are far outweighed by the differences in the capital-raw material ratios and labour-raw material ratios. See Figure 7; differences in quality of the product and factors of production are adjusted for by differences in the prices.

[52] This figure is based on the fact that the work period was half a day (11 actual working hours) in the indigenous sugar industry and that there were two shifts of 11 hours each in the modern sugar industry.

technology that was most common in the 1930s and by identifying the amount of cane that could be processed by that technology and the number of labourers employed.

We start with estimating the labour coefficient in the manufacture of *gur*. By the 1920s, 'iron-made three-roller bullock-driven mills were used widely in all the important *gur* producing areas of India'.[53] In addition, it is said that at the end of the 1930s, about 90 per cent of cane used in the manufacture of *gur* was crushed by iron-made three-roller bullock-driven mills.[54] Thus, it can be considered that the iron-made three-roller bullock-driven mill was the most typical type of crushing technology during the 1930s. The efficiency of this technology depended on the brand of mill (see Table 3), but on average, these mills crushed 2.5 to 3.5 *maunds* (93.3 to 130.6 kilograms) per hour with a juice extraction ratio of between 60 and 65 per cent. These are test results, however. In fact, the farmers loosened the rollers to reduce the strain on their cattle so that the average crushing capacity was reduced to 2.5 *maunds* (93.3 kilograms) and the actual extraction ratio was reduced to 55 per cent.[55] It is these actual figures that are used in our estimation. In the average *gur* factory 1.0 ton of cane (93.3 kilograms multiplied by 11 hours) was crushed daily to yield 550 kilograms of juice. The single-pan type of boiling technology was the most common type used in the production of *gur*. With 25 main brands of single-pan boilers, it was possible to boil up to 17.6 *maunds* (656.5 kilograms) of juice a day.[56] Such an amount of cane juice can be obtained from 1.2 tons of cane when it is assumed that the juice extraction capacity of the mill is 55 per cent. Hence, the cane-processing capacity of the mill and that of boiling pans roughly matched each other. In this chapter we will use the lower limit of the cane crushing capacity, which is 1.0 ton of cane per day.

It was said that 'generally, five men are required to work at a *bel*'.[57]

[53] *Report 1920*, p.213.
[54] *Marketing*, p.51.
[55] *Marketing*, p.51.
[56] Roy, *Monograph*, p.71.
[57] *Marketing*, p.58.

According to the Roy survey,[58] a *gur* factory which crushed 25 *maunds* (932.5 kilograms) of cane a day using a bullock-driven crusher employed four labourers in addition to one *halwai* (sugar boiler). Two people were involved in crushing operations—one drove the bullocks used for turning the crusher and two fed cane between the rollers, and the remaining one was required remove the bagasse (remains after crushing) and spread it out to dry in the sun. It is assumed in this paper that five workers were required by a *gur* manufacturer of an average size.

From these facts, we estimated a labour coefficient of 5 ($^5/_{1.0}$) for the typical *gur* manufacturing process of the 1930s.

The labour coefficient in the khand industry. As pointed out previously, the manufacture of *khand* comprised two stages: (1) the *rab* making process at the *bel*, and (2) the refining process undertaken at the *khandsari*. Let us first discuss them in turn.

(1) *Rab* manufacturing. As has already been explained, *rab* could be manufactured under three forms of organisation. They are (a) the cane growers make the *rab*; (b) the *rab* is made at the *khandsal*'s *bel* from cane juice extracted by the cane growers, and (c) the *rab* is made directly from sugar-cane at the *khandsal*'s *bel*. Taking an average for the four years between 1935/36 and 1939/40, each of the above forms accounted for 33.8 per cent, 55.4 per cent and 10.8 per cent respectively of the *rab* manufacture for the production of *khand*.[59] Hence, we take type (b) to be the typical form of manufacture.

The cane-processing capacity of the *bel* is estimated as follows. An average Rohilkhand *bel* consumed 100 *kardas* (242 tons) of juice in a season extending over 100 days, which gives a consumption rate of one *karda* per day.[60] We are told that 'the crushing work commences at about 4 am. By the sunset, the crushing work is stopped. The hours of work (for *rab*) extend from midday till midnight'.[61] This quote implies

[58] Roy, *Monograph*, p.71.

[59] *Marketing*, p.100.

[60] The weight of one *karda* differs in each district of the Rohilkhand Division. One *karda* is equivalent at 62.5 *maunds* in Bareilly, 65.62 *maunds* in Philibhit and 66.66 *maunds* in Shahjahanpur. Katiha, *Sugar in the Fields*, p.36. In this chapter we take one *karda* to be equal to 65 *maunds* (2.42 tons).

[61] Khatiha, *Sugar in the Fields*, p.48.

that the operation period for a *bel* was almost half a day (11 hours). Thus, taking the extraction ratio of the mills as 55 per cent, the processing capacity of a *bel* that consumed one karda per day amounted to 4.4 tons of sugar-cane per day.[62]

The labour coefficient for the crushing stage at a Rohilkhand *bel* was equal to that in *gur* manufacture. Since crushing capacity was 1.0 ton per day using two people to operate the crusher, 8.8 labour days were required to crush 4.4 tons of sugar-cane. The labour coefficient in the crushing stage is estimated to be 2 ($^{8.8}/_{4.4}$).

An average *bel* employed 11 to 12 paid workers,[63] excluding those employed in the crushing stage. The job composition of these workers was as follows. The *munshi* (one person) was the accountant who weighed and recorded the amount of sugar-cane while the *darga* (one person) undertook general supervision of the *bel*. Both of these workers were usually Hindus, since their job required honesty and reliability, and most of the *khandsaris* were owned by Hindus. Juice extracted by the farmers was transported to the *bel* by a *bhisti* (one person). The boiling was undertaken by a *halwai* or a *karigar* (juice-boiler) and two assistants. These workers were generally Muslims with experience in the job, and were seasonal migrants from the environs of Agra. In addition, six or seven people were required to feed the furnaces (*jhonkias*), and lay out the wet bagasse to dry. These workers were generally of the Chamar caste (untouchables) who came from nearby villages. In this chapter, the number of people employed is taken to be 12.

The labour coefficient in the manufacture of *rab*, undertaken at the *bel* and excluding the crushing stage, is therefore 2.7 ($^{12}/_{4.4}$). When the crushing stage is included, the labour coefficient becomes 4.7. The labour coefficient for the manufacture of *rab* is slightly lower than that for *gur* manufacture, reflecting the economies of scale that resulted from the use of the Rohilkhand *bel*.

(2) *Rab* refining. We now turn to the estimation of the labour coeffi-

[62] Since the *rab*-cane juice ratio is 20 per cent, the daily *rab* production of a *Rohilkhand bel* is 13 *maunds* (484.9 kilograms). The average yearly quantity of *rab* prepared in one *bel* is 1300 *maunds*.

[63] Katiha, *Sugar in the Fields*, p.46 and *Marketing*, p.294.

cient in the *khandsari* stage. The use of centrifuges had become common by the end of the 1930s. Approximately 85 per cent of all *rab* used in the manufacture of *khand* was cured in centrifuges.[64] The early centrifuges were turned by hand, but even and uniform spinning could not be ensured. Thus, the electric power centrifuge became widespread.

In 1937 in Shahajahanpur district of Rohilkhand Division, the centre of *khand* production, 60 out of 70 centrifuges were turned by electricity.[65] Centrifuges powered by diesel engines were also used. In this chapter the typical technology for the *khandsari* stage is taken to be the power centrifuge. In relation to the processing of *rab*, Katiha states that 'on average, electric centrifuges take 10 to 12 minutes in warm months and 12 to 16 minutes in the cold season for machining a charge of 30 to 35 seers (30.0 to 32.6 kilograms) of *rab*'.[66] In one refining operation of 17 minutes, 32.5 seers (30.3 kilograms) of *rab* were processed, giving a processing capacity of 106.8 kilograms per hour. The ratio of *rab* to cane was 11 per cent, so that 106.8 kilograms of *rab* were equivalent to 970.9 kilograms of sugar-cane.[67] These are also test results, and we are told that the actual processing capacity was 80 per cent of the test results.[68] In this chapter, the actual processing capacity of the centrifuge is taken to be 776.7 kilograms of sugar-cane per hour, or 8.5 tons per day. Finally, it is assumed that an average *khandsari* with three sets of *bels* treated 4,500 *maunds* (167.9 tons) of *rab* each season.[69]

An average *khandsari* using electric centrifuges employed one sugarman experienced in the operation of centrifuges, one centrifuge assistant and one supervisor (*munshi*). Four unskilled labourers were hired to smash the earthenware pots containing the *rab*, remove the *rab*

[64] *Marketing*, p.100.

[65] *ITB Report 1938*, p.92.

[66] *One seer* is equivalent to 923.5 grams; 40 seers = 1 *maund*. Katiha, *Sugar in the Fields*, pp.61-2.

[67] Katiha, *Sugar in the Fields*, p.274.

[68] Srivastava, *The Open Pan System*, p.48.

[69] The average *khandsari* had three sets of *bels*. The average yearly *rab* production in one *bel* was 1,300 *maunds*. The remaining 600 *maunds* of *rab* was purchased from farmers who had produced it themselves. Hence a *khandsal* who had an average size *khandrari* purchased the remaining 600 *maunds* in the local markets.

and fill the centrifuge. Generally seven people were employed in total.[70] Thus, the labour coefficient of the centrifuge stage is calculated to be 0.8 ($^7/_{8.5}$).

The sugar produced by the centrifuge (called *pachani*) was then dried by spreading it over gunny mats (*patta*), where it was trodden by labourers (known as *pataha*) with their bare feet. One drying process took seven hours and three *pataha*s to dry 16 *maund*s (596.8 kilograms).[71] Since the recovery ratio in the production of the first *khand* was 4.4 per cent, 16 *maund*s of *pachani* was equivalent to 13.6 tons of sugar-cane. When these figures are converted to the amount used for 11 hours of operation, a labour coefficient of 0.1 is obtained. From the above calculations, the labour coefficient for the first *khand* is estimated at 5.6 (2.0 + 2.7 + 0.8 + 0.1).

The second *khand* is made from molasses that still contains some sucrose. There are no data available for estimating the labour coefficient in the centrifuge stage of the second *khand* production. By regarding the molasses as cane juice, we assume that the second *khand* is manufactured with the same labour coefficient as the first *khand*. 60 kilograms of molasses can be obtained from one ton of cane, since the recovery of the first molasses in the first centrifuge process averages about 60 per cent.[72] Thus the labour coefficient of the second *khand* is estimated at 0.4.[73] Therefore a labour coefficient of 6.0 is obtained for the complete manufacture of the second *khand*.

The estimation of labour absorption. According to the ITB Report 1931, the cane-crushing capacity of the average modern sugar factory around 1930 was 500 tons using a two-shift system, each shift lasting for 11 hours. The total number of employees was 675, consisting of 500 seasonal workers, 70 unskilled annual labourers, 75 technicians and

[70] Katiha, *Sugar in the Fields*, p.63.
[71] Ibid.
[72] *Marketing*, p.64.
[73] It required 3.6 days of labour to produce first *khand* from the 550 kilograms of cane juice that were obtained from one ton of cane. If the 66 kilograms of molasses is assumed to be equivalent to 66 kilograms of cane juice, (3.6 x 6.6)/550 = 0.4 days of labour required to produce the second *khand*.

30 management staff.[74] Hence the labour coefficient for the modern sugar industry is 1.4 (675/500).

The highest labour coefficient in the Indian sugar industry is for the *khand* industry, with a coefficient of 6.0, followed by 5.0 for the *gur* manufacturing industry. The modern sugar industry has the lowest coefficient of 1.4. Now, we will attempt to calculate the number of employees in each type of sugar industry for the mid-1930s (taking an average for the figures for the five years between 1933/34 and 1937/38), during which period the modern sugar industry experienced some growth. In this period, the average yearly amount of sugar-cane consumed in the production of sugar was 48,840,000 tons, of which 76.2 per cent was used for *gur*, 6.2 per cent for *khand*, and 17.6 per cent for the production of white sugar. The calculation of the labour absorption for each type of sugar industry is given below. It should be noted here that the annual number of working days in the modern sugar industry is taken to be 120 days, which is equal to the number of days the modern sugar refinery operated over the year. This sugar manufacturing period also more or less coincides with agricultural slack season in northern India.

We estimate that the *gur* manufacturing industry absorbed 1,549,000 people (86.0 per cent of the total employed in the sugar industry) for 185,930,000 working days, the *khand* industry, 152,000 people (8.4 per cent) for 18,220,000 working days, and the modern sugar industry employed 100,000 people for 12,060,000 working days, out of a total of 1,801,000 people employed in the sugar industry. If we suppose that all the sugar-cane went into the production of white sugar in the modern sugar industry, employment would have been reduced by 1,232,000 people (or 68.4 per cent) to 570,000 people. If all refined sugar made by centrifuging were *khand*, the number of workers employed in the *khand* manufacturing industry would have been 583,000, which is more than double the actual figure of 252,000.

Thus we can say that labour absorption in the indigenous sugar industry was three to four times that in the modern sugar industry. The indigenous sugar industry not only maintained its competitiveness, but

[74] *ITB Report 1938*, pp.192-5.

played an extremely important role in the Indian economy—an economy with abundant surplus labour. It was a labour-absorptive industry based on appropriate technology, and it continues to play this role to this day.

V. *Concluding remarks*

It is sufficiently clear from our research that the indigenous sugar industry had a great significance for the Indian economy. Firstly, from the point of view of the survival principle, the fact that the indigenous sugar industry managed to exist alongside the modern sugar industry for so long is a testament to the competitiveness of the former. For example, the 1932 Sugar Industry Protection Act aided the modern sugar industry by excluding imported white sugar from the home market. The Act had only a limited effect on the modern sugar industry, however, since the modern white sugar industry substituted only for the imported sugar, not the indigenous sugar industry (section II). Moreover, the competitiveness of the indigenous sugar industry continues to this day. More than half of the sugar production today is still carried out by the indigenous sector, which is quite exceptional in international terms.

This Indian situation should be accounted for in terms of both the modern sugar industry's stagnation or lack of competitiveness, and the competitive strength of the indigenous sugar industry. The weakness of the modern sugar industry should probably be explained by reference to its weak entrepreneurial spirit, the peculiarities of Indian agriculture or the constraints placed on the colonial economy. This chapter has focused on the capacity of the indigenous sugar industry for using technological adaptations as a means of enhancing its competitiveness. More specifically, it focused on the technological advances that were induced by the establishment of the modern sugar industry and on the use of appropriate technology (section III). We concentrated on these aspects because the technological innovations of the indigenous sugar industry contain several good examples of appropriate technology that fulfil the necessary condition of being competitive. This is the second significant finding of our study.

Thirdly, the significance of the advances in the appropriate adaptation of technology achieved by the indigenous sugar industry should be seen in the industry's role of absorbing labour. In this chapter we calculated labour absorption by estimating the labour coefficients for each branch of sugar industry. The most important aim was to try and demonstrate the importance of appropriate technology in the pre-independence indigenous sugar industry, in spite of the lack of sufficient production statistics. The indigenous sugar industry was a rural industry situated in Indian villages, where there was abundant surplus labour. We concluded that the indigenous sugar industry created as many as 1,600,000 employment opportunities by managing to compete with the modern sugar industry (this can be compared with the domination of the market by the modern sugar industry that occurred in other countries).

Even today, the indigenous sugar industry has become no less important. The problem of surplus labour in the rural sector has worsened and great expectations are placed on the prosperity of rural industry to provide linkages in India's industrialisation. The amount of labour required to process a unit of raw material is falling in both the modern and indigenous sugar industries, but there is still a large differential in the labour-raw material ratio between the two industries. Moreover, this differential exists because the indigenous sugar industry continues to be sufficiently competitive. Thus, the indigenous sugar industry should continue to attract our attention, since it remains a typical rural industry situated in villages and rural towns, that absorbs both its raw material and, moreover, labour from the rural sector.

Our study suffered from limitations in the availability of data. Hence, we were only able to ratify the competitiveness of the indigenous sugar industry using the ex-post perspective of the survival principle. It would be desirable in future to conduct a more rigorous study using detailed production cost data. Although such a study is difficult for the period of the 1920s or 1930s, it should be possible to conduct one for the contemporary sugar industry. We were only able to indicate possible factors leading to the stagnation of the modern sugar industry by referring to circumstantial evidence. In future we would like to undertake a broader in-depth study of the reasons for this situation, which is peculiar to the Indian sugar industry. It may be possible to derive more meaningful insights by comparing India's experience with the experience of other sugar-producing countries. We were unable to discuss the role played by the sugar industry research institutes which contributed to the advance of several technological innovations, though to a limited extent. A more comprehensive discussion on this point and on the lack of organisations established for technical education and for the diffusion of technology is required. Finally, in taking the indigenous sugar industry as a typical example of rural industry, it will probably be necessary to undertake an analysis of (a) the complementary linkages with other agricultural activities, and (b) the flows of labour and capital between agriculture and industry. These are all topics for future research.

Table 1: *Percentage of cane used in the production of the different types of sugar* (percentages)

	White Sugar	Khand	Gur
1921-25	1.2	19.6	66.6
1926-30	2.4	15.9	67.6
1930/31	3.9	14.6	66.4
1931/32	4.2	12.4	67.8
1932/33	6.8	11.2	65.9
1933/34	9.4	7.3	66.7
1934/35	12.1	5.5	65.3
1935/36	16.0	4.0	62.8
1936/37	17.4	4.6	61.7
1937/38	17.7	4.8	61.2
1938/39	19.2	4.6	57.3

Source: *Marketing*, p.69.
Note: Just under 20per cent of total sugar cane production was used for ratoon and for direct consumption other than sugar production.

Table 2: *The trend in import tariffs on sugar*

Period	Type of Duty	Rate of Duty
Before 1916	Ad valorem	5%
1916-21	Ad valorem	10%
1921-2	Ad valorem	15%
1922-5	Ad valorem	25%
1925-30	Specific (per cwt)	Rs.4 to 8 annas
1930-1	Specific (per cwt)	Rs.6
1931 (revised)	Specific (per cwt)	Rs.7 to 4 annas
1931-37	Specific (per cwt)	Rs.9 to 1 anna
1937-9	Specific (per cwt)	Rs.9 to 4 annas
1939-40	Specific (per cwt)	Rs.9 to 12 annas

Source: As for Table 1. *Note*: 1 cwt (one hundred weight) is 50.8 kg.

Table 3: *Cane crushing capacities of the main brands of iron-made, bullock-driven three-roller mills*

Name	Crushing capacity per hour (mds)	Juice extraction rate (%)
1. Sultan	2- 3	66-70
2. Kumar	2- 2.5	58-62
3. Karamat	1.75- 2.25	65-68
4. Hathee	4- 4.5	60-65
5. Delhi type	1.75- 2.25	62-65
6. Collectory	2- 3	50-60
7. Renwick	3- 3.5	50-55
8. Batala	2.- 3.6	53-60
9. Charkhari	2.4 - 3.6	53-60
10. Sataria	2.5	60-65
11. Attaria	2.5	60-65
12. Vakil	3.4	60-62
13. Ban 3-4	60-62	

Source: S.C. Roy, *Monograph on the Gur Industry of India*, Kanpur, 1951, p.251.

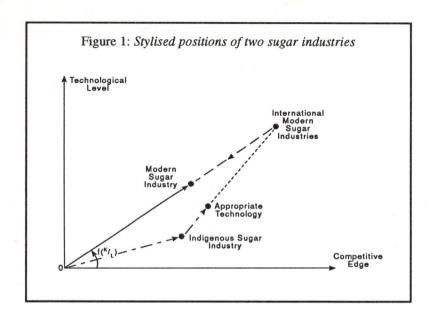

Figure 1: *Stylised positions of two sugar industries*

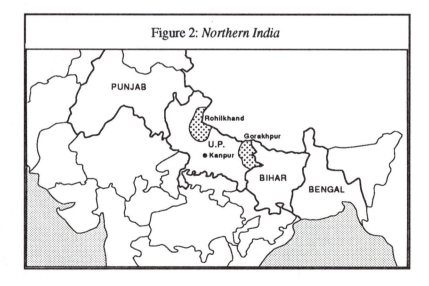

Figure 2: *Northern India*

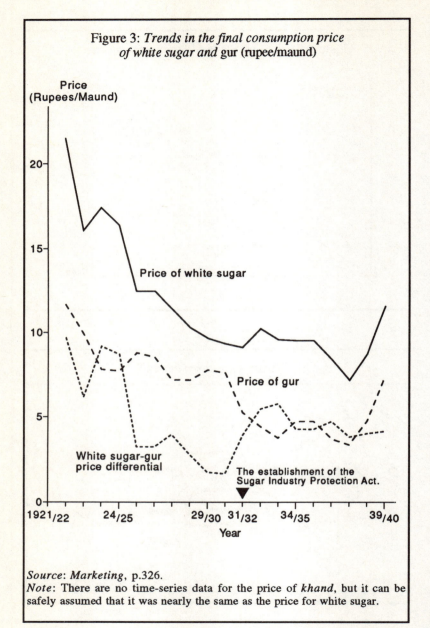

Figure 3: *Trends in the final consumption price of white sugar and* gur (rupee/maund)

Source: *Marketing*, p.326.
Note: There are no time-series data for the price of *khand*, but it can be safely assumed that it was nearly the same as the price for white sugar.

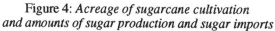

Figure 4: *Acreage of sugarcane cultivation and amounts of sugar production and sugar imports*

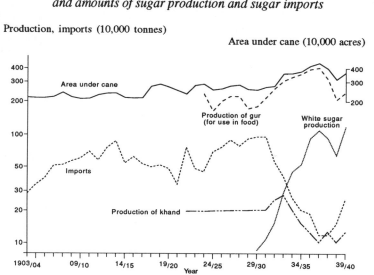

Sources: Area under cane: H.C. Prinsen Geering, *The World's Sugar Cane Industry Past and Present* (Manchester, 1912), p.47 (to 1911-12); Indian Tariff Board,*Written Evidence recorded during Enquiry on the Sugar Industry*, vol.6 (Calcuuta, 1932), p.37 (1912/12 to 1928/9); M.P. Gandhi, *The Indian Sugar Industry, its Past, Present and Future* (Calcutta, 1934), p.50 (1928/9 to 1933/4); *Marketing*, p.293 (1933/4). Production of *gur*: Gandhi, *Sugar Industry*, p.30 (to 1932/3), and *Indian Sugar Industry 1944* (Bombay) (after 1932/3). Production of *khand*: Gandhi, *Sugar Industry*, p. 57 (to 1931/2) and *Indian Sugar Industry* , p.1 (after 1931/2). Production of white sugar: *Indian Tariff Board Report*, 1931, p.20 (to 1929/30), *Indian Sugar Industry 1931*, p.20, and ibid., p.12 (after 1929/30). Import figures: *Marketing*, p.310. *Note*: The production of white sugar by refining *gur* and by the modern sugar industry totalled less than 100,000 tonnes per year before 1929/30. The production figures for *gur* and *khand* are estimated.

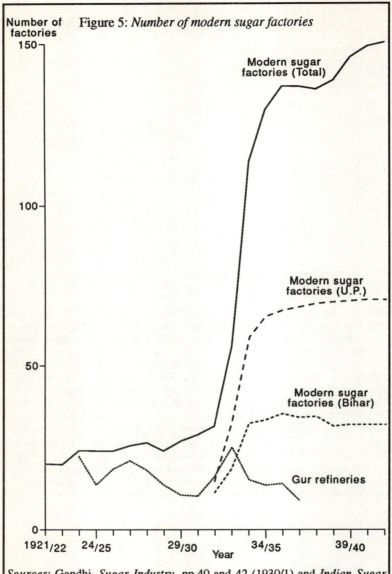

Figure 5: *Number of modern sugar factories*

Sources: Gandhi, *Sugar Industry,* pp.40 and 42 (1930/1) and *Indian Sugar Industry 1944*, p.1 (other years).

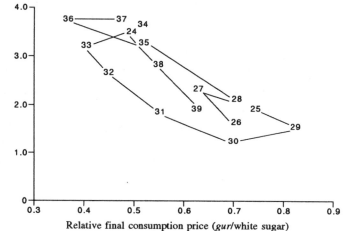

Figure 6: *Substitutability between* gur *and white sugar*

The ratio of per capita consumption
of *gur* over white sugar

Relative final consumption price (*gur*/white sugar)

Sources: Consumption: *Indian Sugar Industry 1934* and *1944*. Price: *Marketing*, p.326.

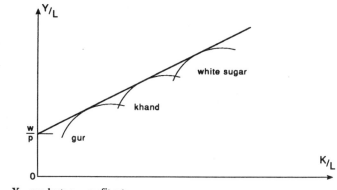

Figure 7: *Differences in capital-labour ratios by raw material unit*

Y = product; r = profit rate;
K = capital; w = wage rate
Note: See note 51 above.

$$\bar{p}\left(\frac{\bar{Y}}{R}\right) = \bar{r}\frac{K}{R}\!\downarrow + \bar{w}\frac{L}{R}\!\uparrow$$

Chapter 7.2

TECHNOLOGY OF THE INDIAN SUGAR INDUSTRY FROM AN INTERNATIONAL PERSPECTIVE

In Chapter 7.1, we made a comparison between the modern and the indigenous sugar industries of India, in order to evaluate their competitive edges in the domestic market as a result of respective technological improvements. However, since cane-sugar has been a world commodity from very early days, there are many comparable cases in other tropical and subtropical regions, and the technological level of the Indian sugar industry can be examined in comparison with those of other countries. The Japanese experience (including that of Taiwan under Japanese rule), for example, as well as those of Java, Mauritius and Cuba, may show some contrasts with the technological development of the Indian sugar industry. The aim of this chapter is to suggest that the technological level of the modern sugar industry in India was less advanced than its equivalents in other countries, whereas that of the indigenous sugar industry was quite similar to comparable industries in other countries.

I. *The modern sugar industry*

The modern sugar industry in India was developed fully only in the 1930s. Furthermore, this development was almost completely due to protection provided by the Sugar Industry Act of 1931. In other words, the rapid expansion of the modern sugar industry merely resulted from the imposition of import duties which allowed domestic white sugar to replace sugar from Java and Mauritius, where modern factories manufactured white sugar more efficiently, with up-to-date equipment, using scientific chemical control. Some technological causes for the weak competitive edge of the Indian modern sugar industry can, therefore, be confirmed by a comparison with these newly-rising cane-sugar industries.

Although it is difficult to exactly judge the technological efficiency of Indian modern factories in the nineteenth century, a drawing pro-

356

vided by P.R. Cola may give us an idea of an 'ideal' sugar factory in the 1860s.[1] The design in his book adopts the modern system, comprising steam clarifiers, a concretor,[2] a vacuum pan, centrifuges, and so on; the total cost for all the equipment would have been more than £5,000. Though we have no idea to what extent these plans were implemented in India, the three factories in Shahjahanpur mentioned in the book may have been equipped with some modern machines.[3]

The Commercial Products of India (1908) by George Watt does not refer to any modern sugar factories, except for F.J.V. Minchin's Aska factory in northern Madras, which used the diffusion method. This method of cane-sugar production in the 1870s is more precisely discussed in G. Martineau's Sugar: Cane and Beet.[4] As for the circumstances after the turn of the century, we can get a little more information from ITB Reports, Reports of the Committee on Rehabilitation and Modernisation of Sugar Factories in India (1965), Martineau's Sugar: Cane and Beet and Geerligs' The World's Cane Sugar Industry. In particular the last two discuss intensively the technological aspects from a global comparative viewpoint. R.G. Padhye's monograph also provides more exact evidence of the technological efficiency in modern Indian sugar factories around 1920.[5]

Judging from the above information, we may point out the following drawbacks that modern Indian sugar factories faced prior to the 1930s. First, the machines installed in Indian sugar factories were neither fully modern nor well-equipped, when compared with other cane-sugar producing countries. For example, while the two- or three-mill system of a

[1] P.R.Cola (ed.), How to Develope [sic] Productive Industry in India and the East (London, 1867), pp. 146-66.
[2] The concretor, which was patented by a Briton, Alfred Fryer, in 1865, was an evaporator of the direct fire-heated type. It is interesting that the latest British innovation was immediately adopted in this plan.
[3] Cola, Productive Industry, p.163. They are the E. Horsman, the Astragan Sugar Co., and Messrs. Caren and Co.
[4] George Martineau, Sugar: cane and beet, an object lesson (London, 1910).
[5] For ITB Reports and Padhye's monograph, see footnotes 6 and 9 in Chapter 7.1. Ministry of Food and Agriculture, Report of the Committee on Rehabilitation and Modernisation of Sugar Factories in India (1965) (Delhi, 1968). H.C. Prinsen Geerligs, The World's Cane Sugar Industry: past and present (Manchester, 1912).

three-roller type was popular in India, five mills and four mills were common in Cuba, Java and Taiwan. This fact identifies the double inefficiency of a low extraction rate from low sucrose-content sugar-cane. Multi-effect evaporators used in India were usually the double- or triple-effect ones, whereas triple- or quadruple-effect evaporators were common in other countries such as Cuba, Java, Mauritius or Taiwan. In addition, in India many of the vacuum (boiling) pans were not of the more efficient calandria type, but of the coil type.

Secondly, use of either the sulphitation method or the carbonation method as a decolorisation process is indispensable for the production of plantation white sugar. The latter is more sophisticated, though more expensive. In India the former method was popular, whereas the latter prevailed in Java. We do not share the view that the weak demand for white sugar in India resulted from a distaste for bone charcoal (a decolorising agent used in the refinery) in Hindu society, since bone charcoal was not used in the modern Indian sugar industry. Also, the development of the modern sugar industry after 1931 disproves the above view.

Thirdly, in comparison with Java, Cuba, Mauritius and Taiwan, R&D activities lagged behind in India. Indian sugar factories lacked a sufficient understanding of the significance of chemical control in the manufacturing process, as we can judge from the size of the technical staff. By contrast, in the case of Taiwanese modern sugar factories,[6] many factories competed, sending their engineers to Java in the 1910s to learn the Javanese carbonation method. Earlier they had tried to learn advanced technology from the Hawaiian sugar industry, but found the technology inappropriate to Taiwanese conditions. This switch for the industry, serving as a model, led to a remarkable reduction in sucrose loss in the Taiwanese sugar industry in the 1920s. Also in the 1920s, each factory made great efforts to train their field-staff to understand the significance of chemical control. These efforts resulted in: (1) a reduction in the time-lag between cutting and crushing the sugar-cane,

---

[6] A lot of information is available on the modern sugar industry in Taiwan, though mostly in Japanese. An easily available handbook is Hiroshi Higuchi (ed.), *Togyo Jiten* (A Sourcebook on the Sugar Industry) (Tokyo, 1959).

and (2) a popularisation of the use of a refractometer to check the sucrose content in crude sugar-cane in the field. In addition, the Japanese colonial government not only promoted the improvement of the irrigation system, but also encouraged research on new sugar-cane varieties in sugar experimental stations. We may suppose that most of the information on sugar machinery and other technical knowledge in India came from Great Britain. It is worthwhile investigating whether such British-biased information was really the most up-to-date, since Britain had only a sugar-refining industry and a not beet-sugar manufacturing industry, in contrast to other European countries.

Fourthly, most cane-suppliers in India were small independent cultivators. Hence, it was very difficult to preserve a steady and adequate cane-supply. In Mauritius the difficulties were surmounted, despite the fact that the majority of cane-suppliers were small-scale. Taiwan overcame the problem by introducing a zoning system, and cane-cultivators were encouraged to form co-operative unions of cane-suppliers. Cultivators in each zone could freely choose whether to grow sugar-cane, rice or sweet potato, judging from the price offered by the sugar factory prior to the new crop season. This price-mechanism and zoning, which India also tried to follow after the 1940s, eventually led to rapid technological development in the Taiwanese sugar industry.

It was not possible to improve the poor management of cane procurement in India, which largely depended on subcontractors or middlemen. This was partly due to the passive attitudes of (British) factory managers. In general the centralised factory system was not so strongly pursued in the British colonies. Cane-suppliers in Jamaica as well as Barbados also included small landholders.

In Chapter 7.1, the weaker competitive edge of the modern sugar industry was compared with the competiveness of the indigenous sugar industry in India. We should keep in mind the above drawbacks for the modern sugar industry, in comparison with other cane-sugar producing countries, when we emphasise the strong competitiveness (in the restricted sense of domestic competition) of the indigenous sugar industry.

II. *The Indian indigenous sugar industry compared with Asian cases*

As discussed in Chapter 7.1, the Indian indigenous sugar industry had enhanced its competitiveness, either through absorption of the imported sugar technology, or through indigenous efforts for the development of sugar technology as was described in Hadi's work.[7] A question that arises along with the competitiveness of the indigenous industry vis-à-vis the modern industry in India, is how the technological level of the former can be appraised in an international context, especially in an Asian perspective, since it has been well documented that indigenous sugar industries flourished in the many regions of Asia.

Indigenous sugar industries in some Asian countries. The history of the Chinese sugar industry can be traced to the period before the Common Era.[8] Marco Polo, who visited China from 1270 to 1295, witnessed the thriving sugar manufacturers in south China. Tian-gong kai-wu (c.1637), for example, gives an account of the sugar technology of those days. In the wake of the increase in sugar exports to Japan and western European countries in the first half of the seventeenth century, some technological improvements were made in sugar technology. The diffusion of bullock-driven, two-roller wooden (or stone) mills working horizontally is a good instance. Refined sugar was obtained by a method similar to the *khanchi* technique of India. The cane juice was boiled in the three-pan system using lime as a clarificant. Massecuite was then put into vessels with perforated bottoms, which were then covered with a layer of moist clay or grass. The moisture from the clay or grass gradually washed the molasses and carried it away through the holes, leaving the sugar crystals. However, the indigenous Chinese sugar industry declined, following the influx of imported white sugar from Hong Kong and Java in the early years of this century.

The Chinese indigenous sugar technology was probably introduced into other Asian countries. The *khanchi* method, for example, may have originated in China, since *khand* used to be called *chini* (China's) in Hindi; this term now means factory white sugar.

---

[7] S.M. Hadi, *The Indian Sugar Industry* (Bophal, 1929). See also S.C. Roy, *Monograph on the Gur Industry in India* (Kanpur, 1951).

[8] Geerligs, *The World's Cane Sugar Industry*.

In the Philippines at the beginning of the twentieth century, according to Geerligs,[9] 1,075 little factories were found producing a *gur* type of sugar, using a multi-pan system which had a structure more or less similar to a Rohilkhand *bel*. Steam mills were installed in 528 factories, and 470 had carabao (water-buffalo) mills. The majority of the factories crushed 50 to 60 tons of cane a day, which meant that they operated on a larger scale than the contemporary Indian *gur* manufacturers. The sugar recovery ratio was said to be 9 to 11.5 per cent, which was almost the same as that of the Indian *gur* industry.

At the time when Taiwan was occupied by Japan in 1895, there were about a thousand small indigenous sugar factories which had buffalo-driven mills. The greater part of their product was a soft, brown, finely-grained sugar, while refined white sugar accounted for only a small portion of sugar production. The technological level of the indigenous white sugar industry at that time was said to be almost the same as that of the Japanese,[10] that is to say, the white sugar recovery-ratio in the two countries reached around 5 per cent, similar to the ratio of the *khand* industry using a *khanchi* method.

As we have briefly observed, the technological level of the indigenous sugar industries in Asian countries has not been inferior to that of India, taking into consideration: (1) that for unrefined sugar-manufacturing, a multi-pan system was in use in Asian countries in contrast to a single pan system in India; (2) that bullock-driven roller mills were introduced in many regions of Asia long before they were introduced in India, and (3) that no significant differences in the sugar recovery ratio of either unrefined or white sugars were found between India and the other Asian countries mentioned above.

Comparison between Japanese and Indian industries. Sugar-cane and sugar manufacturing technology were introduced to the southern islands of Japan in 1610. They produced *kurozato* (similar to Indian

[9] For the history of the sugar industry in China, refer to Tai Koku Ki, *Chugoku Kansho Togyo no Tenkai* (Development of the Chinese Cane-sugar Industry) (Tokyo, 1967).

[10] For the indigenous sugar industry in Japan, we referred to Seizabro Shinobu, *Kindai Nippon Sangyoshi Josetsu* (History of the Japanese Modern Industry) (Tokyo, 1942) and Nippon Togyo Kyokai, *Kindai Nippon Togyoshi* (Modern Sugar Industry in Japan) (Tokyo, 1962).

gur) and supplied it to Japan proper. Sugar technology was then transferred to Japan proper in 1726. After some trial and error an indigenous
white sugar industry was established at the end of the seventeenth
century. In the first decade of the eighteenth century, the amount of
indigenous white sugar produced exceeded the total of both *kurozato*
and imported sugar in Japan.

There was a wide variety of indigenous white sugars, of which the
highest quality was known as *wasambon*. The technology used in manufacturing this variety shares many similarities with *khand* technology.
The process of *wasambon* manufacture was divided into two stages: (1)
the *shime-goya* (the crushing shed) for making *shiroshita-to* (massecuite, similar to Indian *rab*), and (2) the *saku-ya* (the squeezing shed)
for refining shiroshita-to into *wasambon*.

It is said that in the early stages of development, cane-growers, who
were in general small landholders, produced *shiroshita-to* and refined it
into *wasambon* themselves. Later there were several changes in the
organisational forms, similar to those that have been observed in the
khand industry in India. First of all, cane-growers turned into mere
cane-suppliers in the 1830s, especially in the areas where the industry
prospered. The *shime-goya* and the *saku-ya* were further differentiated
into two distinct operations at the end of the Tokugawa era (1603-
1867), especially in the 1860s when the indigenous white sugar industry reached its peak of prosperity. This organisational form corresponds
to the type-A of *khand* manufacturing that was common in the 1930s as
was mentioned in section IV of Chapter 7.1.

The crucial technology at the *saku-ya* concerned the type of mills.
Wooden two-roller mills driven by bullocks were introduced from
China in the seventeenth century and their use became common at the
beginning of the nineteenth century. Then wooden three-roller mills
and stone (granite) three-roller mills were invented. In the white-sugar
zone, stone three-roller mills, whose juice extraction ratio was 40 per
cent, were reported to be widely used. Iron mills, whose extraction ratio
was more than 50 per cent, were introduced at the end of the nineteenth
century. However they did not become dominant in the indigenous
white sugar zone: they could not work smoothly because the juice

extracted was easily frozen. Therefore at the zenith of the *wasambon* industry it can be safely assumed that stone three-roller mills were common. There were technological advances in boiling pans as well. In manufacturing *shiroshita-to* a two-pan system was used in the white sugar zone. Later an improved type of three-pan system, which had a similar structure to the Rohilkhand *bel*, was introduced at the beginning of the Meiji era (1868-1912). From what we have briefly observed, the technology in the manufacture of *shiroshita-to* can be said to have almost reached the level of *rab* manufacturing using the Rohilkhand *bel*.

The *saku-ya* managed two processes: (1) *oshi* (squeezing), by which the molasses was squeezed out of the *shiroshita-to* (rab) in an *oshibune*, and (2) *togi* (polishing), in which half-refined *shiroshita-to* processed in the *oshi* process was hand-kneaded to complete the refining into *wasambon*. This corresponds to the *khanchi* process of *khand* manufacturing, though a different technique was used.

The *saku-ya* was originally combined with the *shime-goya*, but in the closing days of the Tokugawa regime, it became differentiated from the *shime-goya*. That is to say, the *saku-ya* refined *shiroshita-to* purchased from the *shime-goya*. The owners of the *saku-ya* were generally landlords, as was the case with the *khandsals* in India. It is reported that a *saku-ya* having more than 100 *oshibunes* appeared at the time of the Meiji Restoration (1868). This implies the emergence of modern manufacturers out of the village household industries. The technology of the *saku-ya*, however, remained almost at the level of a *khanchi* technique, with a sugar recovery ratio in the production of the first *wasambon* of 3.5 per cent.

The most notable technological progress in the indigenous white sugar industry in Japan was marked by the establishment of the Sanuki-Shido Factory in 1878. It was equipped with an imported steam iron mill and a centrifuge. The extraction ratio of the steam mill was as high as 69.9 per cent. However boiling was done with open multiple pans (four pans). This factory proved successful in that production costs were said to be 30 per cent below those of the former production methods, and the white sugar it produced was of the same quality as imported white sugar. This type of production organisation may correspond to the type (c) of *khand* manufacturing in section IV of Chapter 7.1, and it can be currently observed in India.

Despite the facts (1) that the technological level and the managerial form of the Japanese indigenous sugar industry in the second half of the nineteenth century had reached the level of the Indian indigenous sugar industry of the 1920s and the 1930s, and (2) that sugar demand increased tremendously after the Meiji Restoration,[11] the production of indigenous white sugar declined after the Restoration due to the importation of white sugar from Java and Hong Kong, and later on a massive scale from Taiwan.

In this brief survey we have demonstrated that no significant differences in sugar technology were found between India and other Asian countries at the time when the modern sugar industry came into existence. The Asian indigenous sugar industries, with the exception of that of India, went into decline shortly after the advent of the modern sugar industry or the influx of modern white sugar. They were unable to improve their competitiveness through the absorption of modern sugar technology. This contrast supports our assertion that the technological level of the modern sugar industry in India was less advanced than that of other Asian countries. By the same token, it can be concluded that the survival of the Indian indigenous sugar industry may be ascribed mainly to the lower competitive edge of the modern sugar industry, rather than to the higher competitiveness of the indigenous sugar industry.

[11] Per capita sugar consumption was no more than 1.2 kg in 1877, about 2.4 kg in 1882 and nearly 6 kg in 1897.

Chapter 8

SITUATING THE MALABAR TENANCY ACT, 1930[1]

Toshie Awaya

Introduction

With some justification, Malabar District (now part of Kerala state) has
been treated as exceptional by several southern India specialists. They
exclude Malabar (and South Kanara) from their discussions in order to
keep their analysis coherent.[2] Similarly those scholars whose interest
lies in Malabar proper tend to discuss its history with little reference to
the Presidency, taking for granted that Malabar is peculiar. One reason
is that, ever since Malabar came under British rule in 1792, the Madras
government recognised the ubiquitous existence there of the landowning class called *janmis*, in contrast to the situation in other parts of
the Presidency, where the raiyats were considered tenants under the
government. Accordingly the raiyatwari system introduced into Malabar differed in several respects from the one found on the east coast.

The tenancy movement in Malabar during British rule also has
attracted attention. Scholars usually begin their discussions with the
intentional or unintentional legal (mis)reconstruction of the landholding
system in Malabar by British officials and courts: as is well known,
they considered that janmis had absolute proprietorship, and that other
existing rights in land were more or less terminable tenancies. For
example,[3] *kanam* and *verumpattam* rights, to which this paper will

[1] I wish to thank David Washbrook, Dilip M. Menon, P. Radhakrishnan and
James Chiriyankandath for their valuable comments on an earlier draft. I also
owe thanks to Haruka Tanagisawa and Tsukasa Mizushima for their advice and
sustaining encouragement. [The wording of this chapter has been revised by
Peter Robb and Kaoru Sugihara.]
[2] C.J. Baker, *The Politics of South India 1920-1937* (New Delhi, 1976), p.xii;
N. Mukherjee, *The Ryotwari System in Madras 1792-1827* (Calcutta, 1962),
p.xvii.
[3] Legal definitions were given to as many as 24 types of tenure in the often
quoted proceedings of the Sadr Adalat in 1856. These proceedings are found in
Sir Charles A. Turner, 'Minute on the Draft Bill relating to Malabar Land

often refer, were understood as lasting only for twelve years or for one year respectively.[4] Modern scholars, like tenancy agitators, point out that such interpretations derived from English analogies which were not applicable to Malabar, and allege that disastrous consequences ensued for tenants' interests, especially those of *kanakkarans* (kanam right-holders). In this view, the tenancy movement originated simply from the deteriorating conditions of tenants oppressed by all-powerful jan-mis.[5] It is also common to find two stages in the history of tenancy movements in Malabar, one from the 1910s to the 1920s led by upper-strata tenants (kanakkarans), and the second from the 1930s with the

---

Tenures' (Madras, 1885), Appendix XVI.

[4] Kanam right was defined as a mixture of a lease and a mortgage. At the end of 12 years a kanam tenant had to renew the lease by paying a fee to the landlord. The rent was the balance after deducting interest on money advanced (kanam) by the tenant to the landlord. A landlord could resume the leased land by paying back the advance money plus compensation for any improvements. Verumpattam was a simple lease. There are many treatises on Malabar law (including the land tenures and the Marumakkattayam system), for example, L. Moore, *Malabar Law and Custom* (3rd ed., Madras, 1905); P.R. Sundara Aiyar, *A Treatise on Malabar and Aliyasanthana Law* (Madras, 1922); K. Sreedhara Variar, *Marumakkathayam and Allied Systems of Law in the Kerala State* (Cochin, 1969).

[5] The first person to advance this view was W. Logan, Collector of Malabar, appointed as a Special Commissioner in 1881 to investigate agrarian conditions there. See Malabar Special Commission, 1881-82, Malabar Land Tenures, vol.I, Report (Madras, 1882) (hereafter MLT). It is not too much to say that his report largely determined the course and content of the following tenancy controversy. Its influence on the later studies was also not inconsiderable. The following is a select list of studies on Malabar agrarian history: K.N. Panikkar, *Against Lord and State: religion and peasant uprisings in Malabar 1836-1921* (Delhi, 1989); P. Radhakrishnan, *Peasant Struggles, Land Reform and Social Changes: Malabar 1836-1982* (New Delhi, 1989); A.K. Potuval, *Keralattile Karshaka Prastanam* (Trivandrum, 1976), and *Karshaka Samara Katha* (Trivandrum, 1978); K.K.N. Kurup, *William Logan: a study in the agrarian relations of Malabar* (Calicut, 1981); Prakash Karat, 'Agrarian relations in Malabar, 1925-1948', *Social Scientist* 2, 2-3 (1973), and 'Peasant movement in Malabar, 1934-1940', *Social Scientist* 5, 2 (1976); K. Gopalankutty, 'Movements for tenancy reform in Malabar: a comparative study of two movements, 1920-1939', in D.N. Panigrahi (ed.), *Economy, Society and Politics in Modern India* (New Delhi, 1985). V.V. Kunhi Krishnan, *Tenancy Legislation in Malabar (1880-1970): an historical analysis* (New Delhi, 1993). Dilip M. Menon's recent publication is an attempt to reinterpret Malabar modern history by using unconventional concepts, *Caste, Nationalism and Communism in South India: Malabar 1900-1948* (Cambridge, 1994).

emergence of lower-class tenants (*verumpattakkarans*, verumpattam right-holders). The latter stage, under leftist leadership, is regarded as having been the more radical. In this context the Malabar Tenancy Act of 1930 is treated as a turning-point.

There is no doubt that the British re-interpretation of various land-rights in legal terms was truly an innovation. But the usual assumption, that it simply strengthened the janmis' dominance, is a rather hazardous one, especially when we remember the paucity of studies of agrarian conditions in Malabar immediately prior to British rule.[6] We need first to examine how effectively the legal definitions were translated into practical economic transactions. Further, we need to test the simplistic premises that there was a clear-cut division among the agricultural population, into janmis, kanakkarans, verumpattakarans and so on, and that the tenancy movements involved a specific class stimulated mainly by economic discontent or aspirations.

The purpose of this paper is to re-examine these premises and to throw light upon the various socio-political currents which combined to the Malabar Tenancy Act of 1930. In the first five sections, we explore the meaning of the legal reconstruction of agrarian structure from a hitherto neglected angle. We contend that it formed a discursive sphere of the tenancy movement rather than deciding the economic balance of power between the landlords (janmis) and tenants (kudiyans). In the later sections, we try to show that the Tenancy Act of 1930 was not only the result of economic friction between janmis and kudiyans, but also the culmination of diverse forces working in and outside Malabar. Among the factors examined are the symptoms of a disintegration of *marumakkattayam* (the matrilineal system of inheritance). Additional factors upon which we touch include growing aspirations among the lower classes, and the advent of electoral politics after the Montagu-Chelmsford reforms of 1919. In a word, this paper is an attempt to locate the 1930 Act in a broader perspective.

---

[6] An important exception is K.N. Ganesh, 'Ownership and control of land in medieval Kerala: jenmam-kanam relations during the sixteenth to eighteenth centuries', *Indian Economic and Social History Review* 28, 3 (1991). Commercialisation of the kanam right is demonstrated in this article, an aspect which has not been adequately estimated hitherto.

Categories of the janmis and the kudiyans

It is quite common to set the janmis against other sectors of the agricultural population ('kudiyan' was a general term for them), as if there
was a clear-cut division. Even if we could suppose a relatively sharp
divide in the early nineteenth century, it did not exist when the tenancy
problem was taken up in the latter half of the century. Many statements
show that one person could occupy multiple positions. For instance, a
special Settlement Officer during the resettlement operations in 1930
reported that 'many a man, with whom I have opened a conversation
with the "Are you a janmi or a kudiyan?" formula, has told me that he
has so many seers or paras of janmam right and so many more on some
form of subtenure, or that he is the janmi of his paddy land and a
kanamdar or kuzhikanamdar of his house and garden'.[7]. Thomas, a
former Malabar Collector, also drew attention to a common phenomenon in which 'the average middle-class man who is ably represented
in the legal profession is usually in part a janmi in addition to a
kanamdar and may in addition hold land on simple lease'.[8]

There *were* some prominent janmis who possessed excessively large
amounts of land. In the early 1880s the Logan, Special Commission,
reported the existence of 1,079 principal janmis as shown in Table 1.
What is significant in this Table is a high proportion (more than 40 per
cent) of chieftains and Nambudiri Brahmans who composed at best of
15 per cent (including the Nayar community) of the population. The
proportion would be even higher if we took into account the fact that
landed properties belonging to religious institutions were normally
under the management of those chieftains and Nambudiris as trustees.
Again Innes, the Malabar Collector, observed in his note of 1915 that
about one-fifth of the total land revenue was paid by 86 top janmis
paying over Rs.3,000. They comprised 22 Rajas, 23 Nambudiris, 29

---

[7] Letter from A.R. MacEwen, 5 March 1930, Proceedings of the Board of
Revenue (hereafter PBR) no.80 (Land Revenue and Settlement) 17 October
1930, p.14. See also statements given by the witnesses to the Malabar Tenancy
Committee of 1927-28, Report of the Malabar Tenancy Committee (hereafter
RMTC), vol.II, Evidence.

[8] G.O.no.366, Law (General), 5 February 1926, p.14.

Nayars, 5 Moosads and Variyars (two of the temple service-castes generally called Ambalavasis), 2 Mappillas, 2 Tamil Brahmans, 1 Tiyar and 1 Goundan.[9] On the other hand the existence of a few big janmis means that there were also numerous small janmis, who probably closely resembled kudiyans. It is interesting that the qualification required to become a member of the Janmi Sabha was payment by one's family of an assessment over Rs.500.[10] In other words, the so-called janmi class was far from a homogeneous entity.

This was also true of the kanakkarans. Not a few of them possessed extensive lands on easy terms and were not engaged in actual cultivation. In Guruvayur amsam, reported in Slater's *Some South Indian Villages*, only three out of 385 rich kanakkarans were cultivating the whole of their land (the three were two Mappillas and one Tiyar); 1,200 out of the 1,400 acres held by kanakkarans were leased to sub-tenants.[11] Rich kanakkarans were in many cases Nayars, sharing a common caste with the janmis. The borderline between minor janmis and rich kudiyans was likely to have been fairly blurred.

Under such circumstances it could be argued that the legal definitions given afresh under British rule were very important for constructing the basic discourse of later tenancy agitations affecting the relations between janmis and kudiyans. The following comment by Washbrook seems to be relevant here: 'Although, by every social definition, there was a vast army of tenants in Madras, they were unable to fuse into an organisation because there was no legal-cum-political point

[9] Note by C.A. Innes on the Malabar Tenancy Legislation (hereafter Innes' Note) in G.O.no.3021 (Confdl.) Revenue, 26 September 1917, p.25. The total number mentioned here amounts only to 85. It is likely that one European janmi is not counted.

[10] *Janmi* 4, 3, p.67. The Janmi Sabha was founded in 1904. Its early history and rules can be found in K.C. Manavikraman Raja, *Janmi Sabha* (Kottakkal, 1910); the author was a member of one branch of the Zamorin family and secretary of the Sabha. The membership totalled 181 in 1910, of whom 87 were Nambudiris. Former Rajas and Naduvazhi local chiefs (janmis) were also predominant. The Malayalam monthly magazine *Janmi*, started in 1908 by Mannarkkattu Muppil Nayar (a Naduvazhi janmi), was handed over to the Sabha the following year.

[11] G. Slater, *Some South Indian Villages* (London, 1918), p.149.

of reference common to them'.[12] To paraphrase, the 'misconstruction' of various landed interests by British officials and courts provided the tenancy agitators with the legal-cum-political point of reference in the case of Malabar. The very fact that there were a conspicuous few among the janmis of a distinct character—that is, a high proportion of former chieftains and Nambudiris—was an additional useful reference point for the tenancy agitators. The 'janmis vs. kudiyans' schema was sustained by relying on the image of those prominent janmis, ignoring or simplifying a more complex reality. The category 'kudiyan' was itself an amalgamation of various types of tenants and could be construed only against the existence of such janmis.

Janmis' landed property and its management

Neither the legal definitions of janmis nor the various types of kudiyans afford enough information to assess the relative economic and social positions of each group, unless we look into their actual operations. In the following analysis our main focus is on the big janmis. This can be justified if we bear in mind that they were a prime target of the tenancy agitators, and have been supposed to be most powerful.

The first point to notice, is that the holdings of those big janmis were scattered over the country, as indicated by the third column of Table 1.[13] Naturally they leased out most of the lands, especially in the case of the former local chiefs or Nambudiris, who considered it beneath their dignity to cultivate themselves. Table 2 shows the number of kanam holdings under several janmis, who were described by K. Prabhakaran Tampan as big and middle-class janmis.[14]

Now let us take a closer look at the management of two big estates,

[12] D.A. Washbrook and C.J. Baker, *South India: political institutions and political change 1880-1940* (Delhi, 1975), p.11.
[13] RMTC, vol.1, Report, pp.152-61.
[14] We sometimes come across assertions that those who had kanam tenants under them could truly be called janmis; see RMTC, vol.I, p.48; G.O.366, p.46; RMTC, vol.II, Evidence (Madras, 1940), p.9. Whether this was a generally recognised concept is beyond the scope of this paper. When a kind of janmis' association was set up at the time of Logan's investigation, which gave rise to some commotion among the janmis, those joining were required to have kanak-karans under them. *Janmi* 6, 11, pp.243-54.

namely the Zamorin and the Kavalappara. These were among several in Malabar managed by the Court of Wards. The Kavalappara *swaroopam* (chiefdom) was virtually an independent power at the end of the eighteenth century.[15] The Zamorin was one of the most illustrious rajas in Kerala, from the thirteenth century. After surrendering to the British, he transformed himself into the biggest janmi in Malabar. Both can be understood as typical in the sense that they represented the power and authority that the tenancy agitators most denounced. Both families were also the most active in fighting tenancy agitations.[16]

Kavalappara estate came under the management of the Court of Wards on the recommendation of the Collector, when a six-year-old girl was left as its only heir in 1872. Wards management continued until 1910. In the other case, the Zamorin himself requested the Madras government to take over the management of his stanam property in 1914.[17] After much correspondences between the Collector and the Board of Revenue, to clear away some legal obstacles, the government agreed in 1915, considering it 'expedient in the public interest', on condition that the duration of its management should be 12 years.[18] Table 3 provides a general description of each estate.

From the above summary we get a general idea of big janmis' households, namely unmanageably vast and scattered landed properties, with various types of religious and charitable institutions required to demonstrate their status. The janmis were seldom directly engaged in land management. When Innes, acting Collector of Malabar, asked the Zamorin for information he was unable to supply it, because he had 'only a vague idea of the extent and location of his lands'. He had to ask Innes to give him duplicate copies of his pattas.[19] It could not have

[15] For information on Kavalappara swaroopam, see K.K.N. Kurup (ed.), *Kavalappara Papers* (Calicut, 1984).

[16] The members of these two families were active in the Janmi Sabha. Kavalappara Muppil Nayar was for a while an M.L.C. representing the janmis.

[17] 'Stanam' means rank or dignity; the senior male member of the family, comprising three branches called Kovilagam, occupied the Zamorin stanam. Other Malabar chieftain families had similar institutions with varying numbers of stanams. The Zamorin family had five, including that of the Zamorin.

[18] G.O.no.897 (Confdl.), 17 April 1915.

[19] Letter from C.A. Innes, 12 October 1914, para.5, in ibid.

been otherwise seeing that the record-keeping of the Zamorin estate was once described as a 'highly mythical form'.[20]

Having no 'home' cultivation until the Court of Wards attempted to start some model farms, both janmis showed little inclination to become improving landlords. During the Court's management, the Collector proposed that surplus income from the Kavalappara estate should be utilised to replace kanam tenants with simple tenants by repaying kanam money, which would have increased the estate's income. Though the government at first approved, it then had second thoughts, finding it 'inexpedient to disturb the existing practice in regard to kanam leases in the estate'.[21] In the end, the only ways of investing the surplus proved to be to buy government bonds or, a little more productively, to improve the estate's market-place.

These conditions remind us of the four characteristics of a typical janmi Logan gave in his famous report, that is, the scattered location of his property, his ignorance of its details, and of the persons who actually cultivated it, and his unconcern so long as his rents came in regularly.[22] It was said that it was not uncommon to find a janmi who had 'never seen his own lands with his eyes in his life time';[23] such negligence was sometimes deplored by the janmis themselves. Instead agents (called *karyastans*) looked after janmis' lands. This was noted by MacEwen in 1930 when he commented on the high proportion of 'rent-collectors' among the agrarian population in Malabar compared with the Madras Presidency as a whole.[24]

The karyastans' power was considerable. They usually demanded

[20] Letter from J.A. Thorne, 18 October 1916, para.6, in *Proceedings of the Court of Wards* (hereafter PCW), no.55, 9 December 1916.

[21] G.O.no.490 (Confdl.no.72) Revenue, 29 May 1905.

[22] MLT, para.218. Such examples to illustrate the big janmis' ignorance of their property could be multiplied. See the answer of Witness 93 who was a member of the Olappamanna family (a famous Nambudiri janmi) in the Report of the Malabar Marriage Commission (Madras, 1891) (hereafter RMMC). Though knowing his family possessed a large landed property, he could not tell how much assessment they were paying to the government.

[23] *Janmi* 4, 2, p.45. Such conditions still remained unimproved in 1929. See *Yogakshemam*, 28 September 1929, p.4.

[24] Letter from A.R. MacEwen, 5 March 1930, para.16 in PBR, no.80, 17 October 1930.

remuneration from the tenants for every transaction, such as renewing a lease and granting receipts of rent. One karyastan was able to save Rs.5,000 to Rs.6,000 over 15 years.[25] Another, Kannattuparuttiyil Krishna Nayar, described in Cherukat's autobiography, became increasingly prosperous after he became a karyastan under a big Nambudiri janmi.[26] Therefore even Logan or Innes (both equally enthusiastic critics of the janmis)[27] admitted that sometimes it was not them but their avaricious karyastans who were to blame for the oppression of the tenants.[28] The Janmi Sabha organised a karyastan examination for several years in attempt to improve their land management; but, as admitted in the Janmi Magazine, the number of applicants showed that this attempt did not bring the expected results.[29]

The financial conditions of the two estates were unsatisfactory, despite or because of their vast landed properties. When the Court of Wards took over the management of Kavalappara in 1872, the estate had only Rs.10 in ready money but debts of over Rs.70,000, more than three times the estate's estimated income. The Court had to begin its management by obtaining a loan of Rs.65,000 from the government.[30] It then cut the number of estate servants to 46. Rent collections improved in comparison with those of other estates, and yet there were overdue balances in the current account amounting to 3.4 to 23.4 per cent, and in the arrears account of 7.6 to 37 per cent. The Zamorin estate was reportedly was free from debt, but had to be helped through its initial difficulties by loans of over Rs.130,000, raised from several sources. Its unrecovered balances on current rents were less than 20 per cent in the first year of Court of Wards administration, owing mainly to incomplete accounts, but in the later period remained between 22 and 39 per cent of the demand. The arrears could not be cut below 50 per

[25] MTC, vol.I, p.144.
[26] Cherukat, *Jivitappata* (Pattambi, 1974), p.61.
[27] For instance, Innes declared that the janmis' 'character as landlords is decidedly bad', Innes' Note, para.42.
[28] MLT, para.231, Innes' Note, para.43.
[29] *Janmi* 8, 12, p.271.
[30] Punnattur, another estate taken up by the Court, was in the same condition, with Rs.100 cash in hand and a large debt (over six times its income), PCW, no.30, 18 July 1911.

374 LOCAL AGRARIAN SOCIETIES IN COLONIAL INDIA

cent even with rigorous management by successive estate collectors, including writing off longstanding arrears as unrecoverable.[31]

It is true that, apart from their landed properties, the janmis were supposed to derive power and influence over their tenants from their social and religious status, especially in the case of big janmis composed of rajas and Nambudiris. The Zamorin was well aware of this. He insisted on receiving *tirmulkalchas* (customary presents from tenants or subordinates in general) directly, even when he asked the government to manage his estate for him.[32] His predecessors had resorted to the courts to punish one tahsildar (apparently a Tiyar) who was insolent enough to refer to the Zamorin in an official communication as Raja instead of Tamburan.[33] Similarly, even in the early twentieth century, the Kavalappara exacted Rs.43 as fines from tenants who failed to celebrate 'varam' at Kavalappara temple.[34] The patronage of religious or charitable institutions was an essential to public status. However the costs were sometimes a burden economically. The Zamorin's devaswams were not 'good milch cows for the trustee [the Zamorin], but are actually a drain on the secular funds'.[35] In course of time such patronage also became less effective. Exactions to support it, provoked much resentment and provided a rallying point for the advocates of tenancy reform. They could point to the jamnis' social 'oppression' as one of their justifications.

The janmis and the revenue administration

Thus the big janmis' economic and social position was somewhat vulnerable. We should also consider the peculiar revenue administration in Malabar, a raiyatwari system which deviated from that in other districts of the Madras Presidency. This subject has been noted elsewhere, but

[31] After fasli 1331 (1921-2) the accounts relating to devaswam property were separated. These figures mentioned here are concerned with the main body of estate property.

[32] G.O.no.897 (Confdl.), Revenue, 17 April 1915, p.11.

[33] *Decrees in the Appeal Suits determined in the Court of Sudr Adalut*, 2 July 1859, no.180 of 1858.

[34] Administration Report from fasli 1313, Statement in PCW, no.122, 23 November 1904.

[35] PCW, no.39, 28 November 1923, p.6.

also rather neglected or misrepresented. These considerations are important to an understanding of relations between janmis, kudiyans and government in Malabar.

A general settlement was not introduced into Malabar until 1900 (except in Wynad where the first settlement was in 1886-87). The result of this omission was, in view of the settlement officer in the 1890s, that 'the Malabar Revenue accounts are practically useless, and it would be difficult to find two Revenue officials who could give a satisfactory explanation as to the existing assessment on any given land'.[36] In a similar vein, McWatters, Collector of Malabar, had remarked in 1881:

> We have no land register in Malabar, we have no regular register of gardens, and we have no register of dry lands. In short, we have nothing about them. The rice fields cannot generally be identified by means of the only register we have got, and we have consequently no control over the apportionment of the revenue. It is believed that the poor man with the worst land pays the highest assessment.[37]

In addition, in Malabar, pattas were very commonly issued not to janmis, who were supposed to be raiyats under the raiyatwari system, but to tenants (mostly kanakkarans). Therefore in many cases the total assessment paid by a janmi's patta-holding tenants was greater than that which he paid himself; this is apparent in the answers of witnesses to the Malabar Marriage Commission of 1891 (Table 4).

This 'anomalous system' in Malabar was referred to in a High Court judgment in 1889.[38] The Collector had issued a 'cowle' to an applicant for cultivating waste lands; later the land was sold by the court because the cowle-holder had defaulted on his revenue-payments, but the Madras High Court decided that the janmi's interest was not transferred by this enforced sale. In another case the Zamorin appealed against a compulsory sale of lands on his estate, again caused by a tenant defaulting on the revenue. A patta had been issued in the tenant's name.

[36] Letter from C. Stuart (Special Settlement Officer, Malabar and South Kanara), 27 June 1892 in PBR, no.106-A, 15 March 1893

[37] Cited in H. Morbly's letter, 24 May 1894, para.14, in PBR, no.883, 29 Aug.1900.

[38] I.L.R.13 Mad.89; *Madras Legislative Council Proceedings* (hereafter MLCP) vol.23, 1885, p.241.

The High Court held that under the raiyatwari system a patta-holder should be presumed to be the landholder responsible for paying revenue and subject to the provisions of the Land Revenue Recovery Act of 1864.[39] An attempt was made to rectify this 'anomaly' in the Registration Act of 1896, which launched a troublesome operation to record every piece of janmis' land, as it was intended that the impending revenue settlement of 1900 was to be made with the janmis.[40] Even so the Act provided for an occupant (presumed to be usually a kanakkaran) to be registered along with the janmi, who would thus be able to pass on responsibility for revenue-payment to the tenants.

This ambiguity between legal landowners (janmis) and revenue-payers (tenants) might be understood at first sight to mean that tenants were under a double yoke—of the government and the janmis. Yet at the same time it seems to show tenants' relative independence from the janmis, or at least that janmis were not completely dominant. Evidently, for example, big janmis like the Zamorin and Kavalappara did not have complete control over their waste lands, 'unoccupied dry' land in revenue parlance, where fugitive cultivation was carried on.[41] From the perspective of government finances, these tenants could not be neglected as revenue-payers—and indeed the reason government did not find it necessary to change this anomalous system for a long time was that it worked well enough so far as the revenue collection was concerned. Even the joint registration provision of the 1896 Act was probably added so as to perpetuate long-standing revenue arrangements as much as to show consideration to the janmis, though only the latter reason was mentioned in the Madras Council.[42] It should also be noted that, given that the land revenue demand in Malabar remained stable

[39] I.L.R.7 Mad.405. Charles Turner, one of the Judges giving this decision, had second thoughts a few years later, admitting that he had been ignorant 'of the extent to which the Revenue authorities in Malabar have diverged from the practice in other parts of India', Turner, 'Minute on the Draft Bill', p.96.

[40] On this weary and and long-drawn-out operation of janmam registration, see G.O.no.882A, Revenue, 29 August 1900.

[41] Administration Report for fasli 1314, para.4, in PCW, no.92, 18 October 1905; Letter from J.A.Thorne, para.4. See also the above-mentioned case, I.L.R.13 Mad.89.

[42] MLCP, vol.23, 1895, pp.241-4 and 281-5.

throughout the nineteenth century, and that it was criticised as notoriously low,[43] benefits derived from the rise of grain prices were enjoyed by those tenants paying revenue.[44]

The problem of evictions

It would be a mistake to conclude, from a few indications of weakness in the janmis' economic and social base, that their power was imaginary. Above all, they had a legally recognised right of eviction. Any study of agrarian change in Malabar would be incomplete if it ignored this question. Legal eviction could be effected directly through the courts or by means of a *melchart*, a sort of second mortgage upon lands leased on kanam whereby a third party obtained the right to evict the kanam tenant, after repaying him his kanam advance and paying compensation for any improvements. Melchart was particularly resented because it enabled a janmi to evict a tenant without spending his own money. Even stable and powerful kanakkarans feared the intervention of a more resourceful competitor by this device.

Tenants' advocates emphasised the increase in eviction suits and melcharts to justify curtailing janmis' power, but it is hard to interpret the statistical information available. According to Strange's report of 1852, the average annual number of eviction suits (exclusive of pending cases and those cases that involved Mappillas and Hindus as defendants

[43] *Madras Mail*, 5 January 1887, p.5.

[44] It is difficult to estimate precisely how the increased revenue burden after the settlement, when the total demand was almost doubled, was distributed among the agricultural population. While the Kavalappara estate under the management of the Court tried to induce the kanam tenants to register jointly, the Zamorin seems to have been fairly successful in shifting the responsibility for revenue-payment on to the tenantry by the time the estate came under the Court's control. On the whole, if we believe the word of the President at the seventh Annual Conference of the Janmi Sabha, janmis could ride the crisis without selling their lands because of the rise of agricultural prices. See C.A. Innes, *Madras District Gazetteers, Malabar* (Madras, 1951), p.347; Administration Report for fasli 1313, para.5, in PCW, no.122, 23 November 1904; letter from C.A. Innes, 12 October 1914, para 6 in G.O.897; and *Janmi* 4, 1, p.21. Prabhakaran Tampan's response in the Madras Council is interesting. When asked if he could repeat his statement, that is, that the land revenue in Malabar was relatively mild at the time of resettlement, he replied, 'Why not, Sir, the tenant pays that. What do I care?', *Madras Legislative Council Debates* (hereafter MLCD), vol.30, 1926, p.457.

at the same time) was 1,140.[45] The average for the period from 1877 to 1880 was reported by Logan to be 4,983.[46] The number given by Innes is shown in Table 5. As he does not give the total number of cases filed, it is impossible to compare his figures with Logan's.

In addition, from 1900 to the 1920s, between 2,800 and 3,600 cases of melchart were registered or executed every year, with a tendency for the number to increase.[47] The figures show that evictions in Malabar were not negligible—and without them it is possible that there would have been no tenancy agitation. At the same time, though there is not much statistical evidence for it, the renewal fees demanded by janmis sometimes seem have become enormous and arbitrary.

In addition, from 1900 to the 1920s, between 2,800 and 3,600 cases of melchart were registered or executed every year, with a tendency for the number to increase.[48] The figures show that evictions in Malabar were not negligible—and without them it is possible that there would have been no tenancy agitation. At the same time, though there is not much statistical evidence for it, the renewal fees demanded by janmis sometimes seem have become enormous and arbitrary.

In addition, from 1900 to the 1920s, between 2,800 and 3,600 cases of melchart were registered or executed every year, with a tendency for the number to increase.[49] The figures show that evictions in Malabar were not negligible—and without them it is possible that there would have been no tenancy agitation. At the same time, though there is not much statistical evidence for it, the renewal fees demanded by janmis sometimes seem have become enormous and arbitrary.

On the other hand, it is difficult to say that eviction suits simply prove janmis' oppression. Though the melchart still remained a fearful

[45] *Janmi* 4, 1, p.21. Strange's report, 25 September 1852, para.21, in 'File of Correspondence regarding the relations of landlord and tenant in Malabar, 1852-1856' (Madras, 1881).

[46] MLT, para.147.

[47] Innes' Note, para.29; RMTC, vol.I, para.127; G.O.366, Law (General), 5 February 1926, p.98.

[48] Innes' Note, para.29; RMTC, vol.I, para.127; G.O.366, Law (General), 5 February 1926, p.98.

[49] Innes' Note, para.29; RMTC, vol.I, para.127; G.O.366, Law (General), 5 February 1926, p.98.

weapon, the number of eviction suits seems to have dropped somewhat after the revision of the Compensation Act in 1900. Moreover, the percentage of decrees actually executed indicates that decree-holders quite often did not actually redeem the lands; perhaps they did not find it worth doing so because of the heavy compensation they had to pay to the tenants. The futility of eviction suits because of their disproportionately heavy cost was a theme repeated by the collector of the Zamorin Estate.[50]

More importantly, what is missing in the eviction figures is information as to by whom and against whom these suits were instituted. It is quite possible some were filed by kanam tenants against under-tenants. The Malabar Tenancy Committee of 1927/8 gave the details shown in Table 6, which is supplemented by a statement by a munsiff at Pattambi.

It is significant that kanakkarans also resorted to the munsiff's court in order to evict tenants. Seen this way, the increase of evictions and melchart cases may indicate not one-sided aggression by janmis, but quite possibly the janmis' constrained situation and the existence of people who were rich and ready enough to pay large compensation on their behalf. Another incentive for giving a melchart will be discussed below. For now, suffice it to say that the eviction problem certainly existed, and that the customary relationships between the various agrarian classes must have deteriorated because of a plethora of bitter court struggles. According to Thorne, collector for the Zamorin estate, *kana janma mariyada* 'expresses all the sentimental obligation that bind the janmi and the tenant in their relations to one another'. He reported in the early twentieth century that 'Kana Janma Mariyada...is dying'.[51]

Profits of tenants (kudiyans)

'Kudiyan' is the term which was used during the tenancy agitations of the 1910s and 1920s to refer to tenants in general. Their economic

---

[50] See Administration Report for fasli 1329, para.7, in PCW, no.43, 19 November 1920.

[51] Cited in MTC, 1927-28, vol.I, p.104.

conditions are more difficult to ascertain. The case of the Zamorin estate, though, indicates that there was much scope for some sections to make large profits; this was especially true of kanakkarans, who held lands directly under big janmis. According to an estate collector, in the Palghat area, with its rich paddy fields, more than 60 per cent of the tenants did not themselves cultivate, and even among the remaining 40 per cent, many kept only a small portion in their direct charge subletting the rest. The small number of eviction suits in Palghat (four in all over eight years out of nearly 2,000 rent-payers) also implies their relative prosperity.[52] Thorne had experience of managing the Zamorin Estate, and said that:

The Zamorin, in fact, is the pillar of the middle classes of South Malabar. The old Tarawads [Nayar joint families] and their branches (tavazhis) and the Brahman families, which supply the bulk of the professional and official classes of Malabar, depend largely for their prosperity on the profits they derive from the easy tenures of the estate. And these families sub-let most of their lands. They are middle-men rather than agriculturists.[53]

A statement prepared by Evans, who was Collector of Malabar when tenancy legislation was under consideration in 1915, gives a rough impression of the division of net yield between janmi, kanakk-aran and verumpattakkaran. Out of 50 examples from South Malabar, there are 23 that show the typical three-tier division. Kanakkarans enjoyed from 45 to 80 per cent of yield (average 64), and janmis from 1 to 30 (average 12).[54] Such a situation is exactly what we can see from several cases reported to the Tenancy Committee of 1927/8. For instance, one kanam tenant, a vakil, with a kanam advance of Rs.1,600,

[52] G.O.366, p.70.

[53] Ibid., p.25, 'The Brahman families' should be understood to mean not Nambudiris but Tamil Brahmans, of whom there were many in the Palghat area; unlike the former they were not hesitant about taking up a public job. See also Letter from Estate Collector, 27 November 1927, para.4. He reports that 'These intermediary tenants get for estate lands double to 10 times what they pay the estate for rents', PCW, no.45, 14 December 1927.

[54] Note by F.B. Evans on Innes' Note (hereafter Evans' Note), G.O.no.3021 (Confdl.) Revenue, 26 September 1917. This statement is certainly neither complete, nor without defects. See also Panikkar, Against Lord and State, pp.30-1.

held 50 acres of land cultivated by verumpattakkarans; he received 2,800 paras of paddy as rent (valued at about Rs.2,800). From this he was paying revenue assessment of Rs.300, plus 195 *paras* of paddy and Rs.60 as *michavaram* (rent)—in all about Rs.255—to his janmi, a *devaswam* whose trustee was a *naduvazhi* (a former local chief). The tenant's share thus amounted to Rs.2,245, subject to a renewal fee payable once every 12 years; on the last occasion this had been set at Rs.3,000.[55] Another case, from Palghat, related to a holding of 135 acres. The kanakkaran received 8,000 to 10,000 paras in rent from under-tenants, and paid to his janmi (a *kovilagam*, that is, a chieftain family) 1,000 paras and the revenue assessment of Rs.300. His renewal fee was Rs.8,500.[56] Another kanakkaran in Ponnani with 50 acres held from five or six janmis, received Rs.1,500 from garden lands and 2,000 paras from wet lands. His rent was 300 paras plus Rs.40 (the revenue is not mentioned); the kanam advance was Rs.5,000 and the latest renewal fee Rs.1,000. The rent this person received from his own janmam land (10 to 15 acres) leased out on kanam came to only Rs.10 to Rs.20.[57] Though there were many smaller kanam holdings, we can infer from these cases that investment in a kanam right was often 'more profitable than a janmam purchase', as a High Court vakil from Palghat told the Tenancy Committee.[58] Naturally, for those who thus would 'go in for kanam', it was necessary to deprive the janmis of their power of eviction and to prevent them from giving a melchart to another well-to-do kanakkaran.[59]

Among less-privileged tenantry such as the verumpattakkarans, there was again much variety. One fairly big verumpattakkaran held 15 acres of wet land and 75 acres of garden land, and paid about Rs.1,000 as rent.[60] Another held 7 acres of wet lands and 40 acres of dry lands, paying in rent 450 paras and 450 bundles of straw for the former and

[55] RMTC, vol.II, p.270.
[56] Ibid., p.52.
[57] Ibid., p.543.
[58] Ibid., p.77.
[59] We should not forget that there were many poor kanam tenants. See, for instance, RMTC, vol.1, p.144. In this case the adult male members were working as coolies.
[60] Ibid., p.142.

Rs.208 for the latter. Gross yield from the wet lands was 650 paras, and profits of Rs.450 could be derived from the dry lands by cultivating groundnuts. This tenant had five pairs of cattle and seven or eight farm servants.[61] On the other hand, among the examples recorded by the Tenancy Committee, there are not a few whose rent exceeded their net produce, if we accept the figures that those verumpattakkaran respondents supplied. If true, they indicate a heavy burden of rent,[62] even though the verumpattakkarans retained the straw—a valuable asset, as Evans stressed, eager to prove Malabar rents were not excessive.[63]

To sum up, this fragmentary information provides a fairly complex picture of the so called 'kudiyans'. They formed a highly heterogeneous category, ranging from some kanakkarans with sufficient resources to become vakils, to poor verumpattakkarans for whom straw was an essential item of income. To understand the nature and process of the tenancy agitation, we need to bear this diversity in mind .

The agrarian problem and the marumakkattayam system

Certain contemporary issues, interwoven with the agrarian system but not purely economic in character, played a significant role in creating the background for the tenancy agitations. One relates to *marumakkattayam* (the matrilineal system of inheritance), which started to show some symptoms of malfunction under British rule in the form of family feuds. Therefore in the last quarter of the nineteenth century, some demanded the reform of the system through legislation. The troubles necessarily were reflected in land relations, as so many of the land-owners and tenants were Nayars, the main community following the system.[64] The patrilineal Nambudiri Brahman community—another landed class—was also involved through *sambandham* (conjugal

---

[61] Ibid., p.146.

[62] Ibid., pp.140, 141 and 145. Of course, it is possible that those verumpattakarans were holding as well more profitable leased lands, the produce from which sustained them.

[63] Evans' Note, para.10.

[64] The marumakkattayam system was followed not only by Nayars but by Tiyars and Mappillas of North Malabar, some Ambalavasis (temple service castes), and even the Nambudiris of Payyannur.

relations between a Nambudiri boy and a Nayar girl).[65] The disintegration of the marumakkattayam system in the latter half of the
nineteenth century sharpened the frictions in the agrarian society, and
attitudes towards marumakkattayam helped decide the stance taken on
tenancy reform.

It is fairly certain that most of the problems with the system originated in judicial decisions by the British courts. In 1813, according to
Moore, author of a famous book on Malabar law, the first ruling was
given that partition of a *tarawad* (matrilineal joint family) was not allowed unless all members agreed to it.[66] This often led to an unhealthy
and unnatural swelling of family size. At the same time, British courts
tended to stress the exclusive power of the karanavan (the eldest male
member of the tarawad). Strange (a judge of the Sadr Adalat) had
recommended in his report of 1852 that only karanavans should have
the right to dispose of tarawad property. Similarly Holloway, a Madras
High Court Judge in the 1860s and 1870s, was known as the 'champ-
ion...of the absolute rule of the Karanavan'.[67] His celebrated dictum
was that 'a Malabar family speaks through its head (karanavan), and in
all Courts of Justice, except in antagonism to its head, can speak in no
other way'.[68] That the courts were inclined to keep traditional joint
families undivided and under the karanavan's authority can also be seen
in the finding that the personal (self-acquired) property of an intestate
tarawad member should be inherited not by the appropriate *tavazhi*
(branch of the tarawad) but by the tarawad.[69]

These decisions might appear to have helped strengthen the karanavan's authority. But the courts also ruled that *anandaravans* (junior
members of the tarawads) were entitled to demand maintenance from

[65] These alliances occurred because only the eldest Numbudiri sons were
allowed to marry within their own community. The children born to such cou-
ples belonged to the mother's family. The Nambudiris and the British courts
did not regard sambandham as a formal marriage.

[66] Moore, *Malabar Law and Custom*, p.95; Mayne, *Hindu Law*, 6th ed.,
Preface, p.XIII, p.293, cited in ibid., pp.15-6 (note).

[67] Report of the Travancore Marumakkathayam Committee (Trivandrum,
1908), para.76.

[68] Moore, *Malabar Law and Custom*, p.69; Sundara Aiyar, *Treatise*, p.83.

[69] The leading case was Kallati Kunju Menon *v.* Palat Erracha Menon,
(1864) 2 M.H.C.R.162.

karanavans, and to file a suit to oust a karanavan under certain conditions. Animosity arose between karanavans and anandaravans in the latter half of the nineteenth century. It was often caused by the karanavans' partiality towards their own wives and children—under the marumakkattayam system, a wife and children did not belong to their husband's tarawad, nor were they entitled to be supported by him. Removal or maintenance suits by anandaravans against karanavans seem to have became frequent from the 1870s. Thus family feuds can be said to have been encouraged by the interference of the courts in the functioning of the marumakkattayam system. Such notoriety as Malabar won for its litigiousness came partly from these unsettling conditions with the tarawads—at least this was often said among Nayars, and found expression also in the Report of the Malabar Marriage Commission (1891). There, one witness deplored that 'Nayars are already litigious and wasted a great deal of money in this way'.[70]

More important to tenancy problem is the fact that deteriorating relationships among tarawad members connected with practices such as melcharts and renewals of kanam leases. A karanavan who was eager to benefit his own wife and children, would often resort to giving a melchart as the easiest way of obtaining cash. This was often pointed out by both sides of the debate over tenancy legislation. For instance, M. Narasinga Rao, Subordinate Judge at Calicut, who was against tenancy legislation, argued that recent High Court decisions against the grant of melcharts by karanavans were intended to protect not tenants but rather the janmis' tarawads.[71] On the tenants' side, K. Chantan, retired Deputy Collector, said,

The greatest curse in Malabar is the Marumakkattayam law of inheritance prevailing among the Janmi classes. According to this law the lands of the tarwad are to be managed by the karnavan for the benefit of the members of the tarwad which dose not include his own children. Law requires that he should manage the estate for the benefit of his nephews and nieces. Natural affection demands that he should work for the benefit of his own issue. The karnavan therefore stoops to defraud his wards in the tarwad. This leads him to extract

[70] RMMC, Witness no.13.
[71] G.O.no.366, Law (General), 5 February 1926, p.110.

large renewal fees and to the grant of melcharths.

He reported, as an example, Chirakkal Raja's case in which the new raja leased out by melcharts lands which the former raja had given to his wife and her relatives.[72] The Janmi Sabha, defending themselves against pro-tenant arguments on the basis of the increase of eviction cases, contended that they were only a surrogate for other conflicts, one kind of which was between a karanavan and his anandaravans: the latter's attempts to set aside the encumbrance or alienations of lands by their karanavan took the form of eviction cases.[73] On the other hand, one tenancy supporter claimed that the karanavans of tenants' families raised debts ostensibly for the purpose of paying renewal fees, but in fact for enriching their wife and children.[74] Even more explicitly, K. Krishnan asserted that:

The present day karnavan is quite a different being from his ancestor. On account of his affection for his wife and children, and his anxiety to put them above want, the one aim of his life is to amass fortunes for them.... He tries to make as much out of the few years left for him (since he usually becomes a karanavan in his old age) to enable him to provide a competence for his wife and children. Indolent monied men snatch this opportunity to get a Melcharth of the properties held by his tenants.[75]

These arguments are no doubt difficult to attest in either way owing to lack of concrete examples. Nevertheless, they certainly indicate that the disintegration of the marumakkattayam tarawads was one of the factors that intensified tenancy problems.

The marumakkattayam reform movements

The first move to reform marumakkattayam was initiated by educated Nayars. The direct incentive was a humiliating High Court decision of 1869, declaring that conjugal relations (sambandhams) in relation to *aliya santana*, a matrilineal system followed in South Kanara, could not

---

[72] Ibid., pp.58-9; see also Innes' Note, para.44.
[73] G.O.no.366, p.45.
[74] Ibid., p.50.
[75] 'The Necessity for Legislation' attached to his Tenancy Bill. For this document, see note 89.

be considered as legal marriages but were mere concubinage.[76] This judgment was generally understood to cover the marumakkattayam system as well. Educated Nayars, who most resented this judgment, had already begun to feel uncomfortable at having their social customs exposed to ignorant comment. Thus, one of their objectives was to make sambandham legal in the eyes of the law. This is made plain in the provisions of C. Sankaran Nayar's Marriage Bill (1891), the first of its kind.

Yet a more deep-seated cause underlying the demand for reform was that these educated Nayars had started acquiring separate property of their own. The Malabar Marriage Commission appointed in 1891, with a view to ascertaining public opinion on Sankaran Nayar's Bill., showed that prominent support for the Bill came from government officials and professionals, such as lawyers and teachers, who made up more than half of the Bill's supporters. Of 63 professionals consulted (excluding those who did not observe marumakkattayam customs), 53 were in favour of reform. They were particularly enthusiastic about the provisions allowing personal property to be inherited by wives and children.

It is difficult to ascertain how much of this self-acquired property there was. According to the Malabar Marriage Commission report, it usually comprised just a house and gardens.[77] Table 7 is a summary of the statements made to the Marriage Commission by the witnesses (all Nayars) who were self-employed (other than in agriculture), and whose family economic background could be inferred from the assessment due to their tarawad.[78]

Except for witness no.99, all those shown in Table 7 supported marriage legislation of some sort. The sample is admittedly small, and it is not possible to say whether these reform-minded witnesses came from janmi families or otherwise.[79] But we may infer the emergence of

[76] 4 M.H.C.R. 196.

[77] RMMC, para.77.

[78] Witness no.80 is excluded from this examination on account of his being a Brahman.

[79] However, one case (Witness no.96), which is not included here, is available that conveys a picture of a Nayar tarawad which produced a tahsilder,

an assertive professional class that came not from the extraordinarily rich, such as Raja or Nambudiri, families, but from reasonably well-to-do landed families.[80] No doubt K.P Raman Menon, who was one of the prominent tenancy advocates, was exceptional, in paying assessment of more than Rs.800 on his personal lands and owning 5,000 acres of forest land.[81] But it would have been similar people who could have purchased more stable tenancy rights, even by way of the notorious melcharts, and have defied the janmis' age-old authority. During the heated debate on melcharts, the janmis tried to lay the blame on professional men who were attacking their arbitrary power, by claiming they were the very persons ready to take a melchart.[82] And among them we find the advocates of tenant rights.

By contrast, opposition to the reform of marumakkattayam, especially to any provisions breaking up joint family property, came from the big janmis—that is, the former chieftains and orthodox Nambudiris. There is no doubt that family schisms aroused much anxiety among these janmis. The Janmi Sabha set up a special committee in 1911 to consider the tarawad problem. However, it was difficult for them to accept any roposal for the partition of joint family properties, since their social prestige and economic base depended on their maintenance. The janmis were ready, at most, for legislation to regulate the arbitrary power of the karanavans. Thus we can see that attitudes to marumakkattayam was an another crucial point that formed the contesting camps on the tenancy problem.

Anti-Nambudiri sentiment, expressed especially during the controversy over the Malabar Marriage Bill, also should not be overlooked. One of the respondents to the Marriage Committee, a retired Sub-Judge, stated in his memorial that, 'Even at the present day a Nambu-

---

holding lands on kanam right under a Nambudiri janmi and paying a large assessment (over Rs.600).

[80] It is hard to decide who could be regarded as well-to-do. According to one witness, families paying Rs.50 or more as assessment in their own name or in the name of tenants could be called well-to-do. Witness no.91 of RMMC.

[81] Report of the Malabar Tenancy Committee (of 1939), vol.II, p.157.

[82] *Janmi* 6, 1, pp.31-2. For instance, K.P. Raman Menon was more than once criticised for his acquisition of melcharts. See *Yogakshemam*, 7 February 1923, p.4, and 10 October 1923, p.6.

diri thinks that he has a right to have sexual intercourse with a Nayar lady whatever might be the position of her husband'. He alleged that those who opposed reform included people who still believed a Brahman to be a god on earth.[83] These statements indicate a trend, in which the established measure of social status in terms of marital relations with Nambudiri families, was being challenged and denounced. O. Chandu Menon's famous Malayalam novel *Indulekha* (published in 1889) reflected the aspirations of the educated Nayars and the tarawad problems—though Menon was himself a district munsiff, and in the end became a Sub-Judge. However, though unintentionally, it also severely damaged the prestige of the Nambudiri community as a whole by making a target of derision and contempt of one Nambudiripad (a rich but ignorant and licentious janmi). The novel attained such fame that V.T. Bhattatiripad, the famous Malayalam writer and social reformer (who came from a rather poor Nambudiri family), tells from his own experience that people looked at the Nambudiris through the eyes of *Indulekha*.[84] Such 'revolt' against the social authority enjoyed by Nambudiris was intimately connected with antipathy against Nambudiris as janmis. This was apparent in the declaration that 'No tenant in Walluvanad Taluk, unless he be an educated man or a Government official, dare to perform the funeral obsequies of his deceased father for fear of his Nambudiri landlord's displeasure'.[85] The Nambudiris themselves suspected that tenancy advocates had singled them out as the main target. Naturally, throughout the tenancy agitation, their organ *Yogakshemam* took a stand against the legislation in alliance with chieftain janmis (the model of the janmis for tenancy agitators). The two groups were intimately connected through the practice of sambandham.

Thus, from the Malabar Marriage Act of 1896 onwards,[86] the maru-

[83] RMMC, Memorial no.36.

[84] V.T. Bhattatiripad, *Karmmavipakam* (Trichur, 1988), p.160.

[85] RMMC, Statement no.37.

[86] This Act, which was not compulsory but only permissive, allowed those who got their sambandham registered under the Act to leave half of the separate and self-acquired property to their own wife and children in case of intestacy. However, the fact that there were only 83 registrations under this Act by April 1903 (Moore, *Malabar Law and Custom*, pp.90-1), plainly shows that it was a

makkattayam reform movement and the tenancy agitation proceeded side by side. Not only the timing but the people who led each movement often overlapped—for instance, K.P. Raman Menon and M. Krishnan Nayar, who were both in the forefront of the tenancy agitation of the 1920s. In 1913 the former introduced, into the Madras Legislative Council, two Bills called the Malabar Marriage and Inheritance Bill and the Malabar Partition and Succession Bill.[87] In 1910 Krishnan Nayar introduced the Malabar Succession and Partition Bill.[88] These Bills turned out to be abortive, but, in 1932, two years after the Tenancy Act, the Madras Marumakkattayam Act was passed. K. Madhavan Nayar, who was the first to have introduced such a bill, was a tenant representative in the Council from 1926.

Evolution of the Tenancy Act

It cannot be overemphasised that it was the poor Mappillas of Ernad taluk and those of Walluvanad taluk—the former once being described as 'no better in many cases than an agricultural labourer'—[89] who compelled the Madras government to interfere with the relationships between janmis and tenants and who brought pressure to bear on the janmis through their 'outbreaks'. The tenancy agitations of the 1910s and 1920s, however, were very far from this violent world of 'rural terrorism'.[90] By this time the Malabar tenancy problem had come on to the agenda of a much more sober and open world of electoral politics, where Mappilla outbreaks were used just as a 'trope' to justify tenancy

---

far from satisfactory measure. The main reason the Act failed to become popular was the unfamiliar and—even to the supporters—obnoxious procedures required to legalise sambandhams.

[87] Raman Menon withdrew both Bills and instead in 1916 moved for leave to introduce a new Bill consolidating two Bills with some modifications. MLCP, vol.43, 1915-16, pp.339-41. His former Bills were strongly opposed by the Janmi Sabha. See *Janmi* 6, 8, p.217.

[88] This was dropped because the mover ceased to be a member of the Council. The other Bills ended up in the same way.

[89] G.O.no.366, p.29. The summary power of distraint possessed by landlords in other parts of the Madras Presidency was denied to the Malabar janmis because of the government's fear of outbreaks.

[90] Conrad Wood, *The Moplah Rebellion and its Genesis* (New Delhi, 1987), p.57.

legislation.[91] To understand the origin and evolution of the Tenancy Act of 1930, it is essential to bear this background in mind.

The timing and changes in the original Bills of M. Krishnan Nayar provide a convenient indication of the forces which shaped the Tenancy Act of 1930. Krishnan Nayar first submitted a Tenancy Bill to the public in June 1922. Its main objective was to confer occupancy rights on kanakkaran who had been in possession of lands for 25 years or more. It contained no provisions for verumpattakkarans. Krishnan Nayar dropped this Bill, however, and in December 1922 gave notice of his intention to introduce a revised Bill granting occupancy rights to kanakkarans and other tenants. This notice was again allowed to lapse. Eventually he moved for leave to introduce the same Bill in April 1924. The second Bill was different from the first in that in conferring occupancy rights on all kanakkarans, and on other tenants (including verumpattakkarans) provided they had been in uninterrupted possession of their lands for not less than six years.

Since the first District Conference in 1915 the tenancy problem had played an important role in politicisation in Malabar. Resolutions relating to tenancy legislation were hotly discussed at these conferences. In 1920 the Malabar Kudiyan (Tenants) Association was set up, with K.P. Raman Menon as president.[92] The interests of kanam tenants, especially those of South Malabar, surfaced most prominently during tenancy agitations in the 1910s and 1920s, as many scholars

[91] See 'Statement of Objects and Reasons' attached to Krishnan Nayar's Bill in G.O.no.366, and 'The Necessity for Legislation' in K.Krishnan's draft of a Tenancy Bill. The document is appended to P. Kodanda Rao, *Malabar Tenancy Problem* (Madras, 1924). According to Kodanda Rao, Krishnan's Bill was first published in *Young Men of India*, April 1922, and afterwards revised. The original Bill has not been traced. K. Krishnan's Bill found in RMTC, vol.I, is considerably different, not so much in its contents as in its expressions. Krishnan was an additional member of the Tenancy Committee, appointed in 1927 as a tenant representative, though K. Madhavan Nayar asserted that Malabar tenants were not satisfied with him (MLCD, vol.36, 1927, p.537). The Tenants' Association had decided to boycott the Committee because its composition was partial to the janmis' interest.

[92] For information on the early history of tenants' as well as Congress movements in Malabar, see A.K. Pilla, *Kongrasum Keralavum* (Trivandrum, 1983, first published in 1935), and J. Occanturuttu, 'K. Madhavan Nayar', in K. Madhavan Nayar, *Janmashatabdi Smaranika* (Calicut, 1982).

SITUATING THE MALABAR TENANCY ACT 1930   391

have observed.. Tenancy advocates were often unsympathetic towards
verumpattakkarans, however. R. Sekhara Menon, a lawyer, said that a
verumpattakkaran was 'not much better than a contractor or labourer
agreeing to execute a work for a definite period and gain a living by it'.
He concluded (referring to Palghat taluk and portions of Walluvanad
and Ponnani): '...no permanent right need to be conferred on the ordinary simple tenants of these parts'.[93] The Palghat Bar Association
asserted that, as simple tenants, 'especially of the Palghat taluk in South
Malabar, have generally no abiding or substantial interests at stake in
the land', then 'they have no substantial interest at stake therein and
nothing substantial to lose in case of eviction'.[94] When the Select
Committee on Krishnan Nayar's Bill was in session, the *Madras Mail's*
correspondent reported that 'the activities of the Malabar tenants and
Mr. G. Sanakaran Nair, the Organising Secretary of the Malabar
Tenants' Association, are much in evidence, and not a single week
passes without the holding of meetings in some upcountry village or
other and passing of stereotyped resolutions for not giving compensation to Jenmis and for strengthening the hands of the Kanomdar
[middle man] by all manners of means'.[95] It was evident that the
tenancy issue would become a focal point in the elections after the
Montagu-Chelmsford reforms, which in Malabar, it should be noted,
extended suffrage to the tenants (supposedly, in most cases, kanakkarans) of janmis registered by the Malabar Registration Act of 1896,
who paid land revenue of more than Rs.10.[96] It would be a safe guess
that Krishnan Nayar included all the kanakkarans in his revised Bill
with an eye to these new votes. This would also explain the intriguing
fact that tenants' main supporters in the Councils were themselves from
janmi families. Krishnan Nayar was from a Naduvazhi janmi family
(minor chieftains), though not, it seems, a very prosperous one.[97] He

[93] G.O.no.366, p.52.
[94] Ibid., pp.77-8. For the opinion of K. Gopalan Nayar, Subordinate Judge, in
the same vein, see p.81.
[95] *Madras Mail*, 23 June 1925, p.5.
[96] Indian Statutory Committee, vol.6, Memorandum submitted by the Government of Madras to the Indian Statutory Committee (London, 1930), p.311.
[97] On M. Krishnan Nayar, see K.P. Kesava Menon, *Samakalinaraya Cila
Keraliyar*, vol.I (Kottayam, 1974), pp.52-64.

was legal counsel to the Janmi Sabha and had led a janmi deputation to the Cochin Darbar in 1909 when tenancy regulation was under consideration in that State. Another tenants' advocate was K.P. Raman Menon who had once been an executive member of the Janmi Sabha.[98]

More significant was the conditional inclusion of verumpattakkarans in the 1924 Bill. Non-cultivating kanakkarans could have found this irksome when exacting rent from their verumpattakkarans. Krishnan Nayar's sincerity was surely dubious; one of the officials remarked sarcastically that 'It may be of course that M. Krishnan Nayar knows that so long a period of six years of uninterrupted occupancy of a holding by a simple lessee is exceptional'.[99] Both the British officials and the pro-janmi circle regarded the measure as another device for catching votes. But, too much stress should not be placed on verumpattakkarans' votes, as it seems more likely that it was related to the janmis' counter-propaganda, which depicted the kanakkarans as speculative and parasitical middlemen, and more likely than janmis to rackrent verumpattakkarans.[100] The strategy was to show sympathy towards the 'actual cultivators', and it was to a certain extent effective in enlisting the support of Council Members from other parts of the Presidency, who regarded the kanakkarans as strong enough to take care of themselves. For example, S. Satymurti asserted that 'We hold no brief either for the janmi or for the kanamdar, and we feel that the only class of people in Malabar who deserve the sympathy and help of this hon. House is the class of cultivating tenants'.[101] Similarly, Kaleswara Rao said: 'If it is a case as between Janmis and Kanamdars, I would not bother myself about it. They can settle their disputes between themselves. Both parties are equally strong.'[102]

Such attitudes among non-Malayali members were evident too in several important revisions to Krishnan Nayar's Bill in Select Committee. One proposed to grant permanent occupancy rights to all

[98] The Janmi Sabha had expressed its hope in his election to the pre-Reform Council from the Municipality and Local Board Electorate, *Janmi* 5, 3, p.84.

[99] G.O.366, p.15.

[100] Ibid., pp.26 and 46; *Yogakshemem*, 8 March 1924, pp.4-5.

[101] MLCD, vol.50, 1929, p.647.

[102] Ibid., p.666.

cultivating tenants irrespective of how long they had been in possession, excluding non-cultivating kanakkarans who held land by virtue of melcharts. Another amendment drastically cut the rent payable by cultivators; and a motion was nearly carried to confine permanent occupancy rights to kanakkarans who position dated from before 1853.[103] The Select Committee was also not at all lenient to janmis. For instance, it sought to abolish renewal fees, a change which had not been demanded even by the kanakkarans' advocates. (The fees were later reintroduced to the Bill.) On the other hand, the Committee did not grant occupancy rights to the tenants on lands reclaimed by janmis themselves or belonging to small janmis. Thus, it could be said that its radicalism came partly from its ignorance of affairs in Malabar, but more substantially from its partiality towards the ideal of the 'actual-cultivator'.

By contrast, the Malayali legislators, except K. and C. Krishnan (nominated Tiyar members), were mainly concerned with the kanak-karan interest, irrespective of whether they were actual cultivators or not. Both Saldanha, an outspoken Christian tenancy representative, and K.P. Raman Menon were reluctant to concede anything to the verumpattakkarans—and nothing, according to them, was all the verumpattakkarans demanded.[104] Even K. Madhavan Nayar, who showed some consideration towards the verumpattakkarans, tended to become much more eloquent when the interests of kanakkkarans came under attack. On one occasion, he went so far as to assert 'The whole problem has been the problem of the kanamdars'.[105] A consequence was that the pro-tenancy legislative council members from Malabar had to fight on two fronts, against janmi representatives and against

---

[103] See Dissenting Minute by T.A. Ramalinga Chetti in Report of the Select Committee on the Malabar Tenancy Bill. The particular year of 1853 (some-times 1852 or 1854) was repeatedly referred to by tenancy advocates as the time when the kanakkarans' permanent occupancy right had been denied by the courts. Therefore, it was sometimes assumed by non-Malayali M.L.Cs that kanakkarans who had acquired a kanam right after 1853 were not particularly entitled to legislative protection since they should have been well aware of their tenancy conditions.

[104] MLCD, vol.50, pp.648 and 1157.

[105] Ibid., p.673.

cultivators' supporters from outside Malabar.. Madhavan Nayar once confessed in the Council that he had to take great pains to persuade his Congress colleagues to his way of thinking, and that he did not always succeed.[106]

Bringing verumpattakkarans within the purview of the Bill was not just a pose to get rid of the impression that whole agitation was for kanam tenants. The Malabar Tenants' Association was not the only organisation of its kind, though it was certainly the most articulate at that time. There were several other voices, some of them more radical. Before the turn of the century Manjeri Mappillas had begun to send petitions to the government asking for a Tenancy Act,[107] and in 1919 the Hidayattukal Muslims conveyed their grievances to the Governor of Madras when he visited Manjeri,[108] one of the sites of Mappilla risings. The tenants, mainly Mappillas, in the Zamorin estate in the Kottakkal area started an association called the Zamorin estate Kudiyan Sankatan-ivarana (Grievance Removal) Sangham around 1920. They invited tenancy advocates to their conferences. After attending one of these, where more than 5,000 Mappillas and Hindu tenants gathered, K. Madhavan Nayar recorded his amazement at the ability of the tenant speakers, who had no knowledge of English, in expressing their grievances and their determination to face any difficulties.[109] Their zeal was such that he had to interfere to restrain them when they began boycotting not only the cultivation of lands belonging to janmis who ill-treated tenants, but the janmis themselves.[110]

The Tiyar community of North Malabar was also active in this respect. Their association, the Jnanodaya Yogam, submitted a tenancy bill to the Governor of Madras on his visit to Malabar in 1919. It proposed to confer permanent occupancy rights on the actual cultivators, excluding intermediaries from any protection in the law.[111] There was a

[106] Ibid., p.1157.
[107] RMTC, vol.11, p.604..
[108] Pilla, *Kongrasum Keralavum* ,p.86.
[109] K. Madhavan Nayar, *Malabar Kalapam* (Calicut, 1987), p.91.
[110] Ibid.,. pp.94-5.
[111] C. Govindan Nair, *The Malabar Tenancy Act, 1929* (Madras, 1931), p.49. See also Kollengode Raja's angry reference to this document in his *A Note on the Malabar Tenancy Legislation*, Part I (Madras, 1925), p.36. K. Krishnan was

tenants' association with about 500 members, mainly Tiyars, in Kurumbranad Taluk.[112] Furthermore, in 1925, following criticism of the Malabar Tenants' Association, another tenants' organisation was set up in North Malabar, called the North Malabar Tenants' League. It emphasised the protection of the verumpattam tenants.[113] The objection to the Tenants' Association had been spelt out by a correspondent to the editor of the *Madras Mail*. He criticised the South Malabar tenants for passing a resolution demanding a provision to allow the eviction of a verumpattam tenant who had defaulted on his rent for one year; and he 'hoped that Verumpatadars [sic] will take effective steps to see that their cause is placed before the Select Committee. They cannot afford to entrust the Malabar Tenants' Association to do this especially when they see that Mr. G.S. Nair the Secretary is identifying himself with this agitation.'[114]

Among legislative council members from Malabar the clearest cultivator-oriented stance was taken by the nominated Tiyar members, such as K. Krishnan and A Achutan.[115] 'Mitavadi' C. Krishnan, a defeated Tiyar candidate of the 1920 election, was also articulate in supporting the less-privileged tenants.[116] Pro-tenancy articles combined with expressions of opposition to caste discrimination appeared frequently in the Malayalam magazine *Mitavadi*, owned and edited by C. Krishnan. Tiyars had been active in a caste-reform movement since the beginning of the twentieth century. They were trying to improve their social status mainly by reforming their religious and social customs and spreading education, but the tenancy problem was also taken up. Their community included a vast number of less-privileged tenants, and

a prominent member of the Yogam. Krishnan's Bill was possibly identical, according to Kollengode Raja, with that of the Yogam. See also *Mitavadi* 7, 10, p.483.

[112] RMTC, vol.II, p.355.

[113] *Madras Mail*, 18 July 1925, p.5.

[114] *Madras Mail*, 3 July 1925, p.10. Though one of the initials is lacking, the writer must be a person who was to become one of the joint secretaries of the Tenants' League.

[115] MLCD, vol.51, 1930, p.213.

[116] For his tenancy activities, see K.R. Achutan, *C. Krishnan: Jiviacaritram* (Kottayam, 1971). His brief opinion on Krishnan Nayar's Bill of 1924 can be found in G.O.366, p.55.

therefore it was natural that leaders, such as C. or K. Krishnan, would take pro-cultivator positions, even though they were themselves janmis.[117] The Tiyars' movement thus functioned as a side-attack on the kanakkaran-oriented tenancy agitation.

The Tenancy Act emerged in its final form only after government intervention. The Madras government vetoed Krishnan Nayar's Bill, but then appointed a committee in July 1927 to investigate the matter afresh.[118] The committee prepared a draft bill, which, before being introduced into the legislature, was placed before a conference of janmi and tenant representatives presided over by the Governor.[119] The Bill as modified by this conference was finally passed by the Council in October 1929.

It was rather strange that the Madras government was satisfied with a new Bill that was not very different from the vetoed Bill of 1924. It too conferred permanent occupancy rights on various types of tenants, without providing the janmis with any substantial compensation—one of the issues the government had brought up again and again when opposing Krishnan Nayar's Bill. The government's concern with the actual cultivators, another issue dwelt upon during the former debates, curiously enough also did not find a place in the Act. Retained was a much debated provision requiring a simple-lessee to deposit one year's rent in advance if was required to do so.[120]

To the janmis, the Act was a betrayal indeed, as Sankaran Unni, a janmi representative in the Third Council, protested with much fervour.[121] Every effort was made to appeal to the political sensitivities of the government, reminding it how the janmis's power and influence

[117] MLCD, vol.50, p.78; Achutan, *C. Krishnan*, ch.10; K.R. Bhaskaran, 'C. Krishnan', in G. Priyadarsanan (ed.), *S.N.D.P. Yogam Platinum Jubilee Smarakagrantham* (Quilon, 1978).

[118] G.O.no.2346, Law (General), 29 July 1927.

[119] G.O.no.4457, Law (General), 21 December 1928.

[120] The government insisted that this provision of the deposition replace the right of distraint given to the landlords, again on the ground that the latter would be dangerous, in view of 'the condition of Malabar, particularly in the Ernad and Walluvanad' (referring to the Mappilla population in these area). See MLCD, vol.51, pp.152-5.

[121] Ibid., vol.50, pp.1147-8. See also *Yogakshemam*, 28 September 1929, editorial.

had been and would be at the government's command. The pleading met with little success. Sir Thomas Moir seemed to representing the government's feelings in the blunt statement that 'we have in this Bill recognised the Kanamdar simply because he is there and because he is an important part in the land tenure system of Malabar. It is not because we like him.'[122] In other words, through tenancy agitations the kanak-karans had succeeded in impressing upon the government their irrefutable presence as a political force along with the janmis. That was exactly what some of the tenancy advocates such as Raman Menon had wished for. Congratulating the government for bringing the Bill to a conclusion, he claimed that 'by this Bill the Government as a matter of fact consolidated the landowning classes into one whole'.[123]

Conclusion

This paper tried first to assess the 'dominance' of janmis by looking into estate management and eviction suits in order to understand the real meaning of the Malabar landholding system as interpreted by British officials and courts. Legal definitions were not hastily accepted as a decisive factor in strengthening the janmis' power. We took up janmis who had formerly been chieftains because they became the image of the Malabar janmis as a whole during the tenancy agitations, though the actual conditions were not that simple. There has been a tendency among scholars to emphasise their power and influence, but we suggested some weaknesses in the their economic and social position. Their widely-dispersed pieces of lands could not be easily managed. The British were always ready to sacrifice the strict legal definitions of the janmis' right as landlords to the needs of collecting land revenues regularly or of maintaining law and order, which they were afraid might be disrupted at any moment by Mappilla outbreaks. In addition, the peculiar composition of the janmis—the predominance of the former rajas and Naduvalis along with orthodox Nambudiris—proved not to be necessarily advantageous to the maintenance of their power and influence. It provided a rallying point for the advocates of tenant rights,

[122] Ibid., p.1152.
[123] Ibid., p.1158.

especially when lower castes such as the Tiyars began to try improving their social status by attacking the traditional social order. It was easy for them to denounce janmis' 'oppression' by establishing the image of the janmis as a few powerful chieftains and Nambudiris who monopolised the land.

Viewed in this light, it is reasonable to argue that the fact that the British had singled out the janmis as having the sole right to absolute ownership, helped prepare the ground for the tenancy agitations. The legal re-construction of interests by British officials and courts may then be interpreted not so much as a measure of change in economic positions as a useful reference point when the tenancy agitations took form. In other raiyatwari areas in the Madras Presidency, hardly any tenancy movements seem to have been organised even though there was every evidence to indicate the increasing grouping of tenants as 'raiyats'. In Malabar, the civil suits which were encouraged by the legal re-interpretations of various rights also weakened the traditional ties (so-called *kana janma mariyada*) upon which the authority of the chieftain and Nambudiri janmis largely depended. One result of judicial intervention was the loosening of the 'communal grid', to use E.P. Thompson's phrase, in the agrarian conventions of Malabar.[124]

Social changes going on simultaneously in Malabar were also related to the tenancy agitations; they contributed to shaping the discursive world of the tenancy agitation. First we noticed the swelling complaints against the marumakkattayam system and the demand among the educated for its reform. Hitherto this subject has been dealt with independently, mainly as showing the ascendancy of a new 'enlightened' middle class who found joint families and the domination of karanavans suffocating and outdated. Moreover aspects of the matrilineal joint-family system (such as the indivisible tarawad properties), which seemed to fortify the janmis' power, also might intensify the tenancy problem; we noted instances of friction among those who followed the marumakkattayam system, for example when family feuds led to the grant of melcharts, which intensified agrarian

[124] E.P. Thompson, 'The grid of inheritance: a comment', in Jack Goody *et al.* (eds.), *Family and Inheritance* (Cambridge, 1976).

dissent more generally.

After the Montagu-Chelmsford reforms, electoral considerations influenced the enactment of the Tenancy Act. Reform oriented towards the kanams was modified under pressure from those sympathetic to the actual cultivators. The distinct and often more radical activities of the Mappillas and Tiyars were important here. For example, Tiyar leaders, though most concerned with improving their social status, could not ignore the tenancy question. Most of their fellow Tiyars were less-privileged cultivators.

Certain important subjects, such as the impact on the Malabar economy of world capitalism and the trend of the land market in Malabar, could not be touched upon in this paper. The impact of the depression and that of resettlement, which afflicted Malabar immediately after the enactment of the Tenancy Act, also remain to be explored. The main purpose of this paper, however, was to re-examine a common approach to Malabar agrarian history, and to rescue several elements which had been neglected. The emergence of tenancy agitations could not be properly understood merely by looking into the land-holding system, however peculiar it was. The radicalisation of Malabar politics, including the tenancy movement, would not have been possible without a simultaneous social awakenings on many fronts. What decided any one stance towards agrarian problems was not necessarily economic considerations as such, but often the subtle mixture of varied socio-political values. The process of tenancy legislation was a unique venue for these values, and provides with clues with which they may be explored.

Table 1: *The principal janmis in the 1880s*

	No.	%	Amsams
Religious & charitable institutions	250	23.2	1.3
Rajas	61	5.7	4.9
Nayars, Variyars	339	31.4	1.4
Brahmans	379	35.1	1.3
Tiyars	8	7.0	1.2
Other Hindus	4	0.3	1.0
Mappillas	37	3.4	1.6
European	1	0.2	1.0
Total	1,079	100.0	2.6

Source: Malabar Special Commission, 1881-82, Malabar Land Tenures, vol. 1, Report, paras.232, 241.
Notes: (1) 'Principal janmis' means those janmis who held more than 100 pieces of lands in a single amsam (revenue village). The third column shows the average number of amsams in which they held over 100 pieces of land. (2) Under 'Religious institutions' is included one that was Islamic, while the 'Brahman' total includes 7 Kanara Brahmans and 2 Tamil Brahmans.

Table 2: *Kanam holdings under selected janmis*

Name	Total kanam holdings under each
1. Desamangalam (N)	1,203
2. Vamanjeri (N)	380
3. Olappamanna (N)	943
4. Varikemanjeri (N)	1,496
5. Motapilapalli (N)	425
6. Kizhakkekovilagam (C)	4,804
7. Kavalappara (C)	1,831
8. Kannambra (C)	150
9. Kuthiravattam(C)	2,290
Total	13,522

Source: Madras Legislative Council Proceedings, vol.19, 1924, p.690.
Note: N = Numbudiri; C = chieftain.

Table 3: *Kavalappara and Zamorin estates*

		Zamorin	Kavalappara
Area (acres)	total	155,358 [1]	15,491 [2]
	wet	22,539	4,503
	dry/garden	17,215	
	dry		680
	unassessed	15,604	
	occupied dry		1,620
	unoccupied dry		8,688
Holdings	location	6 taluk [3]	2 taluk [4]
		605 desam	21 amsam
	number	12,062 [1]	9,501 [4]
Tenancy	no. of tenants	9,926 [5]	2,849 [6]
	anubhavan		61 [6]
	kanam		1,786
	verumpattam		1,120
Temples and charitable institutions			
	temples		25 [8]
	cherikkals	23 [7]	
	kalams	10	
	devaswams	28	
	uttuparas/brahmaswams	8	2
	water pandals		5
	others	4	
	educational	1 college	1 elementary high school
Establishment	(for Nambudiris)	1 Vedic school	
	total employees	729 [9]	about 150 [8]
	for devaswams	174	
	for brahmaswams	28	
	for cherikkals	206	
	Zamorin's office	54	
	for household	277	
Assessment (Rs.)		115,000	3,800 [10]

Sources and notes: [1] Figures exclude land in Cochin State, and of the Guruvayur temple where the Zamorin was one of two trustees; CWP, no.45, 14 December 1927. [2] Figures exclude land in Cochin State; ibid., no.122, 23 November 1904. [3] Ibid., no.55, 12 September 1916. [4] BRP, no.2492, 29 August 1879 [5] CWP, no.65, 5 February 1923. [6] Ibid., no.170, 25 Nov.1907; anubhavan = perpetual lease. [7] MLT, vol.1, Report, para.244. [8] BRP, no.1698, 26 August 1872. [9] Letter from the Zamorin,17 August 1914, in G.O.no.897 (Confdl.), Revenue, 17 April 1915. In the 1880s, according to Logan, 839 land-agents, clerks etc., were employed by the estate; MLT, vol.1, para.244. [10] Amount pre-settlement. For the janmis' responsibility for paying the assessment, see below.

Table 4: *Assessment paid by a janmi and by tenants*

Witness no.	Assessment paid (Rs.) by the janmis	by tenants	Numbers of tenants
52	280+[1]		
81	1,000	4,000 to 5,000	2,000
82	500	10,000	500 [2]
86	300		500 [2]
87	200	2,500	
88	300	40	
92	3,000	7,000+	
95	700		40 [2]
97	3,000		400 [3]
98	2,000		400 [4]
101		all	300 or 400

Source: Report of the Malabar Marriage Commission, in Appendix IV.
Notes: [1] Mostly in the name of tenants; [2] with pattas; [3] mostly with pattas; [4] with separate pattas; '+' = (or) more than the sum shown.

Table 5: *Decrees for eviction in Malabar*

Year	Decreed
1887	2,819
1888	2,441
1889	2,627
1890	4,227
1891	4,132 [4,432?]
1892	4,620

	South Malabar			North Malabar			
	Decreed	Executed	%	Decreed	Executed	%	Total
1905	1,426	1,055	74	1,015	646	63.6	2,441
1906	1,402	1,120	79.9	1,148	770	67.1	2,560
1907	1,525	1,202	78.8	976	672	68.9	2,501
1908	1,512	1,144	75.7	801	590	73.7	2,313
1909	1,361	1,131	83.1	1,043	652	62.7	2,404
1910	1,466	1,080	73.7	1,121	782	69.8	2,587
1911	1,530	1,019	66.6	1,244	799	64.2	2,774
1912	1,116	442	39.6	1,513	828	54.8	2,629

Source: Innes' Note, paras. 19, 25.

Table 6: *Eviction suits*

A. *Average number filed annually, 1916-1926*

	By janmis	By melchartdars and others	Total
North Malabar	1,550	793	2,343
South Malabar	1,382	977	2,359
Total	2,932	1,970	4,702

B. *Number instituted at Pattambi Munsiff's Court*

Nature of suits	Number of suits by year				
	1920	1921	1922	1923	1924
Janmi redemption of kanam	219	99	77	98	24
Ditto of panayam (mortgage)	31	14	9	7	
Eviction: landlords against tenants	97	49	36	38	18
Eviction by kanakkarans	136	73	52	76	53
Redemption of sub-mortgages	24	9	12	19	11
Total	1,009	463	558	415	189

Sources: 'A.' Report of the Malabar Tenancy Committee, 1927-8, vol. 1, para. 114. 'B.' G.O. 366, Law (General), 5 February 1926, p.118.
Note: The figures for South Malabar are incomplete because of the Mappilla Rebellion of 1921-22 in which some records were destroyed.

Table 7: *Some cases showing tarawad economic background*

Witness no.	Age	Position in tarawad	Tarawad's assessment (Rs.)	Occupation etc.
39	46	anandaravan	500	Pleader (II)
63	56		300-400	High Court vakil
69	36	anandaravan	200	Malayalam pundit[1]
74	41		600-700	Pleader (I, 15)
77	54	karanavan	500	Pleader (probably II)
79	28		100	Pleader (II)
99	35	anandaravan	150	Pleader (I, 11)
102	35	anandaravan	350 + 80[2]	Pleader (II, 10)[3]
107	37	anandaravan	30	Pleader (II, 12)
108	38	anandaravan	180	Pleader (I, 14)
109	34	anandaravan	100 ~	Pleader (I, 2)
	32	anandaravan	80 +	Pleader (I, 7)
120	45	karanavan	300 ~	Pleader (I, 16)

Notes: [1] In a government college, [2] paid by tenants; [3] a brother was in England for the I.C.S. examination. 'I' shows first- and 'II' second-class pleaders; the Arabic numbers give the years of practice where known.